Theodor Borangiu, André Thomas, and Damien Trentesaux (Eds.)

Service Orientation in Holonic and Multi-Agent Manufacturing Control

Studies in Computational Intelligence, Volume 402

Editor-in-Chief
Prof. Janusz Kacprzyk
Systems Research Institute
Polish Academy of Sciences
ul. Newelska 6
01-447 Warsaw
Poland
E-mail: kacprzyk@ibspan.waw.pl

Further volumes of this series can be found on our homepage: springer.com

Vol. 380. Anthony Brabazon, Michael O'Neill, and Dietmar Maringer (Eds.)
Natural Computing in Computational Finance, 2011
ISBN 978-3-642-23335-7

Vol. 381. Radosław Katarzyniak, Tzu-Fu Chiu, Chao-Fu Hong, and Ngoc Thanh Nguyen (Eds.)
Semantic Methods for Knowledge Management and Communication, 2011
ISBN 978-3-642-23417-0

Vol. 382. F.M.T. Brazier, Kees Nieuwenhuis, Gregor Pavlin, Martijn Warnier, and Costin Badica (Eds.)
Intelligent Distributed Computing V, 2011
ISBN 978-3-642-24012-6

Vol. 383. Takayuki Ito, Minjie Zhang, Valentin Robu, Shaheen Fatima, and Tokuro Matsuo (Eds.)
New Trends in Agent-Based Complex Automated Negotiations, 2012
ISBN 978-3-642-24695-1

Vol. 384. Daphna Weinshall, Jörn Anemüller, and Luc van Gool (Eds.)
Detection and Identification of Rare Audiovisual Cues, 2012
ISBN 978-3-642-24033-1

Vol. 385. Alex Graves
Supervised Sequence Labelling with Recurrent Neural Networks, 2012
ISBN 978-3-642-24796-5

Vol. 386. Marek R. Ogiela and Lakhmi C. Jain (Eds.)
Computational Intelligence Paradigms in Advanced Pattern Classification, 2012
ISBN 978-3-642-24048-5

Vol. 387. David Alejandro Pelta, Natalio Krasnogor, Dan Dumitrescu, Camelia Chira, and Rodica Lung (Eds.)
Nature Inspired Cooperative Strategies for Optimization (NICSO 2011), 2011
ISBN 978-3-642-24093-5

Vol. 388. Tiansi Dong
Recognizing Variable Environments, 2012
ISBN 978-3-642-24057-7

Vol. 389. Patricia Melin
Modular Neural Networks and Type-2 Fuzzy Systems for Pattern Recognition, 2012
ISBN 978-3-642-24138-3

Vol. 390. Robert Bembenik, Lukasz Skonieczny, Henryk Rybiński, and Marek Niezgódka (Eds.)
Intelligent Tools for Building a Scientific Information Platform, 2012
ISBN 978-3-642-24808-5

Vol. 391. Herwig Unger, Kyandoghere Kyamaky, and Janusz Kacprzyk (Eds.)
Autonomous Systems: Developments and Trends, 2012
ISBN 978-3-642-24805-4

Vol. 392. Narendra Chauhan, Machavaram Kartikeyan, and Ankush Mittal
Soft Computing Methods for Microwave and Millimeter-Wave Design Problems, 2012
ISBN 978-3-642-25562-5

Vol. 393. Hung T. Nguyen, Vladik Kreinovich, Berlin Wu, and Gang Xiang
Computing Statistics under Interval and Fuzzy Uncertainty, 2012
ISBN 978-3-642-24904-4

Vol. 394. David A. Elizondo, Agusti Solanas, and Antoni Martínez-Ballesté (Eds.)
Computational Intelligence for Privacy and Security, 2012
ISBN 978-3-642-25236-5

Vol. 395. Srikanta Patnaik and Yeon-Mo Yang (Eds.)
Soft Computing Techniques in Vision Science, 2012
ISBN 978-3-642-25506-9

Vol. 396. Marielba Zacarias and José Valente de Oliveira (Eds.)
Human-Computer Interaction: The Agency Perspective, 2012
ISBN 978-3-642-25690-5

Vol. 397. Elena Nikolaevskaya, Alexandr Khimich, and Tamara Chistyakova
Programming with Multiple Precision, 2012
ISBN 978-3-642-25672-1

Vol. 398. Fabrice Guillet, Gilbert Ritschard, and Djamel Abdelkader Zighed (Eds.)
Advances in Knowledge Discovery and Management, 2012
ISBN 978-3-642-25837-4

Vol. 399. Kurosh Madani, António Dourado Correia, Agostinho Rosa, and Joaquim Filipe (Eds.)
Computational Intelligence, 2012
ISBN 978-3-642-27533-3

Vol. 400. Akira Hirose
Complex-Valued Neural Networks, 2012
ISBN 978-3-642-27631-6

Vol. 401. Piotr Lipiński and Konrad Świrski (Eds.)
Towards Modern Collaborative Knowledge, 2012
ISBN 978-3-642-27445-9

Vol. 402. Theodor Borangiu, André Thomas, and Damien Trentesaux (Eds.)
Service Orientation in Holonic and Multi-Agent Manufacturing Control, 2012
ISBN 978-3-642-27448-0

Theodor Borangiu, André Thomas,
and Damien Trentesaux (Eds.)

Service Orientation in Holonic and Multi-Agent Manufacturing Control

Springer

Editors
Prof. Theodor Borangiu
University Politehnica of Bucharest
Faculty of Automatic Control
and Computers
Bucharest
Romania

Prof. Damien Trentesaux
Université de Valenciennes
PSI / TEMPO Lab
Valenciennes
France

Prof. André Thomas
ENSTIB - Ecole Nationale
Supérieure des Technologies
et Industries du Bois
Centre de Recherche en
Automatique de Nancy
Epinal
France

ISSN 1860-949X
e-ISSN 1860-9503
ISBN 978-3-642-27448-0
e-ISBN 978-3-642-27449-7
DOI 10.1007/978-3-642-27449-7
Springer Heidelberg New York Dordrecht London

Library of Congress Control Number: 2012931563

Printed on acid-free paper

Springer is part of Springer Science+Business Media (www.springer.com)

Preface

The chapters of this book represent papers presented at the first edition of the International Workshop "Service Orientation in Holonic and Multi-Agent Manufacturing Control - SOHOMA'11", held in Paris at the Ecole Nationale Supérieure des Arts et Métiers during June 20–21, 2011.

SOHOMA'11 was organized in the framework of the European FP7 Project no. 264207 ERRIC, the objective of which is to foster innovation in control of sustainable manufacturing and in this context to empower excellence in research in the faculty of Automatic Control and Computer Science of University Politehnica of Bucharest, thus enhancing its national and regional leadership position in selected areas of Intelligent Information Technologies applied to manufacturing.

Service oriented technologies and management have been considered with priority by the scientific and business communities in the past years, indicating a way to create the basis for enterprise agility so that manufacturing enterprises can deliver new, more flexible business processes that exploit the value of service approach from a customer's perspective.

This book gathers contributions from scientists, researchers and industrialists on theory, concepts, methods, frameworks and implementing issues addressing trends in the *service orientation* of control technology and management applied to manufacturing enterprise. The service orientation is emerging at multiple organizational levels in business, and it leverages technology in response to the growing need for greater business integration, flexibility and agility of manufacturing enterprises.

Close related to IT infrastructures of Web Services, the Service Oriented Architecture (SOA) represents a technical architecture, a business modelling concept, a type of infrastructure, an integration source and a new way of viewing units of automation within the enterprise. Business and Process levels information systems integration and interoperability at enterprise level are feasible by considering that the customized product becomes the "active controller" of manufacturing enterprise resources. The product, seen as good by manufacturing resources and as service consumer by information systems, provides consistency between the material and informational flows within the manufacturing enterprise.

Thus, service orientation in the manufacturing domain is not limited to just Web services, or technology and technical infrastructure either; instead, it reflects a new way of thinking about processes that reinforce the value of commoditization, reuse, semantics and information, and create business value. The unifying approach proposed in the bool relies on the methodology of disaggregating siloed, tightly coupled business processes at enterprise level into loosely coupled services and mapping them to IT services, sequencing, synchronizing and automating the execution of processes which encapsulate the software description of such complex business processes related to agile production by means of distributed intelligence and information systems.

The book also demonstrates how Multi-Agent Systems (MAS) and Holonic Manufacturing Execution Systems (HMES) contribute separately or together to global sustainable manufacturing by help of patterns of active product behaviour and interaction with the working environment. Sensing abilities with wireless transmission networks and smart metering can be used to monitor the history and evolution of products, and the behaviour of resources in their environment.

The MAS technology is used to set up and maintain a resource service access record at shop floor level. *Resource service access models* (RSAM) can be configured and then updated in real time to automatically select the resources which will take part in production control: batch planning, operations scheduling, resource allocation, products routing, execution and tracking.

Judiciously defined RSAM may be used in real-time by holonic manufacturing execution systems to solve myopia in resource allocation from the point of view of both production cost and energy efficiency, thus contributing to sustainable manufacturing. On the other hand, HMES use detailed information about resource behaviour, performances and service capabilities to optimize globally (at batch level) or locally (at packet level) production.

Efficient holonic manufacturing execution systems should be able to switch between hierarchical control mode (to provide optimality) and heterarchical control mode (to provide agility and fault-tolerance), triggered by a diversity of changes: (i) resource status (operational / breakdown / recovery) or performance change (due to inflicted penalties or awards granted for high efficiency); (ii) rush orders; (iii) power consumption. Sustainability is addressed in the book with respect to: fault-tolerance, energy efficiency at resource and shop floor level, balancing resource usage and cost efficiency at batch level.

Innovative services will be enablers and drivers of growth of next generation of manufacturing enterprises that are competitive and sustainable. Some of the book's chapters target the development of software infrastructures, based on web services and semantic resources aiming at providing the appropriate support to manage energy-related devices (e.g. renewable energy sources, smart meters, etc.) by considering an environment where energy is generated, stored, distributed, and consumed in a rational and environmentally proper way.

This shall be attained through the comprehensive implementation, instantiation and distribution of computational intelligence at resource level and the development of mechanisms that enable on line control of production such as "intelligent product", "product-driven automation" and "semi-heterarchical scheduling".

A brief description of the book chapters follows.

Chapter 1 presents the Holonic Manufacturing Execution System (HMES) developed at K.U.Leuven, which uses a service-oriented approach to control manufacturing operations in real time. The chapter first explains how manufacturing control emerges from interaction between intelligent products and intelligent resources. Services play a key role in this interaction and form a decoupling point between the generic control system and application-specific elements. To illustrate that this service-oriented approach allows applying the same concepts and principles to various domains, several applications in manufacturing, open-air engineering, robotics and logistics are described. Finally, the chapter describes how supporting services, such as maintenance, can be seamlessly integrated with the core activities of the system.

Chapter 2 presents the new concept of systems characterized by a flow of active entities which run on a guided network. After the general characterization (nature, mobility, service) of an entity, the three basic components of the system are described: active and non active entity and active generalised entity - elements which are structured and implemented as holons. With the three types of basic holons and the generic services offered by them, the structural and dynamic models are formed. The framework is extended with the behavioural model consisting of the planning, scheduling, routing and execution of the operations. The control framework is completed with the implementation architecture based on intelligent embedded devices and a multi-agent programming environment for rapid configuration and integration.

Chapter 3 shows that recent developments in infotronics and communication allow the development of "intelligent" products. This chapter proposes the concept of "activeness" and a modelling approach to increase a system's informational, communicational and decisional capacities in a distributed architecture. The "activeness" concept is then applied in two case studies. The first concerns the "product-driven" control of a real manufacturing cell. The second is related to the advanced diagnosis of complex systems in a railway application.

Chapter 4 introduces a formal model to specify, model and validate hierarchical complex systems described at different levels of analysis. It relies on concepts that have been developed in the multi-agent-based simulation (MABS) literature: *level, influence* and *reaction*. One application of such model is the specification of hierarchical complex systems, in which decisional capacities are dynamically adapted at each level with respect to the emergences/constraints paradigm. In the conclusion, the main perspective of this work is discussed: the definition of a generic meta-model for holonic multi-agent systems (HMAS).

Chapter 5 reveals how the planning and coordination processes in Holonic Manufacturing Execution Systems entail the need of a centralized component, under the form of a staff holon. By using appropriate models (both monochrome and coloured Petri nets) and certain aspects of planning techniques in Artificial Intelligence, a PROSA based implementation is considered. In the obtained holonic scheme, named HAPBA, the solution of the manufacturing goals is attained by holons' cooperation, according to an enhanced form of the Contract Net Protocol, so that certain drawbacks that are possible to appear in a holonic approach are avoided. Theoretical and experimental arguments are supporting the proposed solution.

Chapter 6 In the holonic manufacturing framework, production systems must process orders starting from the basic one submitted by the client, which gives the information only about the type of the products and the quantity (and maybe the delivery date). In order to obtain the final product(s) which has been ordered, the production system must decompose each order into a set of operations based on the capability of the production system. Each entity (in this case robots) can execute a set of operations – offer a service (object assembly, part inspection, etc.), based on its attributes (execution speed, working envelope, dexterity, etc.) and resources (tools magazine, raw materials, components for assembly, vision system, network connectivity, etc.) The chapter presents a case study for intelligent robot services in holonic manufacturing and is focused on how to create an automated system which is capable to decompose the initial order in sets of operations based on the set of services the robots can offer, transform the operations into robot programs and offer high availability services.

Chapter 7 states that integration of intelligent products carrying their own information is more and more present nowadays. In recent years, some authors argued on the usage of such products for the Supply Chain Management Industry. Indeed, a multitude of informational vectors take place in such environments like fixed databases or manufactured products on which one is able to embed significant proportion of data. By considering distributed database systems, one can allocate specific data fragments to the product in order to manage its own evolution. The chapter aims at analyzing the Supply Chain performance according to different strategies of information distribution between manufactured products and fixed databases. The purpose is to determine the key factors which lead to improve information distribution performance in term of time properties.

Chapter 8 Multi-agent societies are often used in manufacturing systems and other large-scale distributed systems. Such systems often need an efficient task redistribution strategy in case of component faults or load variations. This chapter presents a simple, algorithmic approach for such a strategy, requiring low processing and communications resources.

Chapter 9 analysis the myopic behaviour inherent to holonic control architectures. Myopic behaviour is the lack of coherence among local decision-making and system's global goals. So far, holonic architectures use mediator entities to overcome this issue, bringing the holonic paradigms more toward hierarchy than heterarchy. Instead, this chapter explores the recursiveness characteristic of holonic manufacturing systems (HMS) as a possible way to correct myopic behaviour, by distributing decision-making over adjunct entities. The chapter explains the authors' approach and agent-based implementation to solve the allocation problem in a flexible job-shop. Simulation results are compared with a mixed-integer linear program to determine its efficiency in terms of makespan and execution time.

Chapter 10 discusses that current R&D directions sustained by the European Commission and by FP7 and in the near future by FP8 research programs are focusing on the development and standardization of new technologies to sustain the "Future Internet". In this context the development of new Internet-related concepts and technologies oriented towards providing positive benefits for economy has been included in a broad concept of "Future Internet Based Enterprise Systems". The Intelligent Manufacturing Systems (IMS) paradigms lead our Information Society towards New Economy-driven Knowledge Society copying with Global e-Markets new list of System of Systems requirements. The Complex, Adaptive Non-monolithic Dynamic Systems (CADS) has been defined as oriented on Virtual Enterprise (VE) architecture, as well as involved in cross-enterprise methodology like Life Long Product/Service/Organization Management (LPM/PSO) and Business Intelligence (BI). Interoperating efficiently is necessary for the Systemic Integration within Collaborative Concurrent Competitive Enterprises.

Chapter 11 analyses disassembly processes of used manufactured products which are subject to uncertainties. An optimal disassembly level minimizing the costs of these processes and maximizing the end of life components values is hard to establish. In this work, the authors propose a method to find influences and causalities between the main disassembly performance indicators in order to decide the optimal disassembly policy. The proposed model highlights the temporal dependencies between system variables and is validated using the Bayesia Lab software. In the chapter's final part the results of method implementation on a reference case study are presented to demonstrate the performance of this approach.

Chapter 12 proposes the PROSIS model (Product, Resource, Order, Simulation, Isoarchic System), as an alternative to methods fully or partially centralized for decision-making in workshop control. This model is based on the holonic paradigm and operates according to an isoarchic architecture allowing the holonic entities to use ambient services. All holonic entities contribute to the decision-making mechanism. Each holonic entity is seen as the combination of an M_holon composed with the physical part of Holon, with an I_holon, adding to it intelligence and decision-making abilities. Each I_holon is synchronized with its M_holon via infotronic technology of RFID type; it supports the status information of the Holon and access to services, allowing it to take decisions. The I_holon is hosted by an Ambient Control Entity (ACE) network, located, according to SOA, close to each resource. The ACEs provide to the Holons the *ad hoc* services they need.

Chapter 13 presents a generic framework for a Product Driven Control Systems (PDCS) dealing with production planning and control. The framework is based on Viable System Model (VSM) which is introduced for intelligent manufacturing systems. Based on suitable properties as autonomy, self-organization and adaptability, VSM allows modelling and considering these properties for PDCS. An application for a Manufacturing Planning and Control System (MPCS) is proposed. Different points of view are also presented based on this application which are shown and explained in the sense of VSM principles. Finally, a discussion is presented dealing with the main issues of the proposed approach.

Chapter 14 In human-human interaction, para-verbal and non-verbal communication are naturally aligned and synchronized. The difficulty encountered during the coordination between speech and head gestures concerns the conveyed meaning, the way of performing the gesture with respect to speech characteristics, their relative temporal arrangement, and their coordinated organization in a phrasal structure of utterance. The reported research is focused on the mechanism of mapping head gestures and speech prosodic characteristics in a natural human-robot interaction. Prosody patterns and head gestures are aligned separately as a parallel multi-stream HMM model. The mapping between speech and head gestures is based on Coupled Hidden Markov Models (CHMMs), which could be seen as a collection of HMMs, one for the video stream and one for the audio stream. Experimental results with Nao robots are reported.

Chapter 15 shows that the service orientation paradigm is particularly well adapted to distributed manufacturing systems. The difficulty of such systems' production activity control deals with knowledge management. The knowledge is distributed among entities which are able to create, modify or transmit it to other entities. An entity cannot have a full up-to-date access to all the system's data. At shop floor level, a convenient way to implement service-oriented manufacturing systems is to rely on the paradigm of Holonic Manufacturing Systems. The chapter introduces the possibility to specialize a resource holon in order to gather data from the whole holarchy and make these data available to any holon for a decision making. This holon is thus playing the role of a discrete-event observer. Upon positioning the service-oriented architectures, the HMS reference architecture PROSA is used, especially in terms of decision making. After decisions were defined, the problem of online decision making in a HMS is described, and an implementing solution for the observer and the forecasting tools in the architecture is exposed. Finally, two applications are presented, based on an industrial job-shop.

Chapter 16 is devoted to a multi-agent solution for job-shop scheduling and its current migration to a service oriented platform. The coordination mechanism for the agent population and the Decision Control Petri Nets formalism for the behaviour modelling at agent level are introduced. In the framework provided by the Generic Prototyping Approach, the chapter presents the overall orientation of the manufacturing shop control system and identifies the generic prototypes devoted to decision support for resource allocation and activity monitoring at shop floor level. Design specifications are given for the job-shop scheduling particular prototype developed according to the multi-agent paradigm. Finally, in the context of building a Grid computing platform for workflow oriented applications, the feasibility of service oriented implementation of this particular prototype is evaluated.

Chapter 17 With the growing emphasis on the triple bottom line of people, planet and profits, innovative services will be enablers and drivers of growth of next generation of manufacturing enterprises that are competitive and sustainable. This chapter presents desiderata for such services for architecting highly distributed intelligent operation of manufacturing enterprises consisting of production processes, lighting and HVAC systems, renewable energy systems, and energy storage systems to cooperatively achieve in real-time load control, and implicitly emission management. Such distributed architectures will leverage emerging developments in service oriented architectures and in interoperability standards for the smart grid. Research challenges for engineering such systems include understanding their dynamics, ensuring responsiveness, and ensuring humanistic process automation. The work extends recent developments in intelligent product-driven production by combining distributed production scheduling and power consumption scheduling. Preliminary results of an analysis and simulation of the distributed system dynamics are presented.

Chapter 18 describes modern solutions for controlling a complex manufacturing cell consisting of conveyors, robots, machine tools and other devices, using one programmable logical controller. The objective is to design a control system for a Holonic Planned Manufacturing System running into an Intelligent Production Cell. In order to achieve this goal, each product processed in the manufacturing cell has to be identified and its own routing / workstation jobs schedule needs to be implemented without interfering with other products. Two solutions were implemented: one using an offline planner that calculates the necessary data for processing products on pallets, and the other using Intelligent Electronic Devices (IED) attached to the pallets, that can take decisions by themselves in real time. A special array of data structures was used to easily access and manage the pallets processed at the four robotized stations of the cell.

Chapter 19 presents a relatively new way of modelling manufacturing systems using an extension of Petri nets formalism – Hybrid Petri nets. The formalism allows a macroscopic analysis of manufacturing systems evolution, presenting the evolution tendencies. For a better understanding, an analysis algorithm and a case study are included.

Chapter 20 presents a system control frame for implementing the switch between three different production strategies, each with its own planning objective (e.g. makespan, resource load) and perturbation avoidance capabilities. In order to implement such a system, Intelligent Products that possess computing and decision taking abilities are used, the system model and class diagram being presented for all the system agents implemented using the JADE Environment.

Chapter 21 Supply networks are still mainly based on organizations essentially centralized, dedicated and thus fragmented, whose sustainability becomes ever more problematic nowadays. The recently introduced Physical Internet tackles this problem by interconnecting all the logistics services through the encapsulation of the goods in smart modular containers. Within this framework, network adaptation with distributed routing problems take the lead over classical network design with flow assignment problems. Thanks to recent progresses made in the Digital Internet domain, decentralized approaches are foreseen to be applied for solving those problems on the large scale mandated by the Physical Internet. This leads the authors to propose an evolutionist approach to solve the Physical Internet open hub network design problem. The authors model the problem, formally introduce the design approach, analyze empirical results and provide conclusion remarks and opportunities for further research.

Chapter 22 introduces a framework for intelligent search for web services that expose the offer request management functionality for intelligent manufacturing systems. Current research is being done in the area of Semantic Web Service discovery, which involves a search engine that indexes WSDL files, and UDDI repositories. A novel concept of volunteer based search is introduced, in which the search criteria is passed to the manufacturing system to allow a self-assessment against the search criteria. This volunteer based approach is specifically designed for the cases in which the search engine cannot compare the search criteria against the target service, so it cannot determine the matching. The architecture and design of the web service provided by the manufacturing system and of a web service search engine implementing the volunteer based search concept are given. Specific issues arising from this search approach are discussed, such as time availability to decide and respond to an offer. A locking mechanism helping to overcome offer invalidation is described. The advantages of integrating the manufacturing system with SOA and BPEL processes are specified.

Chapter 23 reports on the impact of health information technology on quality of services, work efficiency and related costs of healthcare. The major benefits on quality of health services are considered to be the increased adequacy to standard based care procedures, improved surveillance and decreased medication errors. The key solution to improve the quality of the health services is considered to be the implementation of the Service-oriented Interoperability Paradigm for specifying and standardizing medical services. The chapter demonstrates that Health Level Seven (HL7) specification fits in the context of Service-Oriented Architectures (SOA) supporting enterprise distributed processing. The dynamic model of SOA Interoperability Paradigm and its compatibility with the Reference Model for Open Distributed Processing (RM-ODP) are then analyzed, in order to ensure interoperability between applications based on the HL7 Version 3 Standard. Finally the architecture of a system implementing the principles of the HSSP/HL7 SOA methodology and its specific objectives is given.

Chapter 24 presents a competency based system, developed for an IT project-oriented organization, during the implementation of a conventional Enterprise Resources Planning (ERP) system. The proposed solution is an ontology-based extension of the Human Resources module. Two competence ontologies, one for each relevant profession in the organization are parts of the ontological extension. Even if the proposed solution is based on a loose coupling between the knowledge base and the ERP system, it provides added value for ERP systems, expanding their ability to support all levels of management decision, but especially the executive ones. It is a practical and inexpensive approach for integration the semantic or knowledge oriented applications with the data-oriented or legacy systems.

Chapter 25 describes a new architecture for intelligent CNC machining in the context of a service-oriented manufacturing system involving material processing and robotized assembly operations. The proposed approach is a knowledge-based adaptive machining concept (KBAMC), which encompasses four modules: management, observation, prediction and strategy.

Intelligent Information Technologies are described in the book; they provide advanced control solutions for *Smarter Manufacturing*, and bring into discussion the state-of-the-art in sustainable manufacturing through service orientation for:

- *Dynamic and Green Infrastructure*: the creation of intelligent control infrastructures that reduce costs, are dynamic and secure, and optimize the resource usage for energy and space efficiency.
- *New Intelligence and Smart Work*: efficient data fusion using context-sensitive information processing tools; adopting legacy plant equipment into reconfigurable resource teams to increase agility to business changes.

We hope you will find useful reading this book.

September 2011-10-24

Theodor Borangiu
André Thomas
Damien Trentesaux

Contents

A Service-Oriented Approach for Holonic Manufacturing Control and Beyond 1
Jan Van Belle, Johan Philips, Osman Ali, Bart Saint Germain, Hendrik Van Brussel, and Paul Valckenaers
1 Introduction 1
2 Holonic Manufacturing Execution System 2
3 Application Domains 5
3.1 Manufacturing 6
3.1.1 Car Paint Shop 6
3.1.2 Photographic Foil Facility 7
3.1.3 Machine Tool Shop 8
3.1.4 Heat Treatment Facility 9
3.2 Open-Air Engineering 10
3.3 Robotics 11
3.4 Logistics 13
3.4.1 Chain Conveyor System 13
3.4.2 Cross-Docking Facility 14
4 Supporting Services 15
5 Conclusions 17
References 18

Service Oriented Control Framework for a Holonic System Characterized by a Guided Flow of Entities 21
Theodor Borangiu, Silviu Raileanu, Octavian Stocklosa, Christian Tahon, Thierry Berger, and Damien Trentesaux
1 Introduction 21
2 The System Characterized by a Guided Flow 23
2.1 The Concept and Characterization of Entities 23
Active entity 23
Non-active entities 23
Generalized active entity (GAE) 24
2.2 Control Architecture for an Active Entity System 25
3 Structure of the Holonic Control System 26
3.1 Static and Dynamic Models of the Composing Holons 27
3.2 Static Structure 28
Coordinator Holon (CH) 28
Resource Holon (RH) 28
System Knowledge Holon (SKH) 28

3.3 Dynamic Structure 29
Conduct Holon (COH) 29
Mobility Holon (MH) 29
Services to be Obtained Holon (SOH) 29
Object Service Receiving Holon (OSRH) 29
Flow Holon (FH) 29
3.4 Aggregation Process and Lifecycle of a FH 29
4 Behavioural Model 30
4.1 Planning Process 31
4.2 Resource Scheduling 31
4.3 Service Reception 31
5 Realization 32
6 Conclusion 33
References 33

The Augmentation Concept: How to Make a Product "Active" during Its Life Cycle 35
Yves Sallez
1 Introduction 35
2 The Concept of "Active" Product 36
2.1 Improved Performance of the Pair "Active Products - Support System" 36
2.2 Proposed Model 38
Functional view 38
Evolutionary view 39
3 Application of the Model in a Manufacturing Context 40
3.1 Product Augmentation in a Manufacturing Phase 40
3.2 Real Implementation 41
3.3 Experimental Results 43
4 Application of Our Model in a Condition-Based Maintenance Context 43
4.1 Product Augmentation in Use Phase 43
4.2 Real Implementation 45
4.3 Results 46
5 Conclusion and Prospects 46
References 47

Engineering Hierarchical Complex Systems: An Agent-Based Approach: The Case of Flexible Manufacturing Systems 49
Gildas Morvan, Daniel Dupont, Jean-Baptiste Soyez, and Rochdi Merzouki
1 Introduction 49
2 Two Trends in MABS Research 50
2.1 Multi-level Modelling 50
2.2 The Influences → Reaction Model 51
3 A Generic Meta-model for Multi-level MABS 52
3.1 Specification of the Levels and Their Interactions 52
3.2 Agent Population and Environments 53

3.3 Action Modelling 53
4 Engineering Hierarchical Complex Systems with IRM4MLS 54
4.1 The Emergence/Constraint Paradigm 54
4.2 IRM4MLS Implementation 54
4.3 Conception of Hierarchical Systems 55
4.4 Case Study: AGV Deadlocks in Gradient Field-Based FMS 56
5 Conclusion 57
References 58

HAPBA – A Holonic Adaptive Plan-Based Architecture 61
Doru Panescu and Carlos Pascal
1 Introduction 61
2 Petri Nets Modelling of Holonic Systems; Some Main Planning Issues 62
3 The Need of Holonic Centralized Components – Staff Holons 67
4 Experimental Results and Conclusions 70
References 74

Integrating Intelligent Robot Services in Holonic Manufacturing 75
Florin Daniel Anton, Theodor Borangiu, Silvia Anton, Marco Ceccarelli, and Giuseppe Carbone
1 Introduction 75
2 Decomposing Orders in Operations 77
3 Transforming Operations in Programs 79
4 High Availability Services 81
5 Conclusion 83
References 87

Key Factors for Information Dissemination on Communicating Products and Fixed Databases 89
Sylvain Kubler, William Derigent, André Thomas, and Éric Rondeau
1 Introduction 89
2 The System Characterized by a Guided Flow 91
2.1 General Data Distribution Framework 91
2.2 Distributed Databases through Literature 91
3 Case Study Presentation 93
3.1 Reference Distribution Pattern 93
3.2 Adaptation of the Logistic Process 93
4 DiPA and CoPA Architecture Modelling 95
4.1 Architecture 95
4.2 Estimated "Round Trip Times" via OPNET 95
4.3 Petri Nets: DiPA and CoPA Architectures 96
5 Results and Analysis 98
5.1 Simulation and Results 98
5.2 Key Factor Identification 99
6 Conclusion 101
References 102

A Load Balancing Algorithm for Multi-agent Systems 103
Iulia Ştefan, George Moiş, Szilárd Enyedi, and Liviu Miclea
1 Introduction 103
1.1 Generalities 103
1.2 Load Balancing 104
1.3 Agent Society 105
2 Proposed Solution 106
3 Skill Classes Awareness 108
4 The Load Balancing Algorithm 109
5 Task Agents Transfer 111
6 Distributed Security 112
7 Conclusions and Future Work 112
References 113

A Holonic Approach to Myopic Behavior Correction for the Allocation Process in Flexible-Job Shops Using Recursiveness 115
Gabriel Zambrano Rey, Nassima Aissani, Abdelghani Bekrar, and Damien Trentesaux
1 Introduction 115
2 Myopic Behaviour in Holonic Manufacturing Systems 117
2.1 Recursiveness in HMS 118
3 The Approach 118
3.1 Modelling Holons 119
3.2 Modelling Resources 120
3.3 The Allocation Method 121
3.4 Reactivity to Uncertain Conditions 122
4 Holons and Their Agent-Based Implementation 123
5 FJSP Test Case and Lower Bound Calculation 123
5.1 Mixed-Integer Linear Program (MILP) for FMS 123
6 Results 125
7 Conclusions and Future Work 126
References 127

Integrating e-IMS Platform via Interoperability within Collaborative Enterprises 129
Aurelian Mihai Stanescu, Mihnea Alexandru Moisescu, Ioan Stefan Sacala, and George Burlacu
1 Introduction 129
2 Key Requirement towards the Development of a Methodology for Future Enterprise System of Systems 130
3 From Virtual Enterprise towards Future Enterprise 134
4 Case Study Modules Operator for FInES Supply Chain 138
5 Conclusions 141
References 142

Dynamic Bayesian Network for Decision Aided Disassembly Planning 143
Luminita Duta and Sidali Ad Douche
1 Introduction ... 143
2 State of the Art ... 144
3 Dynamic Bayesian Networks ... 145
4 Proposed Model ... 147
4.1 Notations and Assumptions ... 147
5 Validation and Results ... 148
5.1 Case Study ... 148
5.2 BayesiaLab ... 150
5.3 Implementation ... 151
6 Conclusions ... 153
References ... 154

Service Oriented Architecture for Holonic Isoarchic and Multicriteria Control ... 155
Yves Dubromelle, Fouzia Ounnar, and Patrick Pujo
1 Introduction ... 155
2 Main Characteristics of PROSIS Approach ... 157
2.1 Definition of Isoarchic System ... 157
2.2 Deployment of SOA in Isoarchic System ... 158
2.3 ACE as Support System for Isoarchic SOA ... 159
2.4 Presentation of SOA in PROSIS ... 161
3 Services Provided by ACE ... 162
3.1 Hosting Service ... 162
3.2 Decision Support Service ... 164
4 Conclusion ... 167
References ... 167

Viable System Model Approach for Holonic Product Driven Manufacturing Systems ... 169
Carlos Herrera, Sana Belmokhtar Berraf, and André Thomas
1 Introduction ... 169
2 VSM for Holonic Product-Driven Manufacturing Systems ... 170
2.1 Implementation ... 171
2.2 Coordination ... 171
2.3 Control ... 171
2.4 Intelligence ... 171
2.5 Policy ... 172
3 Application to MPC Systems ... 175
3.1 Part I: Design/Virtual ... 177
3.2 Part II: Design/Physical ... 177
3.3 Part III: Implementation/Physical ... 177
3.4 Part IV: Implementation/Virtual ... 178

4 Discussion ... 178
5 Conclusion ... 180
References ... 180

Speech to Head Gesture Mapping in Multimodal Human-Robot Interaction ... 183
Amir Aly and Adriana Tapus
1 Introduction ... 183
2 Prosodic Features Extraction ... 184
3 Head Pose Estimation ... 185
4 Speech and Head Gesture Segmentation ... 188
4.1 Speech Temporal Segmentation ... 189
4.2 Gestures Temporal Segmentation ... 190
5 Speech to Head Gesture Coupling ... 191
6 Experimental Results ... 193
7 Conclusions ... 195
References ... 195

Myopia of Service Oriented Manufacturing Systems: Benefits of Data Centralization with a Discrete-Event Observer ... 197
Olivier Cardin and Pierre Castagna
1 Introduction ... 197
2 Service Oriented Manufacturing Systems ... 198
2.1 Service Orientation in the Context of Distributed Manufacturing Systems ... 198
2.2 Service Oriented Manufacturing System Example ... 199
3 HMS and SOA ... 200
3.1 PROSA Modelling ... 200
3.2 PROSA in SOA ... 201
4 Centralizing Data, Not Decisions ... 202
4.1 Gathering an Up-to-Date State of a HMS ... 202
4.2 Integration in HMS ... 203
5 Applications ... 204
5.1 Application to Decision DG2 ... 204
5.2 Application to Decision DL1 ... 205
Problem Definition ... 206
Scenarios ... 207
Results and Discussion ... 207
6 Conclusion and Future Works ... 209
References ... 209

A Multi-agent Model for Job-Shop Scheduling ... 211
Gabriel Neagu
1 Introduction ... 211
2 Building Blocks of the Multi-agent Solution ... 212
2.1 MACOR - a Multi-agent Co-ordination Mechanism ... 212

2.2 The DCPN Formalism 214
3 Generic Prototyping in Manufacturing Control 214
3.1 Methodological Framework 214
3.2 DSMC_A Generic Prototype 216
4 Job-Shop Scheduling Particular Prototype 217
4.1 Design Specification of the Control Model 217
4.2 A Case Study: The Job Object Class 219
5 Towards a Service Oriented Implementation 222
5.1 PEGAF Platform 222
5.2 Implementation Feasibility 223
6 Conclusions 224
References 224

Services for Competitive and Sustainable Manufacturing in the Smart Grid 227
Vittaldas V. Prabhu
1 Introduction 227
2 Desiderata of Services 230
3 Simulation-Based Distributed Feedback Control 233
4 Simulation Results 236
5 Conclusions 238
References 239

Different Approaches Regarding the Operational Control of Production in a Flexible Manufacturing Cell 241
Nick Andrei Ivanescu, Mihai Parlea, and Andrei Rosu
1 Introduction 241
2 System Architecture and Production Flow 242
3 First Control Solution and PLC-Based Implementation of Order Holons 243
3.1 Theoretical Backgrounds 243
3.2 Project Structure 244
3.3 The Routing Challenge 246
4 Failure and Perturbations Management 247
4.1 Case 1: Failure / Recovery of a Resource 248
4.2 Case 2: Re-supplying Mechanism 248
5 Second Approach: Using Intelligent Products 249
5.1 Routing a Pallet 251
6 Communication Inside the System 252
7 Practical Results and Conclusions 253
References 253

Using Hybrid Petri Nets for Performance Analysis in Manufacturing Systems **255**
Calin Munteanu, Simona Caramihai, Mihnea Alexandru Moisescu, and Ioan Stefan Sacala
1 Introduction 255
2 Continuous Petri Nets 255
3 Hybrid Petri Nets 259
4 Modelling Manufacturing Systems with Hybrid Petri Nets – Case Study 260
5 Analysis 263
6 Conclusions 264
References 264

A JADE Environment for Product Driven Automation of Holonic Manufacturing **265**
Silviu Raileanu, Mihai Parlea, Theodor Borangiu, and Octavian Stocklosa
1 Introduction 265
2 Using Intelligent Products for Decision Taking in an Industrial Environment 266
2.1 Control System Model 266
2.2 Production Driving Strategies 268
2.3 Using the CNP to Obtain Workstation Offers and Make Reservations 270
2.4 Comparing Robot Offers 271
2.5 Planning the Route 272
2.6 Taking the Production Decision 273
3 Product Driven Automation 273
4 Software System for Implementation Using the JADE Environment 275
4.1 Message Exchange System 275
4.2 System Classes and Implementation 276
5 Conclusions 277
References 277

Physical Internet Enabled Open Hub Network Design for Distributed Networked Operations **279**
Eric Ballot, Olivier Gobet, and Benoit Montreuil
1 Introduction 279
2 A New Logistics Paradigm: The Physical Internet 280
2.1 Motivations for a New Logistics Paradigm 280
2.2 The Physical Internet 281
3 Physical Internet enabled open hub network design 283
3.1 Network Design and Assumptions 283
3.2 Typical Logistics Network Optimizations Problems 284
3.3 Open Hub Network Design for the Physical Internet 284

4 Open Hub Network Design for Physical Internet Proof of Efficiency Purposes ... 286
4.1 A Need of Proof of Efficiency by Simulation of Decentralized Design ... 286
4.2 Open Hub Networks Design by an Ad Hoc Evolutionist Algorithm ... 286
4.3 An Open Hub Network for Food Distribution in France ... 288
5 Conclusion and Future Work ... 290
References ... 291

Volunteer Based Search Engine for Holonic Manufacturing Services ... 293
Cristina Morariu, Octavian Morariu, and Theodor Borangiu
1 Introduction ... 293
2 System Architecture ... 295
3 Holonic Manufacturing System Web Service Design ... 298
4 Volunteer Based Search Engine ... 299
Phase I – Semantic Search ... 299
Phase II – Volunteer Challenge ... 300
5 Conclusions and Future Work ... 303
References ... 305

Impact of Information Technology on the Quality of Health Services ... 307
Radu Dobrescu and Victor Purcarea
1 Introduction ... 307
2 Major Effects of Health Information Technology Implementation ... 308
2.1 Effects on Quality of Medical Care ... 308
2.2 Effects on HIT Efficiency ... 308
2.3 Effects on Costs ... 309
3 The Impact of Service Oriented Architecture on Health Information Systems ... 309
4 Service-Oriented Solutions for Healthcare ... 310
5 The HSSP/HL7 SOA Interoperability Paradigm ... 311
6 State of the Art and Trends of HL7 Implementation ... 313
7 HL7 Integration ... 317
8 Conclusions ... 318
References ... 318

Competency Management System for IT Project-Oriented Organizations ... 321
Constanta-Nicoleta Bodea and Robert Buchmann
1 Introduction ... 321
2 Literature Review ... 324
3 The Proposed Solution for the Competency Management System ... 325
3.1 The Architecture of the Competency Management System ... 327
3.2 The Project Management Competence Ontology ... 328

3.3 The IT Competence Ontology 329
4 The Experimentation of Competency Management System 330
5 Conclusions and Future Work 332
References 333

Knowledge-Based Adaptive Machining Concept for Service Oriented Architectures **335**
Alexandru Dumitrache, Theodor Borangiu, Sylvain Pateloup, and Grigore Gogu
1 Introduction 335
2 Overview of the Knowledge-Based Adaptive Machining Concept 336
3 Management Module 337
4 Observation of the Milling Process 338
5 Adaptive Machining Strategy 341
5.1 Algorithm Overview 342
5.2 Core Algorithm: Advancing with Constant Engagement 343
5.3 Examples of Generated Toolpaths 344
6 The Prediction Module 345
7 Conclusions 346
References 346

Author Index **349**

Subject Index **351**

A Service-Oriented Approach for Holonic Manufacturing Control and Beyond

Jan Van Belle, Johan Philips, Osman Ali, Bart Saint Germain, Hendrik Van Brussel, and Paul Valckenaers

Department of Mechanical Engineering, Katholieke Universiteit Leuven, Leuven, Belgium
{jan.vanbelle,johan.philips}@mech.kuleuven.be,
oali@email.com, {bart.saintgermain,hendrik.vanbrussel,
paul.valckenaers}@mech.kuleuven.be

Abstract. The Holonic Manufacturing Execution System (HMES), developed at K.U.Leuven, utilizes a service-oriented approach to control manufacturing operations in real time. This chapter first explains how manufacturing control emerges from interaction between intelligent products and intelligent resources. Services play a key role in this interaction and form a decoupling point between the generic control system and application-specific elements. To illustrate that this service-oriented approach allows applying the same concepts and principles to various domains, several applications in manufacturing, open-air engineering, robotics and logistics are described. Finally, the chapter describes how supporting services, such as maintenance, can be seamlessly integrated with the core activities of the system.

Keywords: PROSA, MES, intelligent product, delegate MAS, cross-docking.

1 Introduction

The control of manufacturing operations is a complex task. In coordinating the operations and resources, the production manager has to deal with uncertainties (e.g. variable processing times) and unexpected events (e.g. late deliveries or machine breakdowns). Moreover, several (possibly conflicting) objectives have to be taken into account and because of the dynamic nature of manufacturing, 'one-shot optimization' is not sufficient and manufacturing control is a going concern.

The production manager can be supported by a Manufacturing Execution System (MES), which coordinates and controls the manufacturing system in real time. Manufacturing control is the operational level of production planning and control and is concerned with the short-term and detailed assignment of operations to production resources [38].

For several years now, researchers of the Department of Mechanical Engineering at K.U.Leuven are working on a holonic MES or HMES that handles the internal logistics in a manufacturing system. This HMES tries to improve the responsiveness, proactiveness, scalability and flexibility of the system and handles changes and disturbances as business-as-usual [28].

T. Borangiu, A. Thomas, and D. Trentesaux (Eds.): SOHOMA 2011, SCI 402, pp. 1–20.
springerlink.com

In the HMES, the manufacturing control emerges from the interaction between intelligent products and intelligent resources. McFarlane et al. [12] define an intelligent product as follows:

An intelligent product is a physical and information-based representation of a product which:

1. possesses a unique identity;
2. is capable of communicating effectively with its environment;
3. can retain or store data about itself;
4. deploys a language to display its features, production requirements, etc.;
5. is capable of participating in or making decisions relevant to its own destiny.

So, an intelligent product does not only know and communicate its status (1-3), it also makes decisions relevant for its own destiny (4-5). In the HMES, an *intelligent product* searches actively for the services needed to produce its corresponding product and has to decide which services fit its needs best. These services are offered by the *intelligent resources*, corresponding to the factory resources (equipment, staff, etc.).

Services can be described by means of (complex) ontologies, but in many cases it suffices to describe services by their capabilities and constraints. For instance, the service offered by a truck can be described by the transport capability and by constraints like the maximum volume and weight of the load.

Because of this service-oriented approach, it is possible to apply the concepts and principles of the HMES to other application domains than manufacturing. It suffices to have intelligent resources that offer the necessary domain specific services for the intelligent products. The intended control emerges from the cooperation between the intelligent products and the intelligent resources, all equipped with appropriate decision making mechanisms. This will be illustrated by giving an overview of the HMES applications in different domains. This article will also discuss how the organization of supporting services (e.g. maintenance activities) can be easily integrated with the production activities.

The remainder of the article is organized as follows. The next section describes the Holonic Manufacturing Execution System and the interaction between the intelligent products and intelligent resources. In Section 3, an overview is given from several applications of the HMES concepts in various domains. Section 4 describes how supporting services can organized together with the core activities. This will be illustrated with some examples from the described applications. In the last section, conclusions are drawn.

2 Holonic Manufacturing Execution System

The Holonic Manufacturing Execution System developed at K.U.Leuven is based on the Product-Resource-Order-Staff Architecture (PROSA) [36]. This reference architecture is developed in accordance with the holonic manufacturing paradigm (see e.g. [3, 13, 15, 36]) and describes the structure of the software: the components, their responsibilities and their interactions. The HMES is implemented as a multi-agent system (MAS), a detailed description of its software architecture can be found in [39].

Fig.1 shows the main components of PROSA. A resource holon corresponds to a resource in the underlying system and offers the services of that resource to the other holons. A product holon corresponds to a product type. It contains the knowledge on how instances of its type can be produced by the resources (what services are required). An order holon represents a product instance that needs to be produced. It is responsible to handle the required resource allocations to accomplish this. Therefore, the order holon consults its corresponding product holon to find out what services it needs and searches for the proper resources to accomplish it.

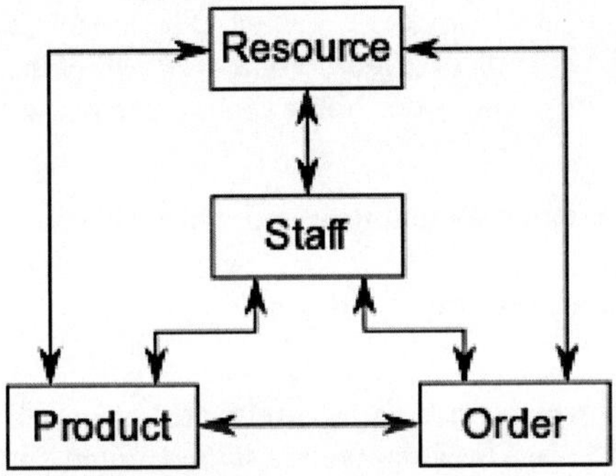

Fig. 1 Overview of the PROSA reference architecture

PROSA separates two different concerns. First, product and resource holons address technical feasibility (making the products in a technically correct and validated manner). Product holons are knowledgeable concerning the capabilities and services of resources that are relevant to them, but they ignore resource capacity and availability aspects. The product holons inform the order holons about (all) technically correct manners to produce their product instance.

Second, order and resource holons address the issue of resource allocation and product flow coordination. Within their solution space, identified by the product holons in cooperation with the resource holons, the order holons arrange the resource allocations and the execution of process, storage and transportation services to get their product instance produced. In view of the complexity of resource allocation, staff holons may assist through advice but the order and resource holons take the final decisions. Staff holons are optional, more information can be found in e.g. [36, 39].

Resources are treated as first class citizens in the overall software system, which implies that all resource entities in the environment which will be used by other holons should have their own separate software component. This is the way to assure that an appropriate reflection of reality is maintained at one single instance. Additionally, placing knowledge of entities used in an application at the entities themselves, rather than spreading this knowledge across multiple components, simplifies software design. This paradigm shift toward explicit resource awareness and allocation has led to insights for handling scalability and flexibility issues most other approaches suffer from. In fact, resource allocation should be a key software design pillar. Explicitly taking resources into account and treating them as separate entities, facilitates resource reusability.

A straightforward implementation of PROSA will handle product routing and resource allocation as if products are cars in (congested) traffic, deciding about the next processing step and resource allocation while trying to guess what the others are about to do. Here, significant margin for improvements remains.

However, PROSA can be accompanied by a bio-inspired coordination mechanism, enabling the holons to achieve improvements by generating the required information: delegate MAS or D-MAS [10, 11, 28]. The PROSA holons create (many) lightweight agents to perform some task on their behalf. The resource and/or product holons use feasibility D-MAS to place *digital road signs* showing technically correct routings to the order holons. The order holons make use of two D-MAS types to:

- Find feasible routings, accounting for resource availability and capacity, capable of fulfilling their task.
- Propagate the order's intentions, informing resource holons about future visits (allocations).

The resource holons maintain an agenda in which future allocations are registered based on the information from the second D-MAS. This agenda is accounted for when lightweight agents from the first D-MAS interact with the resource holon while constructing their candidate routings. The resource's agenda corresponds to a short-term forecast of the load of the resource, while an order's intention is a short-term forecast of its production.

To be able to generate these forecasts, the lightweight agents can make use of the *what-if* functionality of the holons. As the basic holons reflect a part of reality, they have to be able to answer what-if questions about (the dynamic behavior of) their corresponding reality. Therefore, every holon contains a model that supports virtual execution.

In the current implementation, multi-models are used, which allow using different models with different formalisms and different levels of detail [9, 37].

During the development of PROSA and D-MAS, the research team has minimized the introduction of constraints on the applicability of its results. Basically, the technology can be applied to applications with the following characteristics:

- Activities are executed on resources and are subject to technical constraints.
- Virtual execution of these activities in a digital mirror image of the world-of-interest is possible and can be much faster than in reality.
- The socio-economic value of enhanced coordination more than compensates the cost and effort for the virtual execution.

Fig.2 shows a *rephrasing* of PROSA to account for a wider and more abstract or generic application domain. An order holon now corresponds to a task that has to be executed, and does not have to represent a physical product instance.

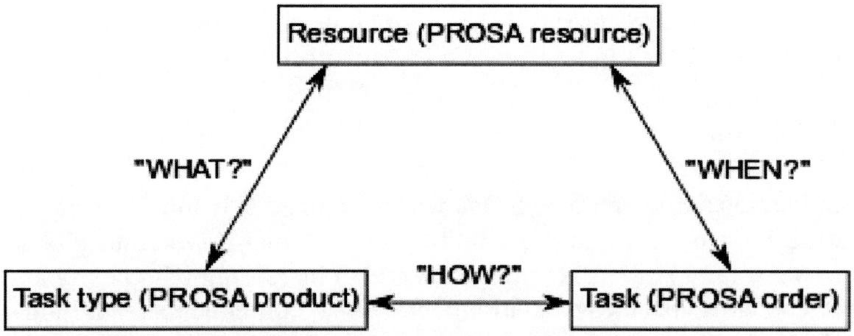

Fig. 2 Rephrasing of PROSA for a generic application domain

Section 1 mentions that the manufacturing control emerges from the interaction between intelligent products and intelligent resources. The intelligent products search for the services required to produce their corresponding products, while the intelligent resources offer these services. An intelligent product corresponds to a PROSA order holon (together with its corresponding PROSA product holon), looking for services to get its product instance completed. Or, after rephrasing for a more generic application domain, looking for services to get its task fulfilled. An intelligent resource corresponds to a PROSA resource holon, offering services to the intelligent products.

According to Valckenaers et al. [25, 29], an intelligent product is the combination of an intelligent agent and an intelligent being. The intelligent being reflects the corresponding product instance or task and is restricted to provide functionality and services for which the corresponding reality provides adequate protection. Any functionality or service that requires decision-making is delegated to the intelligent agent. Similarly, an intelligent resource consists of an intelligent agent and an intelligent being. The intelligent agent takes (local) decisions about the resource, while the intelligent being reflects the corresponding real world resource. So, it is the intelligent being that contains a model to offer the aforementioned what-if functionality of the holons.

The services offered by the resources form a decoupling point between the generic holonic system and application-specific elements. The behaviour of the order and product holons is generally applicable (but with specific plug-ins for decision making), while the intelligent resources offer domain-specific services and contain a model of the domain-specific behavior of the corresponding resources. These models allow the intelligent products to see the effects of a service (e.g. the duration), without having to deal with domain-specific properties.

3 Application Domains

As mentioned above, the HMES concepts are more general applicable than just manufacturing. This section gives an overview from several cases where the concepts and principles of the HMES have been applied. First of all, several manufacturing applications are described, but then applications in the domains of open-air engineering, robotics and logistics are shortly discussed.

3.1 Manufacturing

The original application for which the HMES was developed is manufacturing. As the manufacturing industry became more and more customer-driven and globalized, companies were (and still are) confronted with an increasing complexity and a continuously changing business environment. As a consequence, additional functionality like robustness, scalability, etc. was required, which could not be provided by the conventional production planning and control approaches [6, 13]. So, the HMES was developed as a distributed, flexible and self-organizing manufacturing control approach to deal with these issues.

During the years, the HMES concepts are applied to several manufacturing cases which are now shortly described. In these applications, the intelligent products correspond to physical products, which have to be processed by the available manufacturing resources. The resources are represented by intelligent resources, offering the manufacturing operations as services to the products. The various applications have been tested on a (fine-grained) emulated version of the manufacturing system.

3.1.1 Car Paint Shop

To test the applicability of the HMES for flexible *flow shops*, one of the first applications addressed a car body paint shop. This large paint shop (comprising six floors) paints more than 1000 cars each day and comprises more than 400 manufacturing resources: unidirectional and bi-directional conveyors, turning tables, lifts, processing units for painting, etc. These resources are arranged in a complex topology, in which loops are present. The system has built-in redundancy, i.e. for each processing step multiple resources can be chosen. Similarly for the transportation, more than one routing option is available to move a car body from one processing unit to the next. As the result of a production step is uncertain, the next processing step for a car body will depend on the outcome of the previous one. This means that it is sometimes necessary that a product makes a loop through the paint shop. The main performance measure in this paint shop is throughput. The throughput can be influenced by the batch size, and throughput losses are caused by colour breakdowns on the painting lines and blockages on the transportation system.

The control system is responsible for the routing of the car bodies through the paint shop and has to maintain the required throughput in the face of disruptions. Because of loops in the transport system of this flexible flow shop, the control system also has to deal with deadlocks. Therefore, the intelligent products (corresponding to the car bodies) use a layered decision mechanism to choose their next processing step. The first control layer addresses feasibility. This layer is responsible for deadlock avoidance and assures for instance that a car body is not transported in a direction which lacks the necessary processing capabilities. The second layer handles production goals like maximizing throughput or respecting due dates. A third layer can provide advisory information. These layers are application specific and can be easily replaced if necessary.

The control system is also responsible for the batching of the car bodies for the painting process. Small batch sizes lead to more setups and so a lower throughput. Moreover, as batches are small, there are more defects and so more car bodies have to be repainted, lowering the throughput even more. To deal with this, the intelligent resources corresponding to the painting equipment propagate information about their planned batches (size, colour, time window, etc.). The intelligent products can use this information to decide to join a certain batch.

More information on this application can be found in e.g. [17, 27, 30, 35].

3.1.2 Photographic Foil Facility

A second manufacturing application dealt with the fabrication of photographic products out of large rolls of photographic foil. Such a master roll first has to be split lengthwise. This 'slitting' operation divides the master roll into reels. Depending on the required final product (a stack of sheets or a small roll of photographic foil), these reels have to be cut or rewind and cut. As the material of a master roll is expensive (silver is an important ingredient), losses on the master roll have to be reduced as much as possible. Therefore, several customer orders have to be made out of one master roll, and make-to-stock orders are introduced to fill the remaining gaps. The production floor of this case study is organized as a *flow shop*, but with some flexibility in the routing and storage possibilities (see Fig. 3). Common disturbances include planned and unexpected maintenance, which results in variations in production speeds. More seriously, some disturbances damage the photographic foil and it is possible that a customer order cannot be produced out of this master roll anymore.

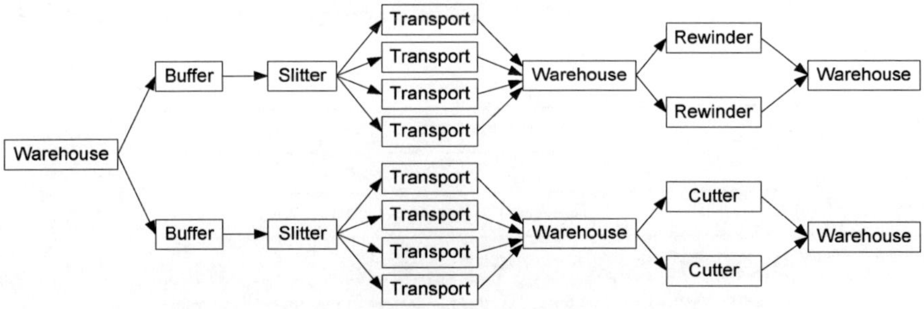

Fig. 3 A schematic representation of the layout of the photographic foil facility

The HMES is responsible for the correct and timely production of the customer orders. This is not straightforward, as the manufacturing processes are in fact disassembly operations in which one product turns into multiple products when it is cut into smaller pieces. The control system also has to deal with disturbances. For instance, when a master roll is (partly) damaged and some orders cannot be produced out of this roll, the control system has to make sure that these orders can be made out of another master roll. Also, as the foil is expensive, the damaged parts should be recycled as much as possible.

The various machines on the factory floor, like slitters, cutters and rewinders (which are also able to cut the rewinded piece), are represented by intelligent resources. There are intelligent products associated with the different physical production parts like master rolls, reels and small rolls. These intelligent products all have a reference to their corresponding 'parent', for instance the intelligent product associated with a reel has a reference to the intelligent product of the master roll from which it is slit. In this way, a change in for instance the due date of a customer order will by 'seen' quickly by all involved agents.

In e.g. [20, 38], more details about this case study can be found.

3.1.3 Machine Tool Shop

Another application addressed a machine tool shop in which components of weaving machines are produced. The shop floor is organized as a *job shop* with a central automated storage and retrieval system (see Fig. 4). This AS/RS consists of a storage area and an automated rail-based transporter to pick up and drop off loads at the various workstations (CNC machine, milling workstation, etc.). The components are transported in containers. Each container contains a variable number of identical components, travelling together until completion. At the workstations, the components of a container are processed one by one and put in another (empty) container. When all components are processed, the transporter is prompted to bring the container to the storage area. The transporter can carry two containers at the same time. So, before moving to a workstation to pick up a container, the transporter can travel to the storage area to take the container that has to be processed next at that workstation. In this way, an additional movement of the transporter is avoided. Most of the processing steps (e.g. sawing, milling, turning, etc.) can be carried out by several alternative workstations, but possibly with different processing times.

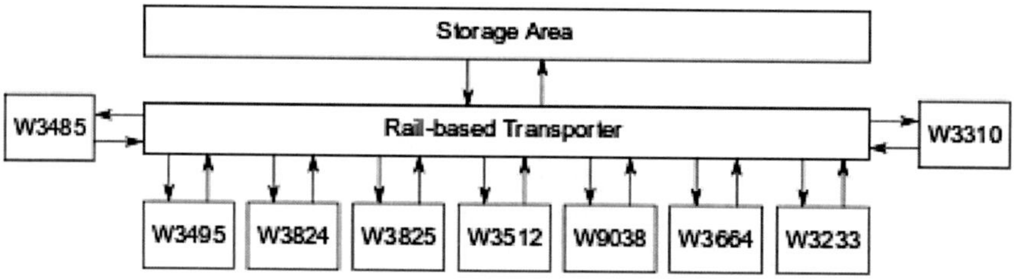

Fig. 4 A schematic representation of the layout of the machine tool shop

The control system has to organize the production by routing the containers - represented by intelligent products - through the machine tool shop. The various (intelligent) resources (workstations, storage area, transporter ...) offer their operations as services to the intelligent products in order to accomplish this. The considered performance criteria are: the increase of throughput, lead-time minimization and the improvement of labour and resource utilization. Another important issue for the control system is the optimization of the use of the transporter. During periods of heavy demand for transportation (rush hour), the transporter is a bottleneck and causes workstations and operators to idle, which is expensive.

More information about this application can be found in e.g. [28, 37, 44].

3.1.4 Heat Treatment Facility

A last manufacturing case study addressed a highly automated heat treatment facility. This facility performs heat treatment of metallic materials. Several processes can be executed: case hardening, vacuum hardening, induction heating, etc. The products demand a certain temperature trajectory inside the furnaces in order to reach the required quality. Also, the time between different processes (for instance between case-hardening and tempering) should not be too long for some products. The various furnaces differ from each other in the range of working temperature and environmental condition (e.g. carbon level). The facility is organized as a *job shop* in which the baskets containing the metallic parts are transported automatically. Fig.5 shows the temperature profile and corresponding resources of the case hardening process.

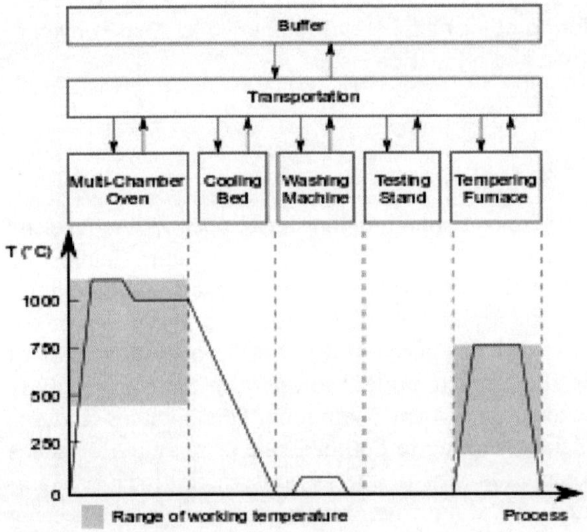

Fig. 5 Case hardening

The control system is responsible for the routing of the metallic parts that have to be treated through the facility and has to make sure that the parts receive the correct treatment in order to obtain the required quality. The intelligent resources correspond to the transportation and heat treatment equipment (e.g. furnaces, washing stations and cooling beds). The services offered by these resources are used by the intelligent products, corresponding to metallic parts that have to be treated in the facility.

Specific for this application is that parts with compatible process temperature trajectories and environmental conditions can be batched. This batching, when properly executed, has significant impact on the performance of this capital-intensive production system. Indeed, a fully loaded furnace and a partially loaded one operate almost at identical cost whereas the output differs significantly. The intelligent products can make use of a delegate MAS to discover batching opportunities or, alternatively, to trigger the build-up of such batches.

This case study also investigated the scalability of the HMES concepts by coordinating manufacturing and transportation activities within networked production. A virtual enterprise was considered, consisting of a network of heat treatment factories. New companies can dynamically join or leave the network and new processes and equipment are introduced as needed. Now the intelligent products have to route their corresponding parts at two levels: the network level and the factory level. At the network level, the intelligent product searches for transportation services between the different factories and heat treatment services (offered by aggregated intelligent resources, offering all services of the resources at a factory). As such a virtual enterprise is a semi-open system, lacking a single command and control centre, the operations have to be organized without the disclosure of sensitive information to other members of the network. Also, a mechanism is required to deal with trust and reputation issues.

More details about this application are in e.g. [21, 22].

3.2 Open-Air Engineering

The concepts of the HMES can also be applied to coordinate open-air engineering processes such as open-pit mining, road construction and agriculture harvesting (see Fig. 6). These processes are usually carried out with high-tech mobile equipment (e.g. excavators, dump trucks, asphalt layers, road graders) that needs to cooperate in order to execute the processes successfully. As the operating costs of the work vehicles are considerable, it is important to optimize their productivity through proper planning and execution of their operations. This involves resource allocation and scheduling decisions, aiming to optimize one or more performance objectives (e.g. minimizing completion time or energy consumption). The dynamics in the open and distributed operating environment of open-air engineering processes make this planning complex.

Current approaches see this problem as a resource constrained project planning problem for which a large number of mathematical and heuristic techniques have been developed. However, most methods that have been developed are ad hoc for specific planning problems as they concentrate on particular aspects of the specified problem.

(a) Open-pit mining (b) Road construction (c) Agriculture harvesting

Fig. 6 Open-air engineering processes

In general, these methods focus on the assignment of process activities to available resources (see for instance [7, 19, 42]). The planning is performed off-line before the process starts. This means that the plan needs to be revised at the time of execution when there are changes in the operating environment. Iteratively switching between planning and execution could address this issue. However, changes in the successive plans often make the planning unusable in practice [8, 23].

For open-air engineering processes, the intelligent resources correspond to the work vehicles, as well as to stationary physical entities (e.g. storage bins for excavated product). These intelligent resources offer domain-specific services such as excavating, harvesting, transporting, etc. Also, these resources contain models of their corresponding reality that encapsulate the domain-specific aspects. These models are used to make short-term forecast, for instance to predict when a storage bin's capacity will be reached. The processes that have to be carried out are the intelligent products, looking for services from the intelligent resources to get their task executed.

Specific for this case is that a product sometimes needs multiple services and resources at the same time (multi-resource allocation). For instance, in open-pit mining, to mine a certain area, the service of an excavator is required, together with the service of a dump truck to transfer the excavated product.

For more details, the interested readers are referred to [2, 26].

3.3 Robotics

Applying the HMES concepts to robotics led to research in multi-robot coordination applications. Consider a set of robots navigating in the same environment, each having its own goal location. The robots should autonomously move around and use range sensors to detect and avoid obstacles. Navigation should be smooth and interference with other robots or humans, also present in the environment, should be minimized.

A possible scenario in which this can be useful is a hospital or retirement home where a limited number of robotic wheelchairs should provide autonomous navigation to patients or inhabitants. These users would request a wheelchair (through some interface) and the robot would then navigate autonomously to the user. After the user is assisted into the wheelchair, a target location is given to which the robot has to navigate. The benefit of this approach is that medical staff is only required when the user wants to mount or dismount the wheelchair. While navigating, the robot autonomously finds its way and is able to avoid obstacles using its range sensors. In this scenario, the need for smooth navigation and low interference is apparent. Minimizing the patient's discomfort is a key criterion for a successful application. In a more industrial context, this application would be useful to allow a set of autonomously guided vehicles (AGVs) to transport goods from one location in a warehouse to another.

Fig.7 shows a suboptimal path execution for this scenario, but in a *robocentric* approach. Each robot executes its own task, assuming the environment is implicitly allocated for its needs and not being aware that other users might be present. Users can either be humans moving around or other robots executing a task. Each room is connected to two narrow corridors and depending on the target, going through one corridor is more efficient than going through the other. This scenario can cause livelocks, provided the number of robots is high. A more common problem however is that the robots do not follow an optimal path to move from one room to another. If two robots enter the same corridor and are not aware of each other's intention, they will most likely replan their route through the other corridors in order to reach their target. This results in a longer travel path.

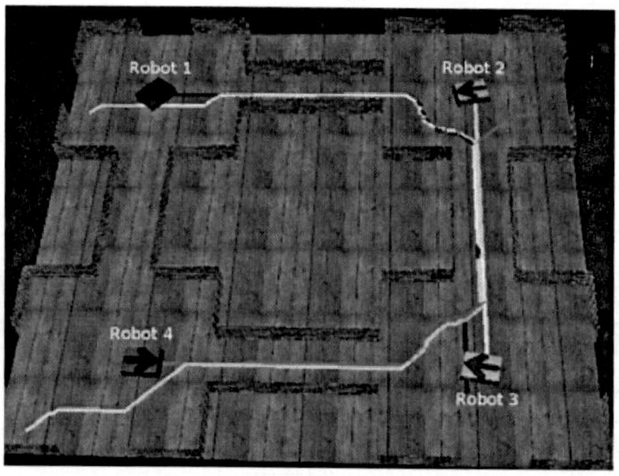

Fig. 7 Example scenario which can lead to suboptimal path execution. Robots are represented by rectangular shapes and their desired paths by a solid line. The configuration is as follows: robot 1 moves to bottom right, robot 2 to bottom left, robot 3 to top left and robot 4 to top right

The main contribution of using the HMES concepts in the context of robotics is that rooms and corridors are represented by intelligent resources, and are thus treated as first class citizens in the overall software system. Most other robot software architectures (such as [1, 16]) on the other hand adopt a functional decomposition and the representation of the environment is spread over the different control systems (each robot maintains its representation). The structural decomposition proposed here improves scalability and flexibility, since explicit resource allocation allows taking other robots' intentions into account. For instance, when several robots with individual capabilities operate in the same environment, implicitly sharing it with each other, interference between robots should be minimized.

Another contribution toward the robotics domain is the introduction of short-term forecasting in multi-robot navigation. The delegate MAS provides a way to adopt the robot's behaviour in such a way it optimally takes into account future tasks or conflicting tasks of other robots. Consider for instance a small corridor, only wide enough for one robot to pass simultaneously. Turning this corridor into an intelligent resource with explicit allocation allows forecasting whether or not it is opportune to navigate through this corridor.

In the robotics context, the intelligent products correspond to the tasks a robotic platform or a fleet of robots need to execute. In the navigation scenario, this boils down to a sequence of navigation operations to move from one location to another. All physical entities relevant to execute a particular task are represented by intelligent resources, e.g. robotic platforms, sensors in the building, doors, corridors, etc. Representing a physical robot by an intelligent resource allows including the robot's available services, such as navigation or manipulation.

PROSA can also be used in robotics at a higher abstraction level to coordinate underlying robot components, provided that PROSA's separation of concerns is adopted. Thus, state-of-the-art robot software is still usable and can benefit from PROSA's advantages. More information on this application can be found in [18].

3.4 Logistics

The concepts and principles used in the HMES can not only be applied for manufacturing, but also in the larger context of the supply chain. Here, we shortly describe two applications in the domain of transport and logistics: the control of a chain conveyor system and a cross-dock.

3.4.1 Chain Conveyor System

Chain conveyors are often used for the internal transport of goods, for instance in distribution centres. In many cases, several chains are connected to each other (by means of diverters) to form a complex transportation network (see Fig. 8). The control system has to decide about the routes that products follow and when these products are transported. Moreover, the control system has to deal with uncertainties and disturbances (e.g. defect charts, delays, jammed chains, etc.). Currently, chain conveyor systems are controlled statically. Routing tables determine the route for each product type. These tables are only adapted when serious changes happen, for instance when the product mix changes seriously. As all products from the same category follow the same route through the system, the control system has no flexibility and cannot react to disturbances.

By applying the HMES concepts, the control can be made more flexible and dynamic. The different components of the chain conveyor system (e.g. the chains and diverters) are represented by intelligent resources which have a model of the behaviour of the corresponding component. Such a model of a chain for instance can forecast at what time a chart will reach a certain position. The intelligent products correspond to the goods that have to be transported by the chain conveyor system, and they search for the necessary services like transporting and

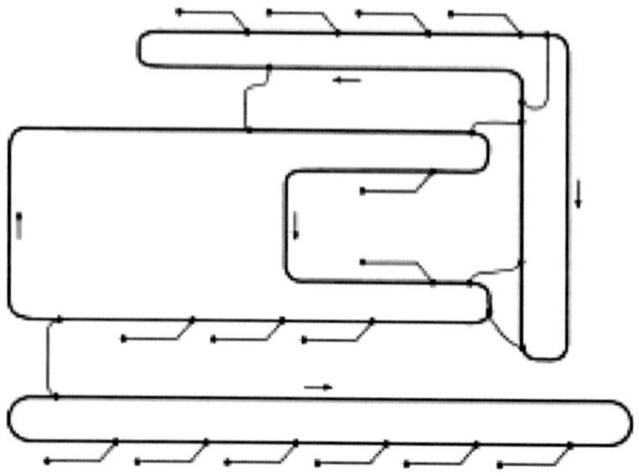

Fig. 8 An example of a chain conveyor network

switching to get their corresponding product at the required destination. In contrast to the static approach, each product can now follow an individual route through the conveyor system and can react to disturbances such as a jammed chain (responsiveness). The short-term forecasts even allow to anticipate certain disturbances (e.g. a congested chain) and to find an alternative route (proactiveness).

More details can be found in [33, 34].

3.4.2 Cross-Docking Facility

Cross-docking is a logistic strategy in which incoming shipments are (almost) directly transferred to outgoing trailers, with little or no storage in between (see Fig. 9). If the shipments are temporally stored, this should be only for a short period of a time. An exact limit is difficult to define, but usually 24 hours is assumed to be the limit. Cross-docking can have several advantages: the consolidation of shipments, shorter delivery lead times, the reduction of costs, etc. However, the organization of the cross-docking operations is a complex and challenging task, certainly because the arrival and departure times of the inbound and outbound trucks need to be synchronized. Moreover, cross-docks operate in an uncertain and dynamic environment, among others due to a tough competition in the transport and logistics sector and ever-increasing traffic.

The current approaches to control a cross-dock are usually planning approaches, in which the plan is made off-line before the operations start. While in practice several operational decisions have to be taken together, these approaches are usually concerned with just one decision, for instance how will the trucks be assigned to the available dock doors (see e.g. [4, 5, 14, 24, 43]), at which location will the goods be temporary stored (e.g. [40]), how will the vehicle routing be performed (e.g. [41]), etc. Moreover, the current approaches usually assume that all necessary information (e.g. the exact content and arrival time of the incoming trucks) is fixed and known on beforehand. Also, the problems are usually assumed to be static, while the control of a cross-dock is inherently dynamic (trucks arrive

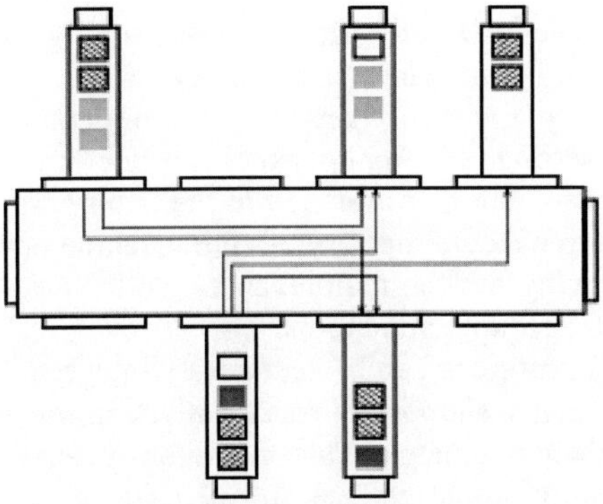

Fig. 9 Material handling at a typical cross-dock

early or late, equipment fails, etc.). Consequently, a more robust and dynamic control approach is desirable.

At present, research has started to apply the concepts and principles used in the HMES to control a cross-docking terminal. For every resource in the world-of-interest, there has to be a corresponding intelligent resource. In the context of cross-docking, this means that for instance all trucks, forklifts and dock doors will have an intelligent resource. This intelligent resource contains a model of the dynamic behavior of the corresponding real-world resource so that what-if questions can be answered. It is also responsible for its own local decisions (e.g. a dock door should decide which truck it handles). All goods that have to be handled in the cross-dock are represented by intelligent products. These intelligent products are responsible for routing their corresponding entities through the cross-docking system. Therefore, they can make use of the available services offered by the intelligent resources, such as loading or unloading, internal transportation, temporary storage, etc.

Also for this application, multi-resource allocation is an issue. For instance, when goods have to be unloaded from a truck, these goods require the unload service from a forklift (and a driver), while at the same time the truck and a dock door have to be available (these resources also have to be allocated, even if they do not perform an active service in this situation).

In [31, 32], more information about this application can be found.

4 Supporting Services

During operations, it might be necessary to introduce services to support core activities of the system. These supporting services have to be carried out by the resources during which these resources are not productive or available for normal tasks. Nevertheless, they are necessary to guarantee the correct functioning of the system. Examples are maintenance in a manufacturing context or the refueling of trucks in a logistics context.

These services can be easily integrated when they are offered by the intelligent resources. As an example, consider machine maintenance in a factory. In a more simplified setting, the machine itself offers a maintenance service, next to its other processing services. In a more advanced setting, this maintenance service is offered by another resource (a technician, consultant, operator ...). In the latter case, two resources are needed at the same time to execute the service (the machine being maintained and e.g. the technician carrying out the maintenance), so the control system has to be able to deal with multi-resource allocation.

However, in both settings, the supporting services can be seamlessly integrated with the core activities. It suffices that for every supporting service that has to be executed, a corresponding intelligent product is created. This intelligent product will search for the intelligent resources required to get its task fulfilled, just as the intelligent products representing the business activities. The intelligent resources do not make a distinction between both types of intelligent products, and will for all products try to allocate a time slot to execute the required service.

In manufacturing, several strategies can be used to create intelligent products in order to perform maintenance. If periodic maintenance is preferred for a machine, the control system can create intelligent products at regular time intervals. Another possibility is that the intelligent resource itself creates intelligent products to execute maintenance tasks, for instance when it notices a degraded performance (in process time or quality). Similarly, when a machine breaks down, the corresponding intelligent resource can create an intelligent product that is responsible for the repair of the machine. It is possible to give these maintenance and repair intelligent products a higher priority to ensure the maintenance or repair is done on time. Moreover, when maintenance is delayed or repair is needed, the intelligent resource will indicate the unavailability or degraded performance of its operations. This induces intelligent products, needing these operations, to wait until maintenance or repair has occurred. Other examples of supporting services in the context of manufacturing are the delivery of additional tools to machines, the delivery of empty containers or palettes to workstations, etc.

In logistics applications, an example of a supporting service is the refueling of trucks. In order to process a set of transportation tasks, a truck needs to refuel regularly. Similarly to maintenance, this refueling can be integrated with the transportation activities by creating an intelligent product that is responsible to perform the refueling task. This intelligent product will then try to allocate a time slot in which the refueling will happen. Again, it is possible to give the intelligent product a higher priority, although this is not necessary. When the refueling intelligent product is not able to schedule the refueling in time, the other intelligent products - representing freight - will notice that the truck will run out of fuel and is not able to reach its destination. Consequently, several intelligent products will find another truck to reach their destination, allowing the refueling to be scheduled in time.

In robotics, similar services can be identified. The machine maintenance service in manufacturing resembles the maintenance of robots and sensors and the battery recharging of mobile robots is the counterpart of the refueling service in logistics. Also resources representing parts of the environment might need maintenance, for instance the floor of corridors needs to be wiped by the cleaning crew at regular times. When an intelligent product is created to reserve a time slot for this task, the mobile robots will notice that the corridor is not accessible during this period.

5 Conclusions

Manufacturing Execution Systems can support production managers to coordinate and control the manufacturing system in real time. The Holonic MES, developed at K.U.Leuven, makes use of a service-oriented approach to organize production. Intelligent products, corresponding to tasks that have to be executed, actively search for required services. These services are offered by intelligent resources, representing physical resources. The active role of intelligent products makes the system flexible and robust against changes. Indeed, when production is disrupted, e.g. by a machine breakdown, intelligent products just have to search an alternative resource to execute the requested service. But their behavior does not change; they keep on searching for correct services, i.e. business-as-usual. Moreover, the control system scales with the underlying reality. The active role of intelligent products makes that intelligent resources can limit themselves to offering services and what-if functionality, which only requires local knowledge. Consequently, intelligent resources remain maintenance-free as long as the corresponding resources remain unchanged. Also, to take a new available resource into account, it suffices to add the corresponding intelligent resource to the control system, no changes have to be made to the rest of the control system.

Because of the service-oriented approach, the concepts and principles of the HMES can be applied to various application domains. This is illustrated with several applications in manufacturing, open-air engineering, robotics and logistics. According to the authors, the same concepts can also be valuable in domains like traffic and power grids. Moreover, these concepts also ease the integration of applications in different domains. For instance, to integrate the manufacturing and logistics of a product, the intelligent product does not only search for manufacturing services, but also for transportation services to move between the production site(s) and sites of partners in the supply chain.

The service-oriented approach also allows the seamless integration of supporting services with the core activities. It suffices that an intelligent product is created for every supporting task that has to be performed. The intelligent resources do not discriminate between requests from intelligent products representing core activities or supporting services, i.e. the handling of a supporting service request is business-as-usual for the intelligent resources.

Acknowledgments. This chapter presents work funded by the Research Council of the K.U.Leuven - *Concerted Research Action on Autonomic Computing in Decentralized Production Systems* (GOA-ACDPS) and the European project *Best Practice in Robotics* (BRICS, FP7-231940).

References

1. Alami, R., Chatila, R., Fleury, S., Ghallab, M., Ingrand, F.: An architecture for autonomy. The International Journal of Robotics Research 17(4), 315–337 (1998)
2. Ali, O.: Operational planning for outdoor engineering processes. Ph.D. thesis, Katholieke Universiteit Leuven (2010)
3. Babiceanu, R.F., Chen, F.F.: Development and applications of holonic manufacturing systems: a survey. Journal of Intelligent Manufacturing 17(1), 111–131 (2006)
4. Bartholdi III, J.J., Gue, K.R.: Reducing labour costs in an LTL crossdocking terminal. Operations Research 48(6), 823–832 (2000)
5. Boysen, N.: Truck scheduling at zero-inventory cross docking terminals. Computers & Operations Research 37(1), 32–41 (2010)
6. Bussmann, S., McFarlane, D.C.: Rationales for holonic manufacturing control. In: Van Brussel, H., Valckenaers, P. (eds.) Proceedings of the 2nd International Workshop on Intelligent Manufacturing Systems, pp. 177–184 (1999)
7. Demeulemeester, E., Herroelen, W.: A branch-and-bound procedure for the multiple resource-constrained project scheduling problem. Management Science 38(12), 1803–1818 (1992)
8. Euwe, M.J., Wortmann, H.: Planning systems in the next century (I). Computers in Industry 34(2), 233–237 (1997)
9. Fishwick, P.A., Zeigler, B.P.: A multimodel methodology for qualitative model engineering. ACM Transactions on Modeling and Computer Simulation 2(1), 52–81 (1992)
10. Holvoet, T., Valckenaers, P.: Beliefs, desires and intentions through the environment. In: Stone, P., Weiss, G. (eds.) Proceedings of the 5th International Joint Conference on Autonomous Agents and Multiagent Systems (AAMAS 2006), pp. 1052–1054 (2006)
11. Holvoet, T., Weyns, D., Valckenaers, P.: Patterns of delegate MAS. In: Proceedings of the 3rd IEEE International Conference on Self-Adaptive and Self-Organizing Systems (SASO 2009), pp. 1–9 (2009)
12. McFarlane, D., Sarma, S., Chirn, J.L., Wong, C.Y., Ashton, K.: Auto ID systems and intelligent manufacturing control. Engineering Applications of Artificial Intelligence 16(4), 365–376 (2003)
13. McFarlane, D.C., Bussmann, S.: Developments in holonic production planning and control. International Journal of Production Planning and Control 11(6), 522–536 (2000)
14. McWilliams, D.L.: Genetic-based scheduling to solve the parcel hub scheduling problem. Computers & Industrial Engineering 56(4), 1607–1616 (2009)
15. Monostori, L., Kádár, B.: Agent based architectures for mastering changes and disturbances in manufacturing. In: Pasqual del Pobil, A., Mira, J., Ali, M. (eds.) Proceedings on Tasks and Methods in Applied Artificial Intelligence, 11th International Conference on Industrial and Engineering Applications of Artificial Intelligence and Expert Systems (IEA-98-AIE), pp. 755–764 (1998)
16. Nesnas, I.A.D., Simmons, R., Gaines, D., Kunz, C., Diaz-Calderon, A., Estlin, T., Madison, R., Guineau, J., McHenry, M., Shu, I.H., Apfelbaum, D.: CLARAty: Challenges and steps toward reusable robotic software. International Journal of Advanced Robotic Systems 3(1), 23–30 (2006)
17. Peeters, P., Van Brussel, H., Valckenaers, P., Wyns, J., Bongaerts, L., Kollingbaum, M., Heikkilä, T.: Pheromone based emergent shop floor control system for flexible flow shops. Artificial Intelligence in Engineering 15(4), 343–352 (2001)

18. Philips, J., Valckenaers, P., Aertbeliën, E., Van Belle, J., Germain, B.S., Bruyninckx, H., Van Brussel, H.: PROSA and Delegate MAS in Robotics. In: Mařík, V., Vrba, P., Leitão, P. (eds.) HoloMAS 2011. LNCS, vol. 6867, pp. 195–204. Springer, Heidelberg (2011)
19. Ruiz-Torres, A.J., Centeno, G.: Scheduling with flexible resources in parallel work centres to minimize maximum completion time. Computers & Operations Research 34(1), 48–69 (2007)
20. Saint Germain, B., Valckenaers, P., Van Brussel, H., Hadeli, Bochmann, O., Zamfirescu, C., Verstraete, P.: Multi-agent manufacturing control: an industrial case study. In: Monostori, L., Kádár, B., Morel, G. (eds.) A Proceedings volume from the 7th IFAC Workshop on Intelligent Manufacturing Systems 2003, pp. 207–212 (2003)
21. Saint Germain, B., Valckenaers, P., Van Brussel, H., Van Belle, J.: Networked manufacturing control: an industrial case. CIRP Journal of Manufacturing Science and Technology (to appear)
22. Saint Germain, B., Valckenaers, P., Verstraete, P., Hadeli, Van Brussel, H.: A multi-agent supply network control framework. Control Engineering Practice 15(11), 1394–1402 (2007)
23. Stoop, P.P.M., Wiers, V.C.S.: The complexity of scheduling in practice. International Journal of Operations & Production Management 16(10), 37–53 (1996)
24. Tsui, L.Y., Chang, C.H.: A microcomputer based decision support tool for assigning dock doors in freight yards. Computers & Industrial Engineering 19(1-4), 309–312 (1990)
25. Valckenaers, P., Saint Germain, B., Verstraete, P., Van Belle, J., Hadeli, Van Brussel, H.: Intelligent products: Agere versus essere. Computers in Industry 60(3), 217–228 (2009)
26. Valckenaers, P., Van Belle, J., Ali, O.: PROSA and delegate MAS for open-air engineering processes. In: Proceedings of the 16th IEEE International Conference on Emerging Technologies and Factory Automation, ETFA 2011 (to appear, 2011)
27. Valckenaers, P., Van Brussel, H.: Deadlock avoidance in flexible flow shops with loops. Journal of Intelligent Manufacturing 14(1), 137–144 (2003)
28. Valckenaers, P., Van Brussel, H.: Holonic manufacturing execution systems. CIRP Annals - Manufacturing Technology 54(1), 427–432 (2005)
29. Valckenaers, P., Van Brussel, H.: Intelligent products: Intelligent beings or agents? In: Azevedo, A. (ed.) Innovation in Manufacturing Networks, 8th IFIP International Conference on Information Technology for Balanced Automation Systems (BASYS 2008), pp. 295–302 (2008)
30. Valckenaers, P., Van Brussel, H., Kollingbaum, M., Bochmann, O.: Multi-agent Coordination and Control Using Stigmergy Applied to Manufacturing Control. In: Luck, M., Mařík, V., Štěpánková, O., Trappl, R. (eds.) ACAI 2001 and EASSS 2001. LNCS (LNAI), vol. 2086, pp. 317–334. Springer, Heidelberg (2001)
31. Van Belle, J., Saint Germain, B., Bahtiar, R., Valckenaers, P., Van Brussel, H., Cattrysse, D.: Towards holonic control for cross-docks. In: Witlox, F.J.A., Weijers, S. (eds.) Bijdragen Vervoerslogistieke Werkdagen 2010, vol. 2, pp. 319–332 (2010)
32. Van Belle, J., Saint Germain, B., Valckenaers, P., Van Brussel, H., Bahtiar, R., Cattrysse, D.: Intelligent products in the supply chain are merging logistic and manufacturing operations. In: Bittanti, S., Cenedese, A., Zampieri, S. (eds.) Preprints of the 18th IFAC World Congress, pp. 1596–1601 (2011)

33. Van Belle, J., Saint Germain, B., Verstraete, P., Valckenaers, P., Ali, O., Van Brussel, H., Cattrysse, D.: A Holonic Chain Conveyor Control System: An Application. In: Mařík, V., Strasser, T., Zoitl, A. (eds.) HoloMAS 2009. LNCS, vol. 5696, pp. 234–243. Springer, Heidelberg (2009)
34. Van Belle, J., Valckenaers, P., Saint Germain, B., Bahtiar, R., Cattrysse, D.: Bio-inspired coordination and control in self-organizing logistic execution systems. In: Proceedings of the 9th IEEE International Conference on Industrial Informatics (INDIN 2011), pp. 713–718 (2011)
35. Van Brussel, H., Valckenaers, P., Wyns, J., Peeters, P., Bongaerts, L.: Holonic manufacturing systems: architectural and manufacturing control issues. In: Proceedings of 2nd CIRP International Seminar on Intelligent Computation in Manufacturing Engineering (ICME 2000), pp. 19–29 (2000)
36. Van Brussel, H., Wyns, J., Valckenaers, P., Bongaerts, L., Peeters, P.: Reference architecture for holonic manufacturing systems: PROSA. Computers in Industry 37(3), 255–274 (1998)
37. Verstraete, P.: Integrating existing scheduling techniques into the Holonic Manufacturing Execution System. Ph.D. thesis, Katholieke Universiteit Leuven (2009)
38. Verstraete, P., Saint Germain, B., Hadeli, K., Valckenaers, P., Van Brussel, H.: On applying the PROSA reference architecture in multi-agent manufacturing control applications. In: Weyns, D., Holvoet, T. (eds.) Multiagent Systems and Software Architecture, Proceedings of the Special Track at Net. ObjectDays (NODe 2006), pp. 31–47 (2006)
39. Verstraete, P., Saint Germain, B., Valckenaers, P., Van Brussel, H., Van Belle, J., Hadeli: Engineering manufacturing control systems using PROSA and delegate MAS. International Journal of Agent-Oriented Software Engineering 2(1), 62–89 (2008)
40. Vis, I.F., Roodbergen, K.J.: Positioning of goods in a cross-docking environment. Computers & Industrial Engineering 54(3), 677–689 (2008)
41. Wen, M., Larsen, J., Clausen, J., Cordeau, J.F., Laporte, G.: Vehicle routing with crossdocking. Journal of the Operational Research Society 60(12), 1708–1718 (2009)
42. Woodworth, B.M., Willie, C.J.: A heuristic algorithm for resource levelling in multi-project, multi-resource scheduling. Decision Sciences 6(3), 525–540 (1975)
43. Yu, W., Egbelu, P.J.: Scheduling of inbound and outbound trucks in cross docking systems with temporary storage. European Journal of Operational Research 184(1), 377–396 (2008)
44. Zamfirescu, C., Valckenaers, P., Van Brussel, H.H., Germain, B.S.: A Case Study for Modular Plant Control. In: Mařík, V., McFarlane, D.C., Valckenaers, P. (eds.) HoloMAS 2003. LNCS (LNAI), vol. 2744, pp. 268–279. Springer, Heidelberg (2003)

Service Oriented Control Framework for a Holonic System Characterized by a Guided Flow of Entities

Theodor Borangiu[1], Silviu Raileanu[1], Octavian Stocklosa[2], Christian Tahon[3], Thierry Berger[3], and Damien Trentesaux[3]

[1] University Politehnica of Bucharest,
Dept. of Automation and Industrial Informatics Romania
{theodor.borangiu,silviu.raileanu}@cimr.pub.ro
[2] East Electric, Bucharest, Romania
octavian.stocklosa@eastelectric.ro
[3] Université Lille Nord de France,
F-59000 Lille, UVHC, TEMPO Lab.
F-59313 Valenciennes, France
{christian.tahon,thierry.berger,
damien.trentesaux}@univ-valenciennes.fr

Abstract. The chapter presents a new concept of systems characterized by a flow of active entities which run on a guided network. After the general characterization (nature, mobility, service) of an entity, the three basic components of the system are described: active and non active entity and active generalised entity, elements which are structured and implemented as holons. With the three types of basic holons and with the generic services offered by them, the structural and dynamic models are formed. The framework is extended with the behavioural model consisting of the planning, scheduling, routing and execution of the operations. The control framework is completed with the implementation architecture based on intelligent embedded devices and a multi-agent programming environment for rapid configuration and integration.

Keywords: flexible manufacturing system, personal rapid transit, control system, service orientation, mobile entity driven automation.

1 Introduction

Generally, scientific contributions tend to offer a standard solution to a single class of problems (e.g.: control of a manufacturing system [7] or of a transportation system [1]). The idea that promoted the research which sustained this publication was to point out the key elements of the target systems that will be automated and, based on these elements, propose a generic control framework. In this case the generic characteristic is a *flow of mobile entities* which run on a *guided network crossing the system*. The entities mentioned above posses embedded

T. Borangiu, A. Thomas, and D. Trentesaux (Eds.): SOHOMA 2011, SCI 402, pp. 21–34.
springerlink.com

intelligence, making them active in the decisional processes associated with their lifecycle (resource allocation, routing and service reception). The set of entities along with the system characterized by a guided flow will be further referred as *system characterized by a guided flow* (SGF). A brief representation of a SGF is given in Fig.1 where the infrastructure is represented by an oriented graph on which the mobile active entities move. The edges represent the direct links between the points (vertices) where services are received. In this structure the mobile entities that move from one point to another following the direct links and that are subject to physical constraints form the guided flow.

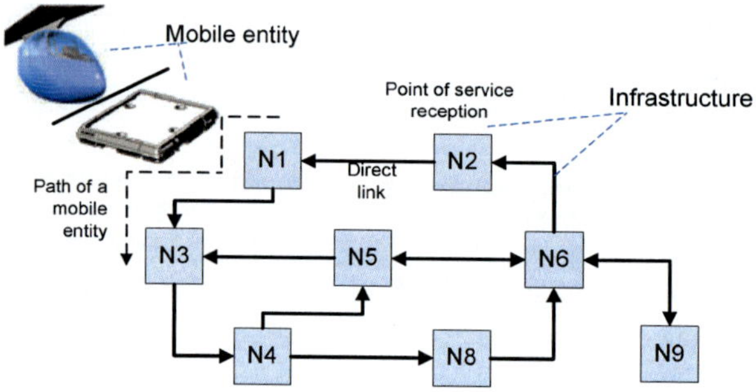

Fig. 1 Functioning principle of a system characterized by a guided flow

Two significant examples of such systems are the flexible manufacturing systems (FMS) [7] that dispose of a conveyor system which automatically transports the products between the workstations and the Personal Rapid Transit / Transportation systems (PRT) [1] which, briefly, offers non-stop, personal and uninterrupted transportation services. Given the common points described above (entities possessing embedded intelligence running on a guided infrastructure), the proposed control framework will deal in the same manner with the three important processes: resource allocation, routing and service reception. Given the fact that the transportation aspect is just one part of the manufacturing domain, this chapter will focus only on the control of manufacturing systems.

In the current industrial context, productions system owners and operators desire a flexible automation solution that should be adaptable to frequent changes rather than one that has an optimized behaviour but is inflexible [9, 15]. In fact, optimality with lots of constraints makes the production system very sensitive to changes in the environment and in client demand. In this context, new control architectures have been proposed in the production domain [4, 7] or solely for the transportation [5], architectures based on the increasing decisional autonomy of products and resources. They implement new decisional activities in order to act according to the system's real-time state and to keep a certain performance level.

2 The System Characterized by a Guided Flow

2.1 The Concept and Characterization of Entities

The systems characterized by a guided flow that are considered in this chapter are composed of elements called *entities*. Thus, in the case of a flexible manufacturing system (FMS) [7] a port-pallet, a processing centre, a conveyor or an industrial robot are entities. For a dedicated transport system (ex. Personal Rapid Transit – PRT [1]) a vehicle, the passengers, the rail or the stations are entities of the system. The different entities present a strong diversity, especially through the functionalities they assure inside the system. In order to characterize them, three parameters are retained: nature, mobility and the associated services.

The **nature** of an entity specifies if it is part of:

- the physical part of the system (machines, conveyor, etc) – *physical entity*
- the informational system (PLC, control PC, etc) – *informational entity*

The **mobility** characterizes the degree of mobility of an entity and is classified into:

- *autonomous entity*: it can move inside the boundaries of the system by its own means;
- *non-autonomous entity*: it can move inside the boundaries of the system by the aid of an autonomous entity;
- *fixed entity*: it cannot move over a fixed period of time.

The entities offer and / or receive **services** which are classified into [8]:

- *space transformation*: physical transformation (ex.: transport system, manipulation), information transformation (ex.: communication);
- *time transformation*: physical transformation (ex.: data storage), information transformation (ex.: data memorization);
- *form transformation*: physical transformation (ex.: machining, assembling), information transformation (ex.: data processing).

Active entity

Certain entities participate to the decisional control process, allocation and routing of different flows, in which case they are called active entities (AE). An AE takes in an autonomous manner control decisions, participates in the monitoring and traceability of the processes of offering and receiving services and assures the monitoring of the current state of the product (ex.: trigger alarms). The active entities constitute the control system.

Non-active entities

The entities that do not directly participate in the decisional processes of allocation, routing and service reception are defined as non-active entity (NAE). A NAE possesses a physical part, receives control commands from an AE and can posses information processing capacities which it uses only for direct resource control.

Generalized active entity (GAE)

A generalized active entity is composed of one (or several) AE and one (or several) NAE interacting as depicted in the Fig.2.

The different entities present a strong diversity especially due to the functions they assure inside the guided flow system. Therefore, in order to better completely characterize the entities, three parameters have been retained:

1. The **nature** of a GAE, abbreviated NAT, which can be physical (NAT=P) or informational (NAT=I);
2. The **mobility** of an entity, abbreviated MOB, which can be autonomous (MOB=MA), non-autonomous (MOB=MNA) or fixed (MOB=F);

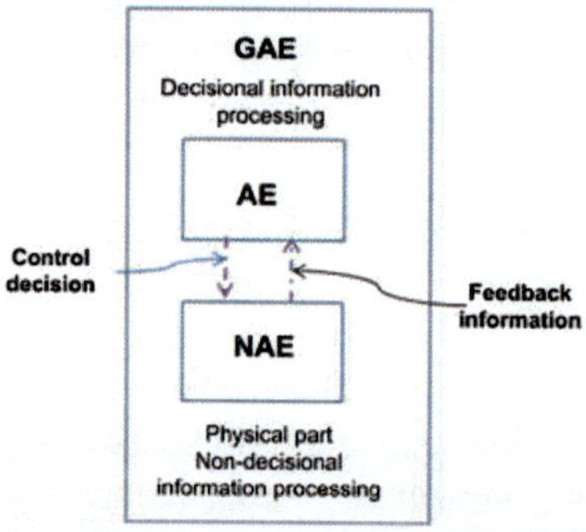

Fig. 2 Generalized active entity, GAE

3. The **services** that the considered entities can offer. In this case an entity can generally offer any of the following three types of services: *spatial transformation* (SERV=ssp for the physical transformation (ex.: transport service), ssi for informational transformation (ex.: communication)), *time transformation* (SERV=stp for the physical transformation (ex.: material storage service), SERV=sti for the informational transformation (ex: data storage)) and *form transformation* (SERV=sfp for physical transformation (ex.: machining, assembling), SERV=sfi for the informational transformation (ex: data processing)). For the informational service of form transformation, sfi, two cases appear: *decisional sfi*, SERV = sfi_D, that generates the high level commands needed for allocation and routing, commands that are elaborated in an autonomous manner by the considered entity and *non-decisional* sfi, SERV = sfi_{ND}, the decisions taken here assure execution of the commands previously issued.

2.2 Control Architecture for an Active Entity System

An active entity system (AES) is composed of a set P of active entities representing the control system responsible for decision making and a set S of non-active entities representing the controlled system responsible for implementing the decisions taken by the active entities (Fig.3). Among the set S we distinguish the mobile NAE (the NAE Flow) participating at the physical flow and receiving services provided by NAE Resources. This structure is recursive: the system P' controls the system S' and S' in its turn composed by a P and an S system.

In an FMS or PRT system the existing flows are physically constituted by certain entities: *pallet-carriers* assuring the movement of products from one workstation to another and *shuttles* which move in order to complete the passengers' transportation orders. The decisional part of these flows resides in the P system and the generalized active entities (GAE), which represent the whole flow and appear on the control architecture as an aggregation of AEs and NAEs which can interact between them through services both at informational and at physical level.

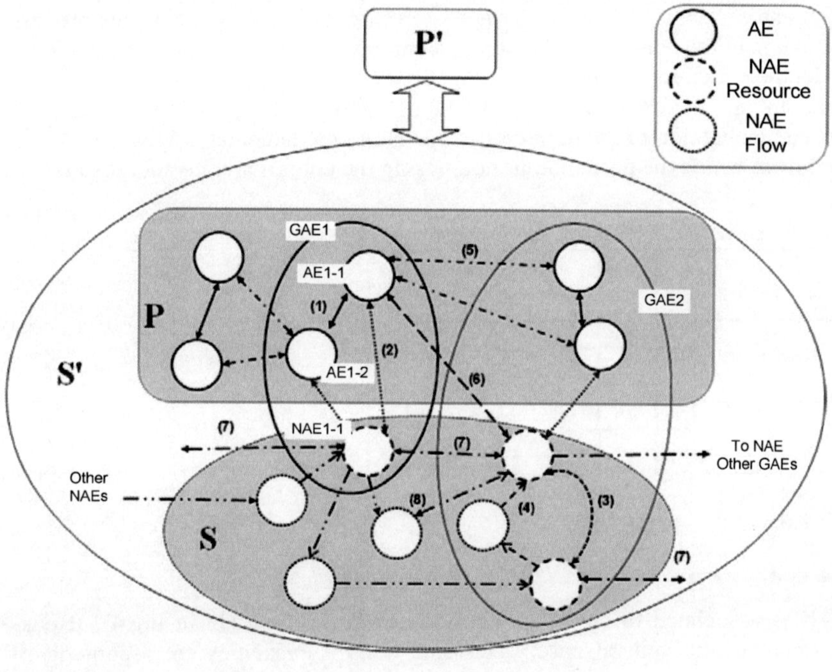

Fig. 3 Structure of an active entity system and the flows inside it

In Fig.3 the following types of flows between the composing entities of the active entity system can be distinguished: (1) informational and physical flows; (2) internal and external flows. These cases are summarized in Table 1.

Table 1 Flows inside an active entity system

		Flow reference	Type of flow
Internal flows	Information flow	(1)	Between AEs of the same GAE
		(2)	Between an AE and a NAE of the same GAE
		(3)	Between NAE of the same GAE
	Physical flow	(4)	Between NAE of the same GAE
External flows	Information flow	(5)	Between AE of different GAEs
		(6)	Between an AE of a GAE and a NAE of a different GAE
		(7)	Between NAEs of different GAEs
	Physical flow	(8)	Between NAEs of different GAEs

3 Structure of the Holonic Control System

The holonification of an active entity system is done by the association of a holon [6] with one or several active and/or non-active entities. The scope of this process is to implement the control architecture: an entity model is used at theoretical level to model the structure and the interactions between the composing parts of the system; the holonic theory associates physical counterparts to the informational entities for the realization of the control system. Three fundamental holons (Fig.4) are defined from which the basic holons composing the control architecture result.

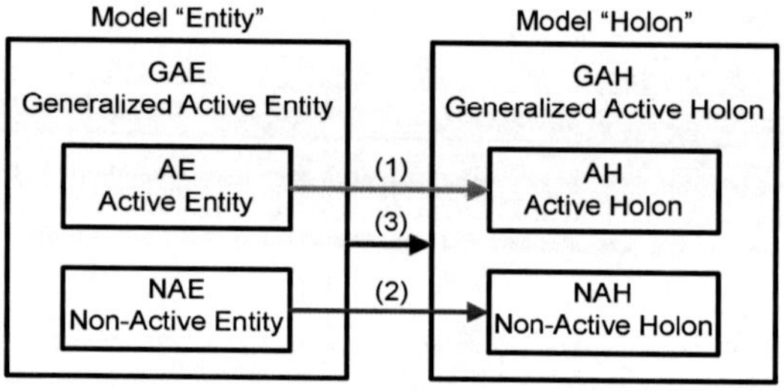

Fig. 4 Model entity – model holon

Active Holon (AH)

The AH is associated to an active entity (association type (1) in Fig.4). It possesses then, in the general case, decisional control capacities corresponding to informational services such as: storing (sti), transfer (ssi) and processing (both decisional (sfi_D) and non-decisional (sfi_{ND})).

Non-Active Holon (NAH)

The NAH is associated to a non active entity (association of type (2) in Fig.4) and doesn't participate to the control process. It possesses a physical part which provides physical space (ssp), time (stp) and form (sfp) transformation services, and also offers informational space (ssi), time (sti) and form (sfi) transformation services, with the observation that its information transformation service is only non decisional (sfi_{ND}).

Generalized Active Holon (GAH)

The GAH is a composite holon consisting of an active holon driving a non-active holon (association type (2) in Fig.4) capable of offering and receiving all the services existing in the system (Table 2).

Table 2 Services associated to a Generalized Active Holon

	Space transformation	Time transformation	Form transformation
Physical	sep	stp	sfp
Informational	sei	sti	sfi_D, sfi_{ND}

Graphically, the association between an AH and a NAH (which forms a GAH) is represented in Fig.5, where the previously presented flows (table 2) are also illustrated.

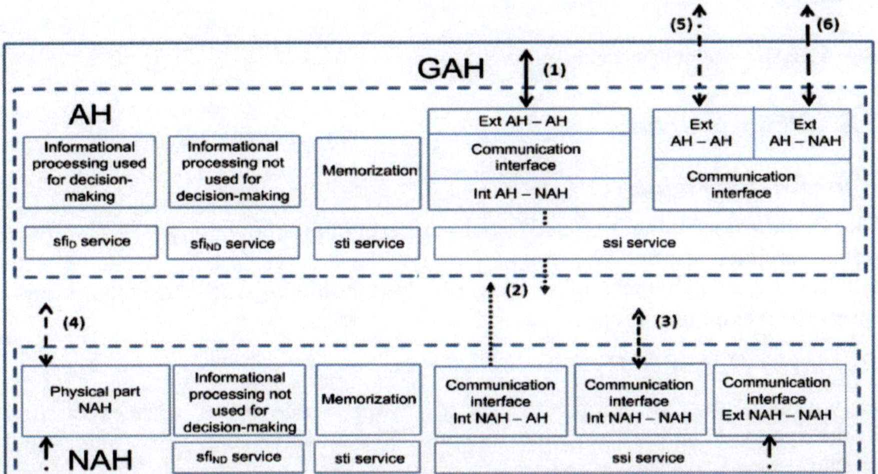

Fig. 5 Model of a GAH

3.1 *Static and Dynamic Models of the Composing Holons*

The holonic model (Fig.6) of a system characterized by an active flow of entities is composed of a *static part* comprising the holons whose structure never changes during the transportation (PRT) / fabrication (FMS) process (coordinator holon, services to be obtained holon and a resource holon) and a *dynamic part* consisting of a single type of holon obtained through the aggregation of a set of base holons. The second part is dynamic because it changes during its lifecycle and it represents the moving parts of the system (flow holon).

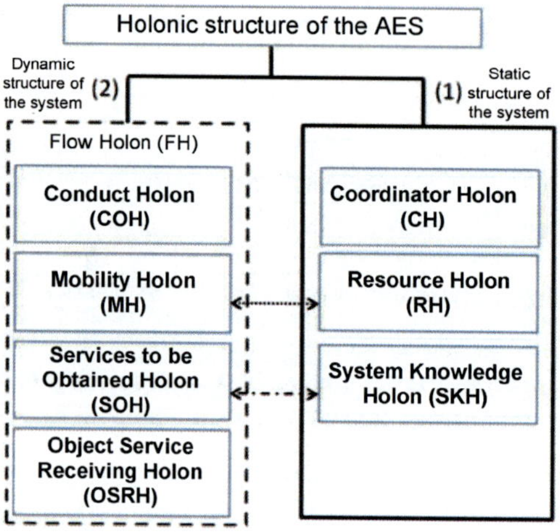

Fig. 6 AESH – Holonic model of an AES

3.2 Static Structure

Coordinator Holon (CH)

The Coordinator Holon is an active holon offering the following types of informational services: sei, sti, sfi_D, sfi_{ND}. Through these services it is in charge with maintaining the coherence of services provided inside an AES and globally optimizes the receipt of services.

Resource Holon (RH)

A type of non-active holon, the RH, offers the whole set of services existing in the system (Table 2), assuring functions like resource reservation and model update.

System Knowledge Holon (SKH)

A type of non-active holon, the SKH contains the information needed to receive a service by a Flow Holon and the set of elements describing the different holons of a FMS (CNC machines, robots, conveyor) or of a PRT (shuttles, stations, rails).

3.3 Dynamic Structure

The Flow Holon is a composite GAH constituted dynamically by four types of holons: *Conduct Holon, Mobility Holon, Services to be Obtained Holon* and *Object of Receiving Services Holon.*

Conduct Holon (COH)

The COH is an active holon in charge with the tasks of: monitoring the service reception, choice of resources and launch, interruption and resumption of services.

Mobility Holon (MH)

The MH is a resource holon which offers transport services to the flow holon.

Services to be Obtained Holon (SOH)

SOH is a generic active holon containing the list of services the flow holon must obtain from the resource holons. It is associated to the technical processes and contains the information concerning the necessary services (technical constraints, quality indexes, resource parameters, etc.).

Object Service Receiving Holon (OSRH)

The OSRH is a generalized active holon that models a passive entity (manufactured piece in the case of FMS and transported person in the case of PRT).

Flow Holon (FH)

The core element of the proposed control framework, the FH represents the materialization of an active entity. As showed in Fig.6, the FH is an aggregation of the following elements: Conduct Holon, Mobility Holon, Services to be Obtained Holon and Object Services Receiving Holon, offering and capable of receiving the whole range of services (see Table 2). The set of existing FHs autonomously deciding where to go, how to go and what to receive represents the inner flow of the AES [17].

3.4 Aggregation Process and Lifecycle of a FH

The FH is initially composed of all or just a subset of the COH, MH, SOH and OSRH holons. This structure is described in Fig.7, which also details the processes of aggregation (the process at the beginning of the fabrication phase of a product (FMS case) in which the four components of a FH are physically put together) and of disaggregation (the process at the end of the fabrication phase of a single product in which the OSRH is physically taken out of the subset; the OSRH, representing the product, is used in further processes and the COH, MH and eventually the SOH are reused for the fabrication of another product). Fig.7 also shows that after the aggregation process, the assembly moves (because of the MH) to the needed RHs in order for the OSRH to receive the services contained in the SOH. The control of the process is assured by the COH.

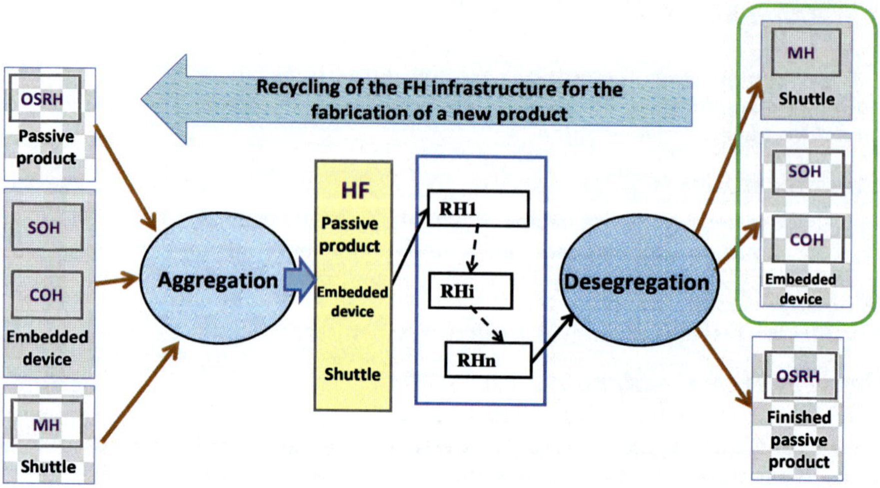

Fig. 7 Lifecycle of a FH, aggregation and desegregation processes

4 Behavioural Model

The behavioural model has three important stages (Fig.8): *process planning* which realizes the sequencing of orders and optimization at global level, *resource scheduling* which realizes the operation attribution and lastly *service reception* which deals with the routing of the FHs and the receipt of the processing operations (just in the FMS case). In parallel with these three stages an update process is taking place for the model describing the way the services of the resources can be accessed (Resource Service Access Model, RSAM [2]). The RSAM is a distributed autonomous entity in charge of collecting resource information during their usage and offering it in a concise manner when taking the decisions of planning and resource allocation. This entity will be initially created using a Graphical User Interface, RHs being manually added to create an initial set of competencies; then it will act as distributed intermediate that will be consulted by the FHs on a regular basis and it will be periodically updated with all RHs status. In this way, each HF owns a list of fixed services to be obtained from the system's resources and based on the real-time updated RSAM it takes a decision concerning the resource and the path to it. Even if the resulting information is computed locally, the final decision is taken through communication with the HRs and HFs [3] in order to optimize the service reception process at the level of HFs in the system [16, 18, 19].

From the three stages enumerated above, only the last two concern the service oriented control framework since they (resource scheduling and service reception) are done in a decentralized manner using the services provided by the holonic AES.

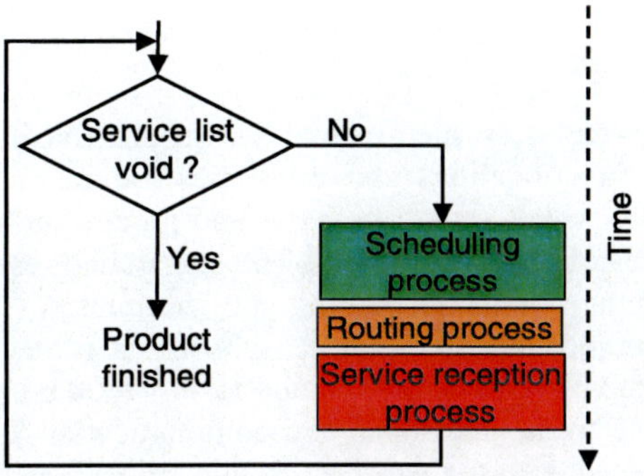

Fig. 8 General product execution process

4.1 Planning Process

The planning process is done at a centralized level and its result is the sequencing of orders. It takes into account the state of all resources over a long period of time in order to globally optimize the way the products are executed.

4.2 Resource Scheduling

After setting the desired objective to a FH through the update of the information within the SOH (e.g.: loading the product recipe in the case of an FMS), the FH has to successively find the RHs needed to acquire the specified services and then to move towards them to obtain these services. The scheduling operation is a heterarchic process, based on the CNP protocol [10] in which each FH seeks through dialogue the needed services within the RHs at its disposal. The selected RH will be chosen based on a multi-criteria decision which takes into account variables such as service time, transportation time and resource queue.

4.3 Service Reception

Once the RH has been selected, it is the routing process that is in charge with the transportation of the FH to the chosen destination (transportation service reception) and afterwards, in the FMS case, with the process of service reception. Also a heterarchical process, service reception with its both aspects, transport and process, consists of an online synchronization between the FH and the RH using the information from the RSAM (Fig.9, inter FH dialogue and FH-RH dialogue) in order to take into account last minute changes in the state of the system (processing and transporting resources online/offline, processing and transporting times variations, etc.).

All the operations offered by the RHs and received by the FHs are stored at this level and are used to **track** the associated process: *fabrication* in the case of FMS and *transport* in the case of PRT.

5 Realization

The design of the control architecture is based on hierarchical and heterarchical control structures, preserving only the functionalities and properties judged as necessary for the proposed control model which should be agile and reconfigurable according to the needs and influences of the environment. The functionalities of the hierarchical systems (global planning, global scheduling and automation – of the routing and processing) are preserved, but the resource scheduling (done offline at the planning level) is now shifted toward the automation level which becomes more autonomous and reactive due to the decisional and communicational capacities (inter HF and HF-HF) and also due to the capacity to dynamically reconfigure the access to services provided by resources. Thus, for controlling the target system, the generic open interface control architecture (Fig.9) is proposed.

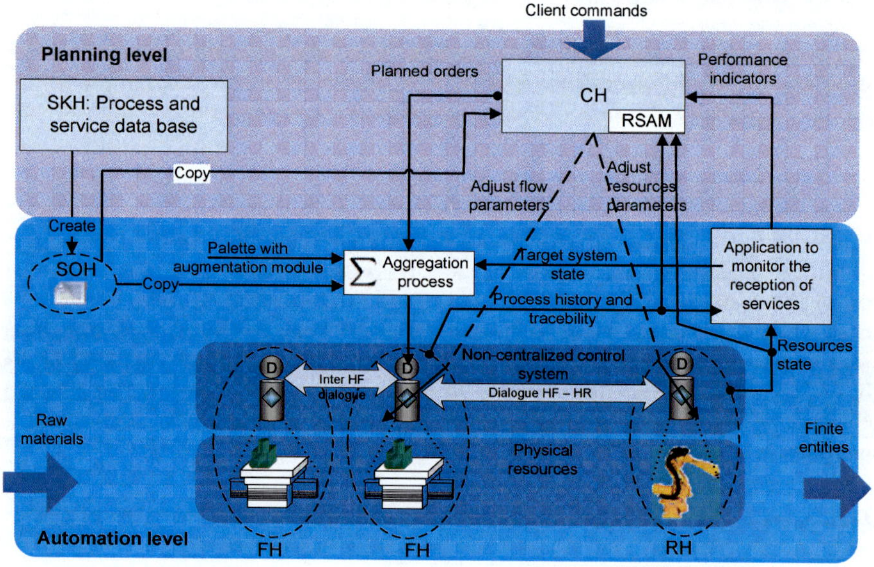

Fig. 9 Generic control architecture and the associated flows (adapted from [11])

We thus propose a two layer control architecture which can be reconfigured in a dynamic way, the choice of the control mode being made according to the technical characteristics of the target system and also according to the performance requirements:

1. A **high level** consisting of a CH and of the SKHs. The CH is in charge with *client and cell GUI*, performing the off-line decisional process of long term *planning*, deciding which strategy to choose, *Execution tracking* and *RSAM operation*. The *SKHs* responsibility is to store the list of services to be obtained for the execution of each FH. This layer is linked to the client by an interface

for order reception and reports, and the client's requests are mapped to SOHs. The SOHs are the input to a centralized application which generates the list of optimally ordered orders which are then associated through an *Aggregation Process* (∑) to the physical *Pallet* and along with the *SOH* (product / transport description) form the FH. The optimization of the orders list is relative to a global cost function, at batch horizon, such as: *makespan* or *resource loading*.

2. A **low level** in charge with *mobile entity driven process automation* composed of two types of entities, FH and RH, which interact in order to *implement the process schedule*.

Dotted arcs between the planning level and automation level represent the indirect influence of FHs and RHs (implicit control through the influence of the mechanism for societal optimization). Other arcs between composing elements represent informational or material flows.

6 Conclusion

The objective of this article was the proposition of a generic control framework for systems characterized by a guided flow (SGF) based on the holonic principles [12, 13, 14].

The main contribution of the generic control framework lies in the adoption of a systemic principle of low coupling and strong cohesion that leads to the implementation of a general active entity using the holonic principles (**Flow Holon**). The Flow Holon is capable, independently, to support the realization of the services offered by the SGF (e.g.: fabrication of a product for the FMS case or transportation of a person for the PRT case). The flow, usually passive, is active in this framework with regard to the control aspect.

Besides the concept of Flow Holon used for the control of a SGF, another important contribution of this article is the orientation towards services of the routing and resource allocation functionalities using the Resource Service Access Model which is permanently updated with the real state of the resources, their availability and quality of provided services.

References

1. Anderson, J.E.: Control of Personal Rapid Transit Systems. JAT 32(1) (1998)
2. Borangiu, T., Răileanu, S., Trentesaux, D., Berger, T.: Open Manufacturing Control with Agile Reconfiguring of Robot Services. In: 19th International Workshop on Robotics in Alpe-Adria-Danube Region, RAAD 2010, Budapest, Hungary, June 23-25 (2010) ISBN: 978-1-4244-6884-3
3. Borangiu, T., Răileanu, S., Anton, F., Parlea, M., Tahon, C., Berger, T., Trentesaux, D.: Product-driven automation in a service oriented manufacturing cell. In: International Conference on Industrial Engineering and Systems Management, IESM 2011, Metz, France (2011)

4. Brussel, H.V., Wyns, J., Valckenaers, P., Bongaerts, L., Peeters, P.: Reference Architecture for Holonic Manufacturing Systems: PROSA. Computers in Industry 37(3), 255–274 (1998); Special issue on manufacturing systems, ISSN: 0166-3615
5. Fischer, K., Schillo, M., Siekmann, J.: Holonic Multiagent Systems: A Foundation for the Organisation of Multiagent Systems. In: Mařík, V., McFarlane, D.C., Valckenaers, P. (eds.) HoloMAS 2003. LNCS (LNAI), vol. 2744, pp. 71–80. Springer, Heidelberg (2003)
6. Koestler, A.: The Ghost in the Machine. Arkana Books, London (1967)
7. Leitão, P.: An Agile and Adaptive Holonic Architecture for Manufacturing, Control. PhD Thesis (2004)
8. Le Moigne, J.L.: La théorie du système général, Théorie de la modélisation, 2nd edn., p. 338. PUF, Paris (1994)
9. Sauer, O.: Automated engineering of manufacturing execution systems – a contribution to "adaptivity" in manufacturing companies. In: Proceedings of DET 2008 5th International Conference on Digital Enterprise Technology Nantes, France, October 22-24 (2008)
10. Smith, R.: The Contract Net Protocol: High-Level Communication and Control in a Distributed Problem Solver. IEEE Transactions on Computers - TC 29(12), 1104–1113 (1980), doi:10.1109/TC.1980.1675516
11. Trentesaux, D.: Les systèmes de pilotage hétérarchiques: innovations réelles ou modèles stériles? Journal Européen des Systèmes Automatisés 9-10, 1165–1202 (2007) ISSN 1269-6935
12. Borangiu, T., Gilbert, P., Ivanescu, N.A., Rosu, A.: An implementing framework for holonic manufacturing control with multiple robot-vision stations. Engineering Applications of AI 22(4-5), 505–521 (2009)
13. Barata, J.: The Cobasa architecture as an answer to shop floor agility, manufacturing the future - concepts, technology. In: Visions, pp. 31–76. Pro Literatur Verlag (2006)
14. Brussel, H.V., et al.: Reference architecture for holonic manufacturing systems: PROSA. Computers in Industry 37(3), 255–274 (1998)
15. Valckenaers, P., Van Brussel, H., Bruyninckx, H., Saint Germain, B., Van Belle, J., Philips, J.: Predicting the unexpected. Computers in Industry (2011), doi:10.1016/l.compind.2011.04.011
16. Demirkan, H., Kauffman, R.J., Vayghan, J.A., Fill, H.-G., Karagiannis, D., Maglio, P.P.: Service-oriented technology and management: Perspectives on research and practice for the coming decade. Electronic Commerce Research and Applications 7, 356–376 (2008), doi:10.1016/j.elerap.2008.07.002
17. Meyer, G., Främling, K., Holmström, J.: Intelligent products: A survey. Computers in Industry (2008), doi:10.1016/j.compind.2008.12.005
18. Barbosa, J., Leitao, P.: Enhancing Service-oriented Multi-agent Systems with Self-organization. In: Proceedings of International Conference on Industrial Engineering and Systems Management IESM 2011, Metz (2011)
19. Leitao, P.: Agent-based distributed manufacturing control: A state-of-the-art survey. Int. Journal of Engineering Applications of AI 22(7), 979–991 (2009)

The Augmentation Concept: How to Make a Product "Active" during Its Life Cycle

Yves Sallez

Univ. Lille Nord de France, F-59000 Lille, France
UVHC, TEMPO Lab, "Production, Services, Information" Team
F-59313 Valenciennes, France
yves.sallez@univ-valenciennes.fr

Abstract. The recent developments in infotronics and communication allow the development of "intelligent" products. This chapter proposes the concept of "activeness" and a modelling approach to increase a system's informational, communicational and decisional capacities in a distributed architecture. The "activeness" concept is then applied in two case studies. The first concerns the "product-driven" control of a real manufacturing cell. The second is related to the advanced diagnosis of complex systems in a railway application. Our results have allowed the origins of some of the defects to be found more easily and quickly than the traditional approaches.

Keywords: closed loop PLM, intelligent product, augmentation, RFID.

1 Introduction

The concept of "active" or "intelligent" products is now widespread. This concept's rapid development is mainly due to the fact that, over the last decade, the increasing growth of embedded technologies (e.g., RFID, smart cards, wireless communication), associated with the concepts of ambient intelligence [8] and machine-to-machine intelligence [5], has allowed the development of products that are fully able to interact with their environment.

In the field of ambient intelligence, in which a product is considered an actor that makes decisions, some researchers have focused their work on the product use phase [8, 9] and others on the manufacturing phase [1, 10, 12]. Providing a complete survey on this topic is beyond the scope of this chapter; a detailed survey of this research field is published by Meyer et al. [11]. The chapter discusses the advantages of making products "active" during their life cycle.

Section 2 introduces the "activeness" concept associated to a product and proposes a modelling approach based on the augmentation concept, available throughout the whole product life cycle. Section 3 describes an initial instantiation of the generic features of our approach in a manufacturing context and a real implementation of active products on a flexible assembly cell. In section 4, another application of the augmentation concept for the embedded diagnosis of complex

T. Borangiu, A. Thomas, and D. Trentesaux (Eds.): SOHOMA 2011, SCI 402, pp. 35–48.
springerlink.com

systems is fully explained, and an on-going industrial railway project is specifically presented. Finally, section 5 provides our concluding remarks and our prospects for future research.

2 The Concept of "Active" Product

The concept of "activeness" is used to increase informational, communicational and decisional capacities of a product. In our opinion, it is interesting to make a product capable of interacting with the different support systems that provide services during the successive phases of its life cycle. This interaction would permit the management of the product's life cycle to be improved.

This last point is clear in the field of closed-loop Product Life cycle Management (PLM), interoperability and traceability. Jun et al. [6] have pointed out some of the relevant specifications that could be applied using an "active product" approach, from product design to product recycling. Unlike traditional PLM, closed-loop PLM focuses on the complete product life cycle, with an emphasis on tracking and managing the information from the whole product life cycle and potentially returning information at each phase in the process. These returns of information or feedback can be given via several forward and backward closed-loops that link the successive life-cycle phases. This feedback allows better decisions to be made throughout the whole life cycle and enhances product traceability.

The concept of "active" product (AP) offers two main benefits:

1. An improved performance of the pair "Active Product - Support System" (AP-SS), for specific phases in the life cycle (detailed in the next section);
2. An improved quality of the information flows between the different phases of life cycle, via the different forward and backwards loops. (This improved quality has already been addressed by the author in [15] and is not dealt with further in this rest of the chapter.)

2.1 Improved Performance of the Pair "Active Products - Support System"

The concept of activeness associated to product contrasts with the traditionally passive product behaviour, which makes the product not able to take the initiative in relation to a support system, and thus is subject to the system's intervention. The minimal activity of an active product is to trigger events: the product is able to identify its state, compare its state with the desired one, and send information (e.g., warnings) when certain conditions are met. Obviously, more complex activities can be considered (e.g., memorization, communication, negotiation, learning), as detailed in the following sections. As shown in Fig.1, for a given phase of its life cycle, the improved performance of the pair AP-SS comes from several interactions.

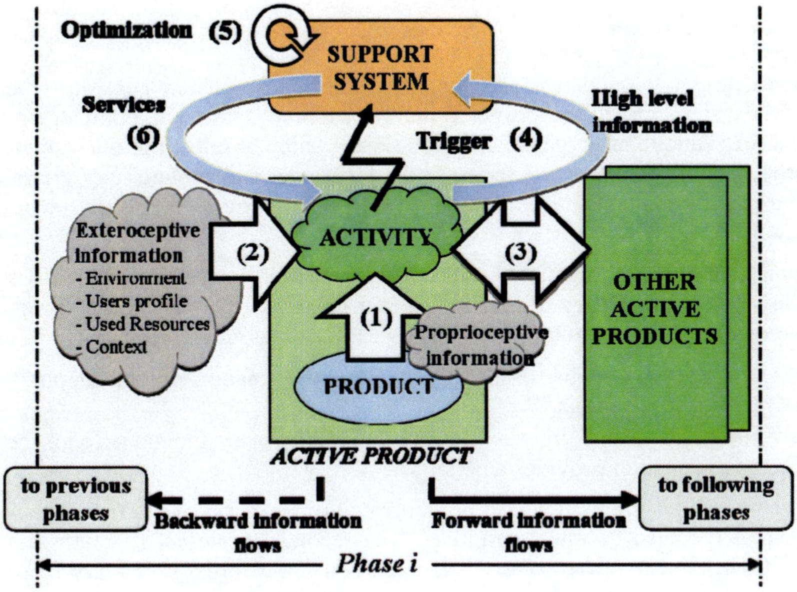

Fig. 1 Illustration of the "activeness" associated to a product for a given phase of its life cycle

To perform its triggering activity or to decision making, the AP can exploit different information sources:

- *proprioceptive information* (1) – This information groups embedded static or dynamic data associated to the product or information obtained via sensors on its physical status (e.g., temperature, internal pressure).
- *exteroceptive information* (2 and 3) – This second information flow is obtained via interaction with its informational environment (e.g., users, resources of the support system) or its physical environment (e.g., temperature, hygrometry) using sensors. The AP can potentially dialogue with the other APs (3). The AP compares then the collected information with those obtained by other APs to reinforce the validity of its self-information. The flow of triggering information back to the support system is enhanced (4): raw data can be transformed in high level information by a preliminary diagnosis and/or by a cross-checking with other information sources.

The flow of “reliable” information allows the support system to improve its performance (5). For example, based on information gathered by the AP, the drift of a resource can be detected and treated earlier, improving its availability. The improved performance of the support system coupled with more accurate information on AP status logically induces a better quality of service for the AP (6).

A model supporting this concept of "activeness" is proposed in the next section.

2.2 Proposed Model

In our modelling approach, we adopt a "systemic" view, which considers the product as a system, called the "target system" Σ^T. Three traditional, complementary views (i.e., functional, organic and evolutionary) are considered in our modelling approach. In this chapter, we focus on the functional and evolutionary views. (Interested readers can consult the reference [15] to obtain more details on the organic view.)

Functional view. From a functional point of view, the target system Σ^T provides services associated to primary functions to a user or users. In complement, secondary functions are also required to:

- allow the target system to be effectively designed, manufactured, supplied, maintained and recycled; and
- improve the performance criteria associated to the primary functions (e.g., the self-diagnosis function provides a better availability to Σ^T).

All primary functions are assumed to be integrated in Σ^T. A set of internal secondary functions can also be integrated in Σ^T. The secondary internal functions are only relevant in the use phase, when Σ^T is operational. The other secondary functions, external, are executed by other systems, called support systems.

In our approach, these functions are then mapped on the "active" product or on the different support systems that intervene during the successive phases). As shown in Fig.2, for each phase of the product life cycle, certain external secondary functions are automatically supported (without human intervention) on a particular support system called the augmentation system $\sum^{S_A}$.

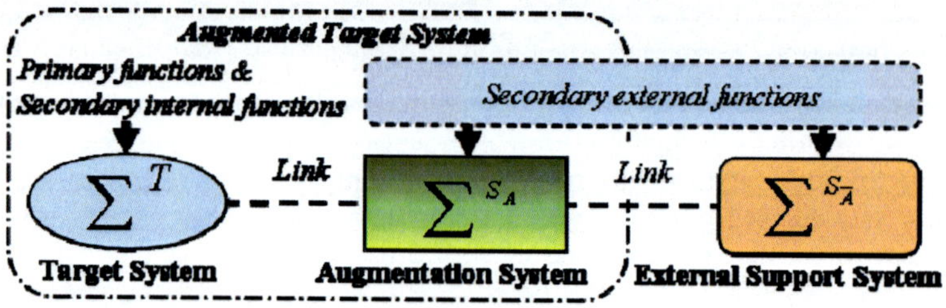

Fig. 2 Augmented target system

The most basic $\sum^{S_A}$ must be able to trigger certain useful tasks that are needed by Σ^T. A set of external secondary functions is supported by $\sum^{S_A}$, and the remaining external secondary functions are supported by the external support system $\sum^{S_{\bar{A}}}$.

An augmented target system, corresponding to the "active" product, is the result of adding $\sum^{S_A}$ to Σ^T.

Σ^T, Σ^{S_A} and $\Sigma^{S_{\bar{A}}}$ are associated via links, which can be informational, physical or temporal. For example, for conditional maintenance, a vibration monitoring system Σ^{S_A} physically linked to an engine Σ^T can send diagnosis data via an informational link to a remote maintenance centre $\Sigma^{S_{\bar{A}}}$.

The notations introduced above allow defining more formally the activeness:

- A passive system/product is a system / product in which all external secondary functions are supported by an external support system $\Sigma^{S_{\bar{A}}}$.
- A system / product is active in a specific phase of its life cycle if it supports one or more external secondary function(s) relevant for this phase, via an augmentation system.

Evolutionary view. The mapping of the functions between the different systems Σ^T, Σ^{S_A} and $\Sigma^{S_{\bar{A}}}$ might evolve with the successive product life cycles (e.g., successive versions of a car or a train). This evolution may be caused by many things, for example, providing new services to the consumers or increasing the traceability of the Σ^T. This evolution may:

- add (or cancel) a primary or secondary function supported by the Σ^T, corresponding to the evolution of the target system, or
- provide a new mapping of the secondary functions between the systems Σ^T, Σ^{S_A} and $\Sigma^{S_{\bar{A}}}$.

Between two successive life cycles, some functions may migrate from one system to another. For example, Fig.3 shows a scenario where a diagnosis function f3 (for predictive maintenance), originally supported in cycle *i*-1 by the system $\Sigma^{S_{\bar{A}}}$, could be supported by the augmentation system Σ^{S_A} in cycle i.

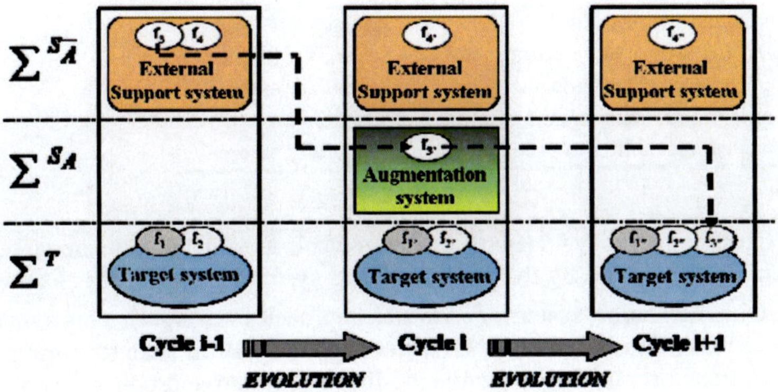

Fig. 3 Evolutionary view

In the end, the function f3 may be integrated into the target system Σ^T for a new generation of products (cycle i+1). This approach takes into account the evolution from a "prototype" to a mature industrialized product.

The next section proposes an initial instantiation of this augmentation concept.

3 Application of the Model in a Manufacturing Context

This section presents an initial instantiation of the model described above for a manufacturing context, via experiments on a real flexible manufacturing cell. This application is clearly situated in the context of product-driven manufacturing control [10, 12, 19]. In this innovative philosophy of control, the "active" product makes decisions concerning its "destiny" related to resource allocation and routing processes.

3.1 *Product Augmentation in a Manufacturing Phase*

As shown in Table 1, in our specific manufacturing application, several functions were considered.

- The two first functions, f_1 and f_2, are internal primary or secondary functions. For example, in the case of an engine, f_1 and f_2 can be respectively responsible for the propulsion task and for alerting the user of any failure. They are not operational (in grey in Fig. 3) because the product is not at its final manufacturing stage.
- Three other external functions $\{f_3, f_4, f_5\}$ are respectively responsible for resource allocation, product routing and product traceability.

Table 1 Description of the functions (manufacturing phase)

Functions	Description	Type
f_1	Perform a specific task	Primary
f_2	Alert the user of any failure	Secondary internal
f_3	Perform resources allocation	Secondary external
f_4	Route product towards the resources	Secondary external
f_5	Manage traceability in production	Secondary external

Unlike the traditional form of manufacturing control, in which the secondary functions are totally supported by the manufacturing system, these functions are supported by an augmentation system Σ^{S_A} in our approach (see Fig.4). This system is associated to the passive product Σ^T at the beginning of the manufacturing phase in order to transform it into an "active product". This active product is considered to be a consumer, with a list of services (i.e., transformation, transport, storage) to obtain from the resources in the manufacturing system. Such a list can be provided by an Enterprise Resource Planning (ERP) system.

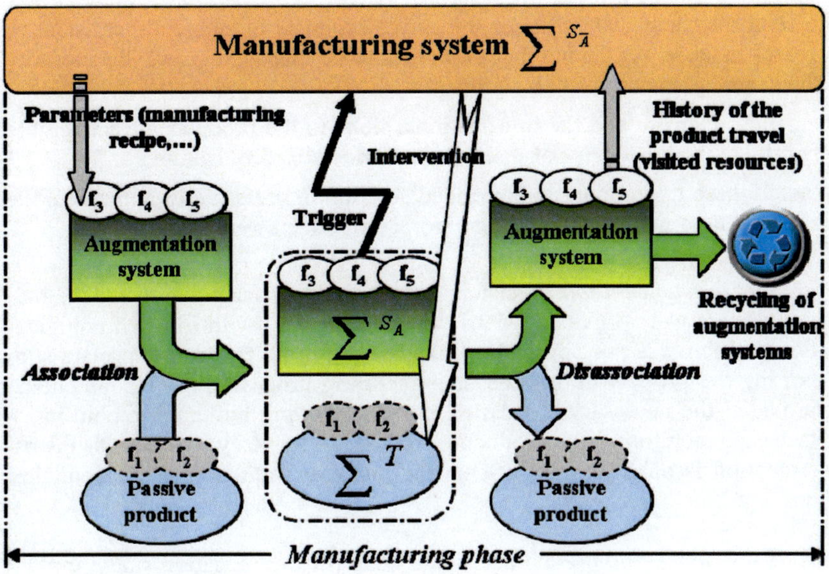

Fig. 4 Augmentation system in a manufacturing context

$\sum^{S_A}$ is then configured with the necessary data (e.g., manufacturing recipe). When the product is at the end of its manufacturing phase, it is disassociated from the augmentation system, which is then recycled for use with other products. The data recorded by the function f5 (e.g., list of visited resources, routing times, resource processing times) are sent to the manufacturing system $\sum^{S_{\bar{A}}}$.

The following sections describe briefly the instrumentation supporting our approach and the experiments.

3.2 Real Implementation

Real implementations were conducted in a flexible assembly cell at the Valenciennes AIP-Primeca Lab., and allowed one to check different interaction mechanisms to support the two functions, f_3 and f_4. In the first implementation [15], the allocation process and the routing process were respectively supported by a traditional contract-net protocol [17] and an embedded Dijkstra algorithm [4] that found the shortest time path toward the chosen resource. In Zbib et al. [21], the two functions exploit the same interaction mechanism, based on potential fields.

To support the interaction mechanisms, the augmentation system must be sufficiently open and flexible. The usual approaches use agents, located on a "distant" computer, to manage the activeness of the product [10]. The synchronization between the distant agent and the real product is generally executed via an RFID tag. The distant implementation leads to cheaper solutions, but has several limitations:

- The central system that manages the active products is inherently resistant to increases in scale. As the number of entities to be managed grows, the capacity of the central element must also grow.
- The central element and the informational links to the products are vulnerable to failure, with a major risk of losing information synchronization.
- To avoid these limitations, an embedded solution in preferred to implement the augmentation system. Two types of control equipment were used:

  - *Product instrumentation*: Decisional capabilities associated to the augmentation system were provided via an embedded eeePC (with IrDA communication), which was not physically linked to the product but to the plate supporting the product during the manufacturing process (Fig.5). This choice provided the necessary processing power, allowing building graphic interfaces for monitoring the production process. Obviously, the current instrumentation is only a "proof-of-concept", used to validate the augmentation concept.

  - *Node instrumentation*: Each decisional node (i.e., the location where the AP makes decision) and each resource was equipped with a Wago 750-841 controller. An integrated Ethernet interface allowed node-to-node interactions to be easily supported. However, since the product cannot directly use the Ethernet technology, it must be connected to the network via an IrDA communication (ESB-301 Clarinet system) at each decision node. (Interested readers can consult the reference [3] for more details.)

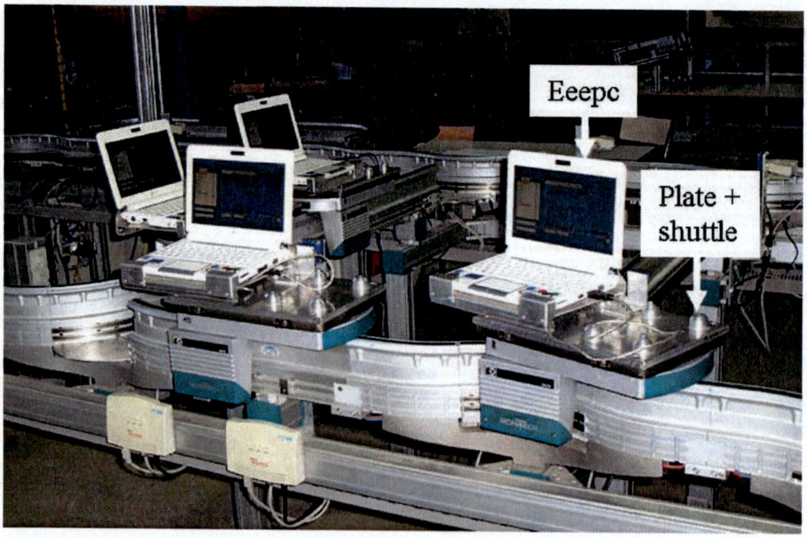

Fig. 5 "Active" product implementation

3.3 Experimental Results

Several scenarios have highlighted the different capabilities of our approach with respect to self-adaptation to perturbations (e.g. curative maintenance of resources, shuttle slowdowns or traffic jams, urgent and / or cancelled orders, re-manufacturing of products after failing quality control). The main benefits are:

1. Making decisions locally without any central point of control makes the system more reactive and more robust against perturbation. For the allocation function f_3, the ability to manage resources perturbations (e.g., breakdown or maintenance operations) was demonstrated in [15] and [21].
2. The efficiency of the routing function f_4 was equally proved in the reference [15] and Zbib's doctoral dissertation [22]. Unlike passive products, which always take the same path, the active product is able to adapt itself to the perturbation and to find an efficient alternative path. The dialogue between APs allows a dynamic routing that increases the system's reactivity when confronting a routing perturbation.

A video available at www.univ-valenciennes.fr/sp/ActiveProduct/ was prepared to show the experiments.

Contract-Net protocol (for allocation) coupled with Dijkstra (for routing) and potential field approach have also been compared. In our specific manufacturing case study, our experiments [22] have showed that potential fields seem to offer at least a 10% improvement over the total production time.

4 Application of Our Model in a Condition-Based Maintenance Context

One of the most challenging problems arising in the use phase is the predictive maintenance of complex systems. Condition-Based Maintenance (CBM) [16, 20] must replace the old prescheduled maintenance systems to identify incipient faults before they become critical and their causes in order to avoid imminent and future failures.

As mentioned in the introduction, new information technologies and the progress in infotronics made it possible to develop intelligent CBM [20]. Approaches based on remote maintenance [2], e-maintenance [13], smart sensors and "Watchdog Agents" [7] then appeared.

This section explores the possible benefits of the augmentation concept for the monitoring and the diagnosis of complex systems (e.g., nuclear plants, planes, trains).

4.1 Product Augmentation in Use Phase

A complex system is composed of several sub-systems in interaction. In addition, the "augmentation" of a complex system passes through the "augmentation" of some sub-systems (e.g., doors, HVAC for a transport system). The example below is related to a complex system monitoring / diagnosis application and the following functions are considered (see Table 2).

- The primary function f_1 deals with the task to be performed, depending of the studied complex system (e.g., door opening / closing for passenger access in a transport vehicle).
- Two external secondary functions $\{f_2, f_3\}$ are respectively responsible for performing an advanced diagnosis of any failures and for keeping a history of the defects in order to optimize the maintenance process.

Table 2 Description of the functions (use phase)

Functions	Description	Type
f_1	Perform principal task	Primary
f_2	Perform an advanced diagnosis of any failures	Secondary external
f_3	Record any failure in the history file	Secondary external

Unlike the traditional form of CBM, in which the secondary functions are totally supported by the maintenance system, our team suggests processing the data (e.g. measured variables) and the diagnosis as near as possible to the equipment considered. As proposed in the Open System Architecture for CBM (OSA-CBM) [18], seven functional levels are considered: (1) Sensor Module, (2) Signal Processing, (3) Condition Monitor, (4) Health Assessment (diagnosis), (5) Prognostics, (6) Decision Support, and (7) Presentation. Our augmentation system supports the four first levels.

As described on Fig.6, the function f2 uses a non-intrusive monitoring of the signals exchanged between the equipment Σ^T and its controller (1). The observed signals are compared to those generated by a real-time execution of a behavioural model of monitored equipment (2). The behavioural model must be calibrated with the data describing the context or the environment of the equipment (3) and the history of maintenance operations conducted on the equipment (e.g. replacement dates of some components) (4).

If a mismatch occurs, a diagnosis phase is triggered, and the results are exploited for curative or predictive maintenance:

- *Curative maintenance*: if the difference between the reality and the behavioural model is high, a failure is detected. A list of possible defective components is then built by analyzing the event cascade.
- *Predictive maintenance*: If the difference is low, a possible drift of the subsystem is detected. This work will be complicated due to the variability of environmental parameters.

To improve the diagnosis, other information sources can also be exploited: the measurements performed via the dedicated sensors on the equipment (5), and the information exchanged with the other "sister" augmentation systems (6). Suppose, for example, that an augmentation system is associated to each similar sub-system (e.g., engines in a plane, doors in a train). If an augmentation system associated to an equipment detects a dysfunction, problems with the internal components can be suspected. But if the same dysfunction is detected simultaneously by all augmentation agents, it means that the problem is not specifically to a sub-system but is probably linked to the context. Thus, the dialogue between augmentation systems prevents the emission of false alarms.

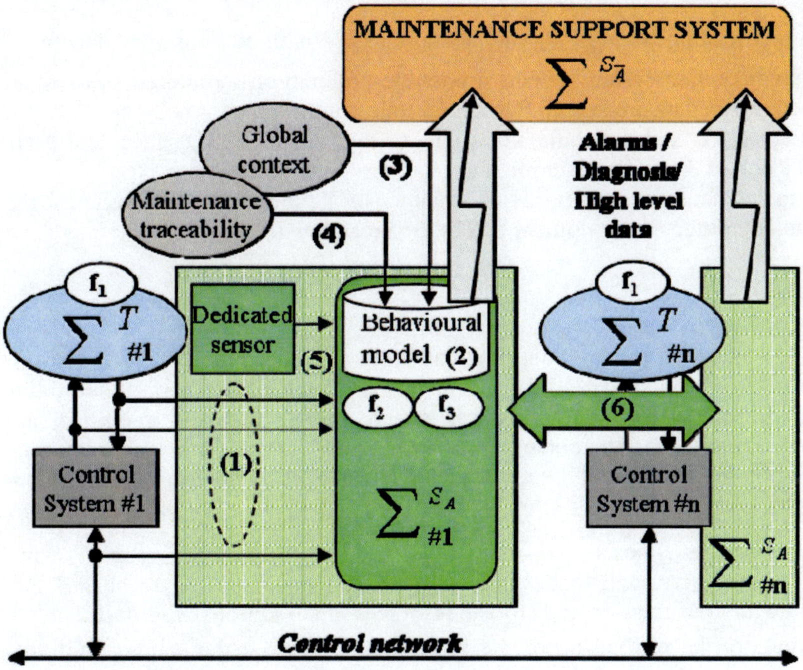

Fig. 6 Active monitoring principle

4.2 Real Implementation

Currently, our team is applying this approach to a major industrial project in the field of railway transportation. This project is being conducted by two research teams and three other industrial companies in the Valenciennes region. For confidentiality reasons, the details on technical aspects are voluntarily limited.

This project is trying to provide a more advanced solution that the existing remote solution [14]. In the existing solution, the data about equipment states (e.g., HVAC, brakes, doors) are collected in real-time from the vehicle and sent to a remote maintenance centre, where they are analyzed, which leads to a diagnosis.

Using this approach implies sending a large amount of data to a central database via General Packet Radio Service (GPRS) or wireless links in the railway stations. This approach suffers from significant limitations, such as the huge amount of data to be analyzed and the lag in detecting the problems.

To resolve these limitations, the goals of the project are to study, develop and implement the following functionalities:

- predictive diagnosis,
- assistance for predictive and curative maintenance,
- dynamic optimization of maintenance plans, which take into account the state of the monitored equipment on an entire fleet.

In this specific application of our augmentation system, three steps take place:

1. An augmentation system detects a possible problem on monitored sub-system and diagnosis data are sent to the central train controller;
2. This controller sends the diagnosis data to the remote maintenance centre via GPRS link in the railway station; and
3. The maintenance centre builds the maintenance schedule to intervene on the potentially defective equipment before the breakdown.

4.3 Results

The project is just beginning, and it is too early to discuss the results. However, the first trials are finished, and the results of these trials have allowed the origins of some defects to be found more easily and quickly than the traditional approaches, by exploiting the context data in real time. Compared to the traditional remote maintenance approach, the following advantages of the augmentation concept are expected:

- The in-situ diagnosis performed by the augmentation system allows maintenance personnel to be more reactive and eliminates "data bursts" that saturate the transmission and storage resources in a traditional approach.
- In addition, the processing of the informational data at the source limits the risks associated with archiving and future asynchronous processing. It reduces delays in the feedback loops, from the occurrence of a problem to its detection and solution.
- Furthermore, the constant local monitoring and the consideration of the context improve the detection of "furtive" failures and those failures that are not reproducible.
- Finally, the dialogue between augmentation systems generates more "robust" alarms.

5 Conclusion and Prospects

After presenting the benefits linked to the "activeness" of a product over its life cycle, the chapter proposed a model based on the augmentation concept. The augmentation approach allowed defining several types of functions: primary & secondary, respectively internal & external, which were mapped onto three systems: the target system, the augmentation system and the external support system.

The experiments have demonstrated that an augmentation system can increase robustness and adaptability in face of perturbations in a manufacturing context. Several experiments on a real flexible cell have shown the interest of product-driven systems. We also met the challenge to improve the monitoring / diagnosis of complex systems. An industrial railway project highlighted the relevant expected benefits.

To improve and to check the developed model for the distribution phase, our short-term research prospects focus on the traceability of blood products for healthcare systems. In a real context, this project directly applies the concepts developed in the field of closed-loop PLM. In fact, the blood product traceability involves several steps:

- collecting the blood and gathering donor information,
- testing the quality of collected products, processing (e.g., transforming the blood into plasma) and packaging in bags,
- distributing the blood to the healthcare network, and
- using the blood in hospitals or healthcare centres.

The aim of this traceability project is to enhance the performance of the blood distribution network and to insure the respect of some constraints (e.g., cold-chain compliance, traceability of the different blood products from the same donor). According the closed-loop PLM concepts, real-time traceability in distribution and in use phases should allow efficient backward loops towards the blood product manufacturing phase. In addition, all donors will be sent statistics on the use made of their blood to stimulate new donations.

Acknowledgments. This research has been financed by the Inter-ministerial Fund (FUI) and the Nord/Pas-de-Calais Region, and sponsored by the i-Trans competitiveness cluster. The author gratefully acknowledges the support of these institutions.

References

1. Bajic, E.: Ambient Networking for intelligent objects management, mobility and services. In: Seminar Institute For Manufacturing, IFM University of Cambridge (2004)
2. Bangemann, T., Rebeuf, X., Reboul, D., Schulze, A., Szymanski, J., Thomesse, J.P., Thron, M., Zerhouni, N.: PROTEUS - Creating distributed maintenance systems through an integration platform. Computers in Industry 57, 539–551 (2006)
3. Berger, T., Sallez, Y., Valli, B., Gibaud, A., Trentesaux, D.: Semi-Heterarchical allocation and routing processes in FMS control: a stigmergic approach. J. Intell. Robot Syst. 58(1), 17–45 (2010)
4. Dijkstra, E.W.: A note on two problems in connexion with graphs. Numer. Math. 1, 269–271 (1959)
5. FING: Machine to Machine. White paper co-directed by Fing, Syntec Informatique et France Télécom, Paris (2006)
6. Jun, H.B., Kiritsis, D., Xirouchakis, P.: Research issues on closed-loop PLM. Computers in Industry 58(8-9), 855–868 (2007)
7. Liao, L., Wang, H., Lee, J.: Reconfigurable Watchdog Agent® for machine health prognostics. International Journal of COMADEM (2008)

8. Lindwer, M., Marculescu, D., Basten, T., Zimmermann, R., Marculescu, R., Jung, S., Cantatore, E.: Ambient intelligence visions and achievements: Linking abstract ideas to real-world concepts. In: Design, Automation and Test in Europe Conference (DATE 2003), Munich (2003)
9. Mattern, F.: From smart devices to smart everyday object. In: SOC 2003 (Smart Objects Conference), Grenoble (2003)
10. McFarlane, D., Sarma, S., Chirn, J.L., Wong, C.Y., Ashton, K.: The intelligent product in manufacturing control and management. In: 15th IFAC World Congress, Barcelona (2002)
11. Meyer, G.G., Främling, K., Holmström, J.: Intelligent Products: A survey. Computers in Industry 60(3), 137–148 (2009)
12. Morel, G., Valckenaers, P., Faure, J.M., Pereira, C.E., Diedrich, C.: Manufacturing plant control challenges and issues. Control Eng. Pract. 15, 1321–1331 (2007)
13. Müller, A., Crespo Marquez, A., Iung, B.: On the concept of e-maintenance: Review and current research. Reliability Eng. Syst. Safe 93, 1165–1187 (2008)
14. Orbita, B.T.: Data monitoring cuts rail running costs. Prof. Eng. 19(20), 50–53 (2006)
15. Sallez, Y., Berger, T., Deneux, D., Trentesaux, D.: The lifecycle of active and intelligent products: The augmentation concept. International Journal of Comput. Integ. M 23(10), 905–924 (2010)
16. Sethiya, S.K.: Condition Based Maintenance (CBM). Secy. to CME/WCR/JBP (2008)
17. Smith, R.G.: The contract net protocol: High level communication and control in a distributed problem solver. IEEE T. Comput. 29(12), 1104–1113 (1980)
18. Thurston, M.G.: An Open Standard for Web-Based Condition-Based Maintenance Systems. In: IEEE System Readiness Technology Conference-Autotestcon, Valley Forge (2001)
19. Trentesaux, D.: Distributed control of production systems. Engineering Applications of Artificial Intelligence 22(7), 971–978 (2009)
20. Vachtsevanos, G.J., Lewis, F.L.: Intelligent fault diagnosis and prognosis for engineering systems. Wiley, New York (2006)
21. Zbib, N., Pach, C., Sallez, Y., Trentesaux, D.: Heterarchical production control in manufacturing systems using the potential fields concept. Journal of Intelligent Manufacturing (2012), doi:10.1007/s10845-010-0467-3
22. Zbib, N.: Allocation et routage dynamique dans un FMS basés sur le concept de champ de potential, PhD Thesis, University of Valenciennes (2010)

Engineering Hierarchical Complex Systems: An Agent-Based Approach. The Case of Flexible Manufacturing Systems

Gildas Morvan[1,2], Daniel Dupont[1,3], Jean-Baptiste Soyez[1,4], and Rochdi Merzouki[1,4]

[1] Univ. Lille Nord de France,
1bis rue Georges Lefèvre 59044 Lille cedex, France
[2] LGI2A, U. Artois, Technoparc Futura 62400 Béthune, France
gildas.morvan@univ-artois.fr
[3] HEI, 13 rue de Toul 59046 Lille Cedex, France
daniel.dupont@hei.fr
[4] LAGIS, EC-Lille, Avenue Paul Langevin
BP 48 59651 Villeneuve D'ascq cedex, France

Abstract. This chapter introduces a formal model to specify, model and validate hierarchical complex systems described at different levels of analysis. It relies on concepts that have been developed in the multi-agent-based simulation (MABS) literature: *level, influence* and *reaction*. One application of such model is the specification of hierarchical complex systems, in which decisional capacities are dynamically adapted at each level with respect to the emergences/constraints paradigm. In the conclusion, we discuss the main perspective of this work: the definition of a generic meta-model for holonic multi-agent systems (HMAS).

Keywords: multi-level multi-agent based simulations, formal models, hierarchical systems.

1 Introduction

Engineering a complex system such as a flexible manufacturing system (FMS) is a challenging problem. The target system is complex, holonic, relies on distributed decisional processes, and must be adaptive, i.e., robust to perturbations and easily reconfigurable.

To solve these problems, proposed solutions,[1] take advantage of system features such as:

- complexity, distributing the control in system components that embody primitive cognitive capacities, e.g., be able to be identified, to communicate, to react to environmental changes,

[1] E.g., heterarchical [3] or semi-heterarchical [25] control, holonic multi-agent systems (HMAS) [2, 6, 35, 34] or intelligent product based concepts [24].

T. Borangiu, A. Thomas, and D. Trentesaux (Eds.): SOHOMA 2011, SCI 402, pp. 49–60.
springerlink.com

- holonic structure, using dedicated meta-models and conception methodologies.

An important tool in the design, simulation and validation of such solutions has been *multi-agent-based simulation* (MABS). This article introduces a formal model to specify, model and validate hierarchical complex systems. It takes inspiration from two trends in MABS research:

- the formalization of interaction models,
- multi-level modelling, where interacting agents are ontologically distributed among multiple layers of organization.

The article is organized as follows:

- in section 2, the two trends of MABS research cited above, multi-level modelling and formal modelling[2], are introduced;
- the section 3 presents a generic formal model for multi-level MABS;
- an abstract implementation of this model, focusing on the specification of hierarchical multi-agent systems (MAS), in which decisional capacities are dynamically adapted at each level with respect to the emergences/constraints paradigm, is proposed in the section 4;
- the conclusion (section 5) summarizes the authors' contributions and perspectives for future work.

2 Two Trends in MABS Research

2.1 *Multi-level Modelling*

A *level* represents a point of view on the system, and its relations to other points of view [16]. While this concept seems important to understand complex systems[3], it generally remains abstract: implementations tend to constraint this definition, in particular the relations between levels. Therefore, a *multi-level model* integrates knowledge on different levels and their relations. Multi-scale are multi-level models characterized by hierarchical relations in levels [7, 12, 19, 23]. A level may represent, according to the context, a spatial-temporal extent, a position in a decision hierarchy, etc. Let us consider these two examples.

1. The system is characterized by processes that have different spatial-temporal extents. Two types of relations can be commonly found in such models:

- *scaling*, i.e., computing macroscopic (respectively microscopic) variables from microscopic (respectively macroscopic) processes;
- *grouping* and *degrouping* (or aggregation and disaggregation) [6,20, 26], i.e., defining a process at a level as a group (respectively part) of processes (respectively a process) at another level.

---

[2] The presentation focuses on the *influences* → *reaction* model (IRM). Other approaches such as IODA [10] or based on DEVS [18] are not described.

[3] Multi-level approaches have proven useful in many domain such as statistics [8], chemistry [9, 11], physics [28], hydrology [26] or biology [33].

2. Levels are characterized by decisional capacities; relations represent the *emergence* of new capacities and the *constraint* over existing capacities [15, 17].

A level is often viewed as a *level of organization*. This concept is closely related to the notion of holon [6]. This aspect is discussed in the section 5.

2.2 The Influences → Reaction Model

The *influences* → *reaction* model (IRM) has been developed to address issues raised by the classical vision of action in Artificial Intelligence as the transformation of a global state [5]:

- simultaneous actions cannot be easily handled,
- the result of an action depends on the agent that performs it but not on other actions,
- the autonomy of agents is not respected.

Basically, it decomposes action in two phases: agents and environment (micro level) produce a set of influences, then the system (at macro level) reacts to influences; e.g., detects and solves influence conflicts such as in the platform Jaak[4]. As [13] notes, "the influences [produced by an agent] do not directly change the environment, but rather represent the desire of an agent to see it changed in some way". Thus, reaction computes the consequences of agent desires and environment dynamics. In recent years, variants of IRM have been developed to handle specific situations [13, 16, 36, 37]. This presentation focuses on the influence reaction model for simulation (IRM4S) [13].

Let $\delta(t) \in \Delta$ be the dynamic state of the system at time *t*:

$$\delta(t)=<\sigma(t),\gamma(t)>, \tag{1}$$

where $\sigma(t) \in \Sigma$ is the set of environmental properties and $\gamma(t) \in \Gamma$ the set of influences, representing system dynamics. The state of an agent $a \in A$ is characterized by its physical state $\phi a \in \Phi a$ with $\Phi a \in \Sigma$ (e.g., its position) and its internal state sa $\in$ Sa (e.g., its beliefs).

The evolution of the system from *t* to *t+dt* is a two-step process:

1. Agents and environment produce a set of influences[5] $\gamma`(t) \in \Gamma`$,
2. The reaction to influences produces the new dynamic state of the system.

An agent $a \in A$ produces influences through a function $\text{Behaviour}_a: \Delta \rightarrow \Gamma`$. This function is decomposed into three functions executed sequentially:

$$p_a(t)=\text{Perception}_a(\delta(t)), \tag{2}$$

---

[4] `http://www.janus-project.org/Jaak`

[5] The sets of producible influence sets and influences produced at t are denoted respectively $\Gamma`$ and $\gamma`$to point out that the latter is temporary and will be used to compute the dynamic state of the system at *t + dt*. (HMAS) [2, 6, 35, 34] or intelligent product based concepts [24].

$$s_a(t+dt)=Memorization_a(p_a(t),s_a(t)) \tag{3}$$

$$\gamma`_a=Decision_a(s_a\ (t+dt)) \tag{4}$$

The environment produces influences through a function $Natural_\omega: \Delta \rightarrow \Gamma`$:

$$\gamma`_\omega=Natural_\omega(\delta(t)) \tag{5}$$

Then the set of influences produced in the system at t is:

$$\gamma`(t)=\{\gamma(t)\cup\gamma`_\omega(t)\cup\bigcup_{a\in A}\gamma`_a(t)\} \tag{6}$$

After influences have been produced, the new dynamic state of the system is computed by a function Reaction : $\Sigma \times \Gamma` \rightarrow \Delta$ such as:

$$\delta(t+dt)=Reaction(\sigma(t),\gamma`(t)) \tag{7}$$

3 A Generic Meta-model for Multi-level MABS

In this section, a generic meta-model for multi-level MABS, called IRM4MLS, is presented[6]. This model has the following interesting properties:

- any valid instance can be simulated [27],
- simulation scheduling is logically distributed by level,
- complexity of simulation algorithm can be optimized according to model structure.

3.1 Specification of the Levels and Their Interactions

A multi-level model is defined by a set of levels L and a specification of the relations between levels[7]. Two kinds of relations are specified in IRM4MLS: an influence relation (agents in a level l are able to produce influences in a level $l` \neq l$) and a perception relation (agents in a level l are able to perceive the dynamic state of a level $l` \neq l$), represented by directed graphs denoted respectively $< L,E_I >$ and $< L,E_P >$, where E_I and E_P are two sets of edges, i.e., ordered pairs of elements of L. Influence and perception relations in a level are systematic and thus not specified in EI and EP (cf. eq. 8 and 9).

E.g., $\forall l,l` \in L^2$, if $E_P=\{ll`\}$ then the agents of l are able to perceive the dynamic states of l and l` while the agents of l` are able to perceive the dynamic state of l`.

The in and out neighbourhood in $< L,E_I >$ (respectively $< L,E_P >$) are denoted N_I and N_I^+ (respectively N_P and N_P^+) and are defined as follows:

$$\forall l\in L, N_I^-(l)\ \ (resp.\ N_I^-(l))=\{l\}\cup\{l`\in L: l`l\in E_I(resp.\ E_P)\}, \tag{8}$$

$$\forall l\in L, N_I^+(l)\ \ (resp.\ N_I(l))=\{l\}\cup\{l`\in L: ll`\in E_I(resp.\ E_P)\}, \tag{9}$$

---

[6] The dynamic aspects of the meta-model, i.e., simulation algorithms, are not described here. An exhaustive presentation can be found in [16].

[7] The notion of level is here similar to the notion of brute space in the MASQ metamodel [29].

E.g., $\forall$l,l`$\in$L2, if l`$\in N_I^+(l)$ then the environment and the agents of l are able to produce influences in the level l`: conversely we have l`$\in N_I(l)$, i.e., l`is influenced by l.

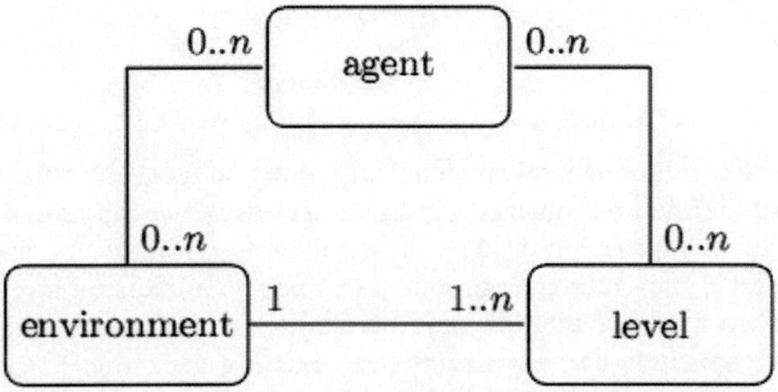

Fig. 1 Main concepts of IRM4MLS (cardinalities are specified in the UML fashion)

3.2 *Agent Population and Environments*

The set of agents in the system at time t is denoted A(t). $\forall$l$\in$L, the set of agents belonging to *l* at *t* is denoted $A_l(t) \subseteq A(t)$. An agent belongs to a level if a subset of its physical state ϕ_a belongs to the state of the level:

$$\forall a \in A(t), \forall l \in L, a \in A_l(t) \; if \; \exists \varphi_a^l(t) \subseteq \varphi_a(t) \mid \varphi_a^l(t) \subseteq \sigma^l(t) \quad (10)$$

Thus, an agent belongs to zero, one, or more levels. As notes [29, p. 815], the physical state of an agent in a level, i.e., its body, is "the manifestation of an agent in the environment and allows others to perceive it." An environment can also belong to multiple levels (see Fig. 1).

3.3 *Action Modelling*

The dynamic state of a level l $\in$ L at time t, denoted $\delta l(t) \in \Delta l$, is a tuple < $\sigma l(t), \gamma l(t)$> where $\sigma l(t) \in \Sigma l$ and $\gamma l(t) \in \Gamma l$. are the sets of environmental properties and influences of *l*.

$$\forall a \in A_l, \; Behaviour_a^l : \prod_{l_P \in N_P^+(l)} \Delta^{l_P} \rightarrow \prod_{l_I \in N_I^+(l)} \Delta^{l_I} \quad (11)$$

Once influences have been produced, interactions between levels do not matter anymore. Thus, the reaction function defined in IRM4S can be reused:

$$Reaction^l : \Sigma^l \text{ x } \Gamma^{l'} \rightarrow \Delta^l \quad (12)$$

where $Reaction^l$ is the reaction function proper to each level.

4 Engineering Hierarchical Complex Systems with IRM4MLS

4.1 The Emergence/Constraint Paradigm

In many MABS, processes are considered on the next 2-level relative hierarchy:

Arrows represent causality relations between levels. Dashing suggests that they are generally not explicitly defined but emerge from interactions between entities. On the contrary, a multi-level approach considers these relations explicitly. In engineering applications, a level may rather represents a position in a decision hierarchy (cf. section 2.1). Two kinds of relation may be distinguished in such systems: *emergence* of new capacities and *constraint* over existing capacities [14]. Let us consider an example in the domain of FMS engineering. In a case study on automated guided vehicle (AGV) control presented in [17] (cf. section 4.4), the model relies on the following relations:

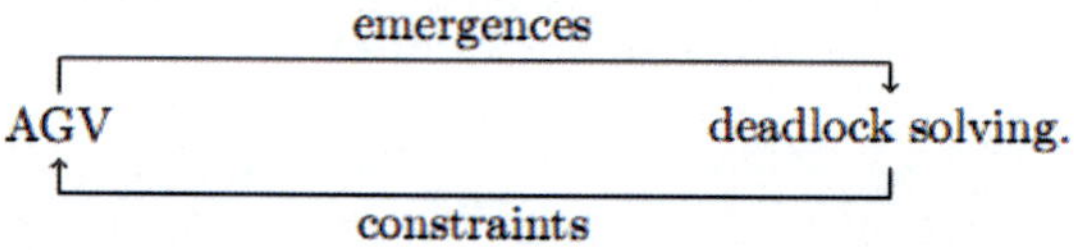

Macro agents (representing a set of "trapped" AGVs) emerge from micro agent interactions when an interaction pattern defined as a deadlock is detected, and then constrains their behaviours to solve it. While the notions of emergence and constraint were informally defined in [17], formal definitions in the context of IRM4MLS are given in the following.

4.2 IRM4MLS Implementation

Let L be a hierarchy and $\{\mu, M\} \subseteq L$ two hierarchically coupled levels, μ referring to the micro level and M to the macro level. Thus, A_μ (respectively A_M) denotes the agents of the micro-level (respectively macro-level). The emergence/constraint paradigm supposes that $E_I \supseteq \{\mu M, M\mu\}$.

$$\forall l \in L, \gamma^{l}`(t) = \{\gamma^{l}(t), \gamma_{\omega}^{M}, \gamma_{\omega}^{\mu}, \bigcup_{a \in A_M} \gamma_{a}^{M}`(t), \bigcup_{a \in A_\mu} \gamma_{a}^{\mu}`(t)\} \quad (13)$$

An *emergence* e at the level M is an influence that has the following properties:

- *e* belongs to the macro-level but not to the micro-level:

$$e \in \Gamma^{M} \; but \; e \notin \Gamma^{\mu} \quad (14)$$

- *e* cannot be produced by the behavior of an agent or the environment of M:

$$\forall t, e \notin \bigcup_{a \in A_M} Behaviour_a^M(\gamma(t)) \cup Natural_\omega^M(\gamma(t)) \quad (15)$$

with $\gamma(t)=<\delta^M(t), \delta^\mu(t)>$.

Emergent influences generally determine the life-cycle (creation, evolution, destruction) of agents at the macro-level.

A *constraint* over an influence *i*, denoted ¬i, is the special kind of influence that has the following properties:

- {i, ¬i} belongs to the micro-level but not to the macro-level:

$$\{i, \neg i\} \subseteq \Gamma^\mu \; but \; \{i, \neg i\} \subseteq \Gamma^M \quad (16)$$

- ¬i cannot be produced by the behavior of an agent or the environment of μ

$$\forall t, \neg i \notin \bigcup_{a \in A_\mu} Behaviour_a^\mu(\gamma(t)) \cup Natural_\omega^\mu(\gamma(t)) \quad (17)$$

with $\gamma(t)=<\delta^M(t), \delta^\mu(t)>$.

- ¬i inhibits i:

$$\text{if } \{i, \neg i\} \subseteq \gamma^{\mu`}(t) \text{ then} \quad (18)$$
$$Reaction^\mu(\sigma^\mu(t), \gamma^{\mu`}(t)) = Reaction^\mu(\sigma^\mu(t), \gamma^{\mu`}(t) \setminus \{i\})$$

4.3 Conception of Hierarchical Systems

The approach described below can be viewed as a semi-heterarchical control one and takes advantage of complexity and hierarchical (not yet holarchical) organization of the system, distributing the control by level. Heterarchical control methods rely on self-organization principles[8] and therefore assume that the system is able to achieve its goals and is easily reconfigurable, i.e., that the normal functioning mode emerges from the interactions between system components (products, machines, simulated entities, etc.) that embody limited cognitive capabilities (cf. introduction). However, the trajectory of such systems may lead to non desired attractors.

The proposed methodology is presented in Fig.2. The system is designed iteratively in a two-step process.

1. From an initial specification of the system, a model of the system in normal functioning mode is defined and verified, i.e., those system components have the necessary cognitive capacities to perform their tasks.
2. From non desired attractors exhibited by the simulation of the model, the control strategy may be designed and validated. However, it is likely that the specification of the system has be modified to do so, e.g., because a new decisional level is needed.

---

[8] Self-organized systems are generally characterized by the use of environment as a communication medium to carry local information as well as positive and negative feedbacks.

The notion of influence is very general and therefore, may have many possible meanings. In this case, let

- $\gamma^{l^{\wedge}}$(t) be the capacities of each agent of a level l at time t, i.e., the tasks they can perform at the moment,
- γ^{l}(t)the actual affectation of tasks to agents; the only cognitive capacity required for agents is to expose services they may provide.

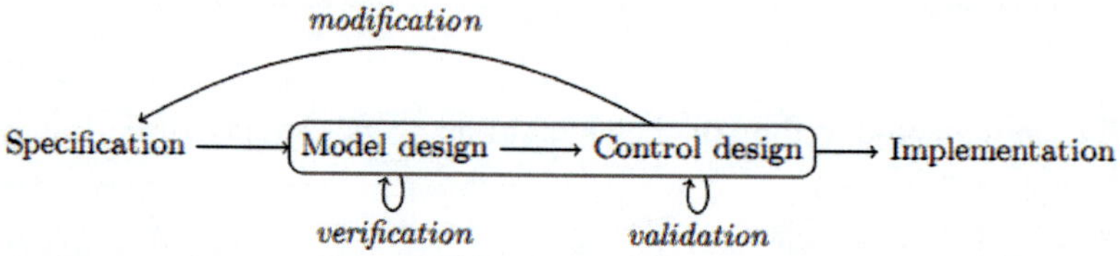

Fig. 2 Engineering methodology

Thus, *Reaction*^{l} is a task assignment algorithm that computes $<\sigma^l(t),\gamma^l(t)>$ from $<\sigma^l(t),\gamma^{l^{\wedge}}(t)>$. Note that the hierarchical nature of the system allows decomposing the specification of the system S by level:

$$S = \{\gamma^l(\delta^l) : \forall l \in L, \forall \delta^l \in \Delta^l\} \tag{19}$$

i.e., task assignments for all functioning modes.

That design should lead to the definition of reaction functions that control goal affectations. If such a function cannot be defined, the system design is not valid and must be redefined. This process is iterated until a solution is found (Fig. 2).

4.4 Case Study: AGV Deadlocks in Gradient Field-Based FMS

The main functionalities of an intelligent transportation system (ITS) are: (1) transport assignment, (2) routing, (3) gathering traffic information, (4) collision avoidance, (5) deadlock avoidance [38].

Gradient field-based approaches, where AGV trajectories are computed from gradient fields, allow to implement efficient ITS in FMS [31, 32]. A dedicated task assignment algorithm is generally used to ensure functionality 1, while functionalities 2-4 rely on AGV and shop self-organization properties. Thus, an AGV has two cognitive capabilities: senses attractive or repulsive force fields and emits a repulsive force field. Similarly, a shop is able to emit attractive fields to require products to process and give back the result to the system. A known problem of gradient field-based approaches is that a group of AGVs may be trapped in local minima that lead to a system deadlock [30, 32, 39]. However, this issue can be easily addressed by hierarchical control methods that compute explicit trajectories[9].

---

[9] Readers interested in general, i.e., not gradient-field based approaches, deadlock avoidance techniques in FMS may refer e.g., to [1, 4, 40].

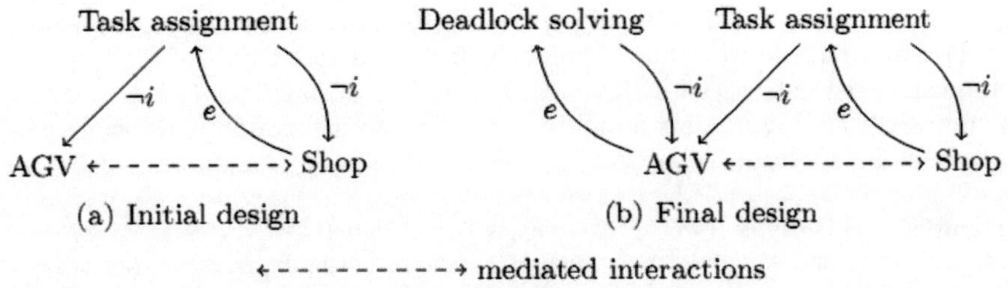

Fig. 3 Decisional levels in the case study on AGV control

The first design of the system is presented in Fig. 3(a): a task assignment algorithm affects goals to AGVs (statically, a signal to maximize) and shops (dynamically, products to process). The deadlock avoidance functionality is not explicitly programmed but is supposed to emerge from mediated interactions between AGVs and shops. Various researches have shown that such a solution may reduce the number of deadlock occurrences but not eliminate it: routing is not deadlock avoidance [39]. A new system architecture is then designed (see Fig. 3(b)): if a deadlock (triggered by an emergence) is observed by a deadlock solving algorithm, constraints over signal sensing and emission are computed to solve it[10].

5 Conclusion

In this article, we have presented a formal model for MABS and its implementation to engineer hierarchical complex systems. Two types of influences have been distinguished in this approach: *emergence*, that basically triggers a new system behaviour when a specific pattern is detected (in the previous short example of gradient field-based FMS, the detection of a deadlock triggers the modification of AGV repulsive signal emission) and *constraint*, that, as its name suggests, constrains decisional capacities of system entities to solve a situation.

The main advantage of this approach lies in the multi-level and simulation capabilities of IRM4MLS to model a system in which decisional capacities are distributed in its components and evolve along time to meet user's goals and to simulate a model whiteout bias and temporal deadlocks[11]. Its main drawback is the strict hierarchical organization in levels.

Holonic multi-agent systems (HMAS) can be viewed as a specific case of multi-level multi-agent-systems (MAS), the most obvious aspect being the loosely hierarchical organization of levels. However, from a methodological perspective, differences remain: thus, most of holonic meta-models focus on organizational

[10] Practical aspects of this approach are discussed in [17]. E.g., AGVs embody the deadlock solving algorithm, becoming multi-level agents. This problem has been an important motivation in the development of IRM4MLS.

[11] Simulation properties of IRM4MLS may be exploited to explore model behaviour using, e.g., the polyagent concept [22, 21]. Such an approach may be used to determine fail probabilities of system components or control strategies.

aspects (cf. e.g., [2, 6, 35, 34]). An important issue towards a generic meta-model for HMAS would be to define a holon with respect to IRM4MLS concepts: a holon cannot be defined with IRM4MLS first class abstractions (level, agent or environment), as it represents a multi-level entity. This situation is the main perspective of this work.

Acknowledgments. Authors would like to thank Daniel Jolly (LGI2A, Université d'Artois, Béthune France) and Alexandre Veremme (HEI, pôle recherche Ingénierie et Sciences du Vivant, Lille France) for their help and support. Jean-Baptiste Soyez is funded by the In-Trade project[12].

References

1. Banaszak, Z., Krogh, B.: Deadlock avoidance in flexible manufacturing systems with concurrently competing process flows. IEEE Transactions on Robotics and Automation 6(6), 724–734 (1990)
2. Bendriss, S., Benabdelhafid, A., Boukachour, J., Boudebous, D.: Meta-modèle de référence holonique pour la gestion de la traçabilité du produit dans la chaîne logistique. In: 5ème Colloque International Conception et Production Intégrées -CPI 2007 (2007)
3. Duffie, N.: Heterarchical control of highly distributed manufacturing systems. International Journal of Computer Integrated Manufacturing 9(4), 270–281 (1996)
4. Ezpeleta, J., Tricas, F., Garcia-Valles, F., Colom, J.: A banker's solution for deadlock avoidance in fms with flexible routing and multiresource states. IEEE Transactions on Robotics and Automation 18(4), 621–625 (2002)
5. Ferber, J., Müller, J.P.: Influences and reaction: a model of situated multiagent systems. In: 2nd International Conference on Multi-agent systems (ICMAS 1996), pp. 72–79 (1996)
6. Gaud, N., Galland, S., Gechter, F., Hilaire, V., Koukam, A.: Holonic multilevel simulation of complex systems: Application to real-time pedestrians simulation in virtual urban environment. Simulation Modelling Practice and Theory 16, 1659–1676 (2008)
7. Gil Quijano, J., Hutzler, G., Louail, T.: Accroche-toi au niveau, j'enlève l'échelle: Éléments d'analyse des aspects multiniveaux dans la simulation à base d'agents. Revue d'Intelligence Artificielle 24(5), 625–648 (2010)
8. Goldstein, H.: Multilevel Statistical Models. Wiley Series in Probability and Statistics, 4th revised edn. Wiley-Blackwell (2010)
9. Horstemeyer, M.: Multiscale Modeling: A Review. In: Practical Aspects of Computational Chemistry Methods, Concepts and Applications, pp. 87–135. Springer, Heidelberg (2010)
10. Kubera, Y., Mathieu, P., Picault, S.: Interaction-oriented agent simulations: From theory to implementation. In: Proceeding of the 2008 conference on ECAI 2008: 18th European Conference on Artificial Intelligence, pp. 383–387. IOS Press (2008)
11. Lucia, A.: Multi-scale methods and complex processes: A survey and look ahead. Computers & Chemical Engineering 34(9), 1467–1475 (2010); Selected Papers from the 7th International Conference on the Foundations of Computer-Aided Process Design (FOCAPD 2009), Breckenridge, Colorado (2009)

[12] `http://www.intrade-nwe.eu`

12. McGregor, S., Fernando, C.: Levels of description: A novel approach to dynamical hierarchies. Artificial Life 11(4), 459–472 (2005)
13. Michel, F.: The IRM4S model: the influence/reaction principle for multiagent based simulation. In: AAMAS 2007: Proceedings of the 6th International Joint Conference on Autonomous Agents and Multiagent Systems, pp. 1–3. ACM, New York (2007)
14. Morin, E.: Method: Towards a Study of Humankind, vol. 1. Peter Lang. Pub. Inc. (1992)
15. Morvan, G., Jolly, D., Veremme, A., Dupont, D., Charabidze, D.: Vers une méthode de modélisation multi-niveaux. In: Actes de la 7ème Conférence de Modélisation et Simulation MOSIM, vol. 1, pp. 167–174 (2008)
16. Morvan, G., Veremme, A., Dupont, D.: IRM4MLS: The Influence Reaction Model for Multi-Level Simulation. In: Bosse, T., Geller, A., Jonker, C.M. (eds.) MABS 2010. LNCS (LNAI), vol. 6532, pp. 16–27. Springer, Heidelberg (2011)
17. Morvan, G., Veremme, A., Dupont, D., Jolly, D.: Modélisation et conception multiniveau de systèmes complexes: stratégie d'agentification des organisations. Journal Européen des Systèmes Automatisés 43, 381–406 (2009)
18. Müller, J.-P.: Towards a Formal Semantics of Event-based Multi-agent Simulations. In: David, N., Sichman, J.S. (eds.) MAPS 2008. LNCS, vol. 5269, pp. 110–126. Springer, Heidelberg (2009)
19. 19 Müller, J.P., Ratzé, C., Gillet, F., Stoffel, K.: Modeling and simulating hierarchies using an agent-based approach. In: Proceedings of the MODSIM 2005 International Congress on Modelling and Simulation, 1631–1638 (2005)
20. Navarro, L., Flacher, F., Corruble, V.: Dynamic level of detail for large scale agent-based urban simulations. In: Proc. of 10th Int. Conf. on Autonomous Agents and Multiagent Systems (AAMAS 2011), pp. 701–708 (2011)
21. Van Dyke Parunak, H.: Pheromones, Probabilities, and Multiple Futures. In: Bosse, T., Geller, A., Jonker, C.M. (eds.) MABS 2010. LNCS (LNAI), vol. 6532, pp. 44–60. Springer, Heidelberg (2011)
22. Van Dyke Parunak, H., Brueckner, S.: Concurrent Modeling of Alternative Worlds with Polyagents. In: Antunes, L., Takadama, K. (eds.) MABS 2006. LNCS (LNAI), vol. 4442, pp. 128–141. Springer, Heidelberg (2007)
23. Ratzé, C., Gillet, F., Müller, J.P., Stoffel, K.: Simulation modelling of ecological hierarchies in constructive dynamical systems. Ecological Complexity 4(1-2), 13–25 (2007)
24. Sallez, Y., Berger, T., Deneux, D., Trentesaux, D.: The lifecycle of active and intelligent products: The augmentation concept. International Journal of Computer Integrated Manufacturing 23(10), 905–924 (2010)
25. Sallez, Y., Berger, T., Raileanu, S., Chaabane, S., Trentesaux, D.: Semi-heterarchical control of FMS: From theory to application. Eng. Appl. Artif. Intell. 23, 1314–1326 (2010)
26. Servat, D., Pierrer, E., Treuil, J., Drogoul, A.: Towards Virtual Experiment Laboratories: How Multi-Agent Simulations Can Cope with Multiple Scales of Analysis and Viewpoints. In: Heudin, J.-C. (ed.) VW 1998. LNCS (LNAI), vol. 1434, pp. 205–217. Springer, Heidelberg (1998)
27. Soyez, J.B., Morvan, G., Merzouki, R., Dupont, D., Kubiak, P.: lisation et simulation multi-agents multi-niveaux. Submitted to Studia Informatica Universalis (2011)
28. Steinhauser, M.: Computational Multiscale Modeling of Fluids and Solids. Springer, Heidelberg (2008)

29. Stratulat, T., Ferber, J., Tranier, J.: Masq: towards an integral approach to interaction. In: AAMAS 2009: Proceedings of The 8th International Conference on Autonomous Agents and Multiagent Systems, pp. 813–820. International Foundation for Autonomous Agents and Multiagent Systems, Richland (2009)
30. Ueda, K., Kitob, T., Fujii, N.: Modeling biological manufacturing systems with bounded-rational agents. CIRP Annals - Manufacturing Technology 55(1), 469–472 (2006)
31. Ueda, K., Markusb, A., Monostori, L., Kalsc, H., Arai, T.: Emergent synthesis methodologies for manufacturing. CIRP Annals - Manufacturing Technology 50(2), 535–551 (2001)
32. Ueda, K., Vaario, J., Ohkura, K.: Modelling of biological manufacturing systems for dynamic reconfiguration. CIRP Annals - Manufacturing Technology 46(1), 343–346 (1997)
33. Uhrmacher, A.M., Ewald, R., John, M., Maus, C., Jeschke, M., Biermann, S.: Combining micro and macro-modeling in devs for computational biology. In: Proceedings of the 39th Conference on Winter Simulation: 40 years! The Best is Yet to Come, WSC 2007, pp. 871–880. IEEE Press, Piscataway (2007)
34. Van Brussel, H.: Holonic manufacturing systems, the vision matching the problem. In: Proc. of First European Conf. on Holonic Manufacturing Systems (2007)
35. Van Brussel, H., Wyns, J., Valckenaers, P., Bongaerts, L., Peeters, P.: Reference architecture for holonic manufacturing systems: PROSA. Computers in Industry 37(3), 255–274 (1998)
36. Weyns, D., Holvoet, T.: Model for Simultaneous Actions in Situated Multi-agent Systems. In: Schillo, M., Klusch, M., Müller, J., Tianfield, H. (eds.) MATES 2003. LNCS (LNAI), vol. 2831, pp. 105–118. Springer, Heidelberg (2003)
37. Weyns, D., Holvoet, T.: A formal model for situated multi-agent systems. Fundamenta Informaticae 63(2-3), 125–158 (2004)
38. Weyns, D., Holvoet, T., Schelfthout, K., Wielemans, J.: Decentralized control of automatic guided vehicles: applying multi-agent systems in practice. In: Companion to the 23rd ACM SIGPLAN Conference on Object-oriented Programming Systems Languages and Applications, OOPSLA Companion 2008, pp. 663–674. ACM, New York (2008)
39. Weyns, D., Boucké, N., Holvoet, T.: Gradient field-based task assignment in an agv transportation system. In: Proceedings of the 5th International Joint Conference on Autonomous Agents and Multiagent Systems, AAMAS 2006, pp. 842–849. ACM, New York (2006)
40. Yoo, J.W., Sim, E., Cao, C., Park, J.W.: An algorithm for deadlock avoidance in an agv system. The International Journal of Advanced Manufacturing Technology 26, 659–668 (2005)

HAPBA – A Holonic Adaptive Plan-Based Architecture

Doru Panescu and Carlos Pascal

"Gheorghe Asachi" Technical University of Iasi
Department of Automatic Control and Applied Informatics
Blvd. Dimitrie Mangeron 27, Iasi, 700050, Romania
{dorup,cpascal}@tuiasi.ro

Abstract. This contribution reveals how the planning and coordination processes in Holonic Manufacturing Execution Systems entail the need of a centralized component, under the form of a staff holon. By using appropriate models (both monochrome and coloured Petri nets) and certain aspects of planning techniques in Artificial Intelligence, a PROSA based implementation is considered. In the obtained holonic scheme, named HAPBA, the solution of the manufacturing goals is attained by holons' cooperation, according to an enhanced form of the Contract Net Protocol, so that certain drawbacks that are possible to appear in a holonic approach are avoided. Theoretical and experimental arguments are supporting the proposed solution.

Keywords: Holonic manufacturing execution systems, planning, Petri nets, contract net protocol, staff holon.

1 Introduction

The PROSA holonic architecture is the starting point for several approaches and implementations [14]. It has the advantage of combining the hierarchical and heterarchical control methods [1], establishing some general patterns to be followed in the design and implementation phases. The proposed architecture is a PROSA implementation; it is named HAPBA (Holonic Adaptive Plan-Based Architecture) in order to suggest the importance of an appropriate solution for the planning phase [10], which should facilitate an adaptive and safe operation for the Holonic Manufacturing Execution System (HMES) [2, 13].

Two main ideas were guiding our approach. First, in various researches on HMESs the agent based techniques are considered as main means for holonic solutions; however, the possibilities of the Artificial Intelligence planning techniques were not entirely used and adapted to the specific requirements of holons, and HAPBA is an attempt in this respect. Then, holons' functioning is mostly based on few clear statements: they act as autonomous, deliberative entities with their social activities determining the organization of the holons' groups (the holarchies), which are supposed to resolve the manufacturing goals. Thus, certain

T. Borangiu, A. Thomas, and D. Trentesaux (Eds.): SOHOMA 2011, SCI 402, pp. 61–74.
springerlink.com

ingredients contribute to the construction of an HMES: the holons' typology established by PROSA, an agent based design and implementation, with the relations between holonic agents set up according to the Contract Net Protocol (CNP) [4, 11]. This chapter reveals some weak points appearing when such a combination is used without certain enhancements, by making use of both theoretical and experimental arguments. These elements added in HAPBA represent the second point to individualize this holonic scheme. The following sections of the chapter explain some main issues of the considered modelling formalism, of the holonic planning and coordination processes, and plead for the necessity of the staff holons.

2 Petri Nets Modelling of Holonic Systems; Some Main Planning Issues

In order to conduct a systematic design and implementation of HMESs, one needs an appropriate modelling tool. As in other holonic investigations, the Petri nets are utilized in developing HAPBA [5, 9]. The modelling aspects concern the principal component of a holon, namely its deliberative part built under the form of a holonic agent. Besides this, a holon is composed of a structural (execution) component that carries out the commands established by the decisional constituent. The holonic structural part can be a proper physical device (only for a resource holon) or a holarchy [8, 9, 10, 14]. Between the holonic decisional and execution components suitable communication interfaces must exist [7, 8].

A distinctive point of the proposed methodology refers to the construction of the Petri net models valid for all the three types of holons – *order*, *product* and *resource* holons, that are supposed to make explicit the two phases of holon's operation: (1) planning which is developed from the moment of receiving a goal until finding a plan able to solve the goal, in the most cases by cooperation, and (2) execution that starts at the moment when a complete plan is validated within the holonic scheme and regards the carrying out and monitoring of the established set of actions. The models have been firstly devised as monochrome Petri nets. The universal model suitable to describe the operation of the holonic agent is presented in Fig. 1.

The two working phases of the holonic agent – planning and execution – are represented by the places P_1 and P_2, respectively. Besides these, the agent can be in an idle (waiting state) denoted by the place P_0. This happens when it has no goal to solve, or it is waiting for feedback from the HMES. The idle state can appear during planning, because when a goal is solved by cooperation the solution is constructed according to the information asynchronously changed between the manager and contractor agents within the CNP; it can appear during execution too, when after sending a command, the agent has to wait its carrying out. The planning/execution processes regarding different goals can be interleaved and this multi-tasking operation of the holonic agent is made possible by passing through the idle state.

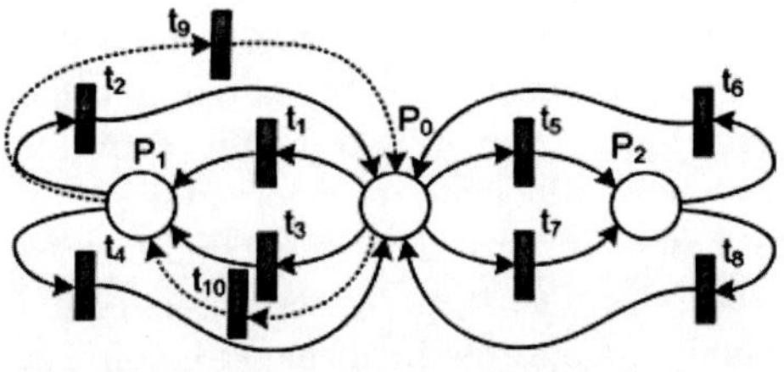

Fig. 1 The Petri net model describing the holonic agent operation

The Petri net of Fig. 1 catches all types of events that determine changes of the holonic agent's state (the elements with dashed lines in Fig. 1 will be explained in Section 3). It is intended for all the cases possible according to the application of the CNP: a holonic agent acts as a manager, as a contractor or it undertakes both roles. The transition t_1 represents the receiving of goals, t_3 regards the receipt of bids, t_2 concerns the sending of goals (in this way t_2 and t_3 reflect the cases when the agent is manager) and t_4 appears when the agent issues bids (meaning it acts as a contractor, for the goals got during t_1). When the agent is manager and is on the top level in a holarchy, the planning phase is ended by sending an internal contract (which allows the start of its own execution phase) and this event is modelled by t_4, too.

Concerning the execution phase, its pre- and post-conditions are modelled by appropriate transitions. t_5 happens when the holonic agent receives a contract: it had proposed a bid, and t_5 means it can start the execution as it was selected by the manager. t_7 models the feedback from contractors (sub-contractors) or from the controlled device regarding the ending (with success or failure) of an entire contract or of part of it (an action). The transition t_6 designates the execution phase finalization; this transition provides the feedback towards the manager holon. t_8 models the events of sending contracts when the execution part of the holon is a holarchy, or commands towards the physical device for a resource holon.

The model of Fig. 1 is only the highest abstraction of holonic agent operation, and it can be further detailed. In HAPBA the reasoning mechanism applied by the holonic agents is of the BDI (Belief Desire Intention) type [15]. From the planning point of view this means that the agent planning phase is not done from scratch, but each holonic agent is endowed with a plan library, with the components selected and used when needed. There is a specific mode of constructing and using the agents' plans in HAPBA. In the agents' libraries a plan is partially specified, as in the planning methodology of the plan space approach [3]. Thus, according to the application of the least commitment principle, the actions of a plan are partially ordered and the plan variables are partially bound (e.g., the actors of plan actions may be not specified). This creates a greater flexibility for the management of goals within the HMES. As an example, the following expression represents a plan with two actions (a_1 and a_2) whose order is not established:

$$a_0 < a_1 \leq a_2 < a_\infty \qquad (1)$$

The usual notation of planning theory is utilized, with a_0 and a_∞ representing the dummy actions for the initial state from which the plan is to be started, and the goal state, respectively. There are two advantages obtained by using this form for the planning phase of the HMES. First, for holons which are managers in the CNP it allows the establishing of the contractor holons to fulfil the actions in a distributed manner, according to the bidding process results. It means the actors of the actions a_1 and a_2 are firstly seen as un-bound variables by the manager holon that is using the plan (1) – for example it can be a product holon, their values being determined by the result of the CNP. Though in a centralized planning approach the problem of keeping the consistency of the variable binding is known to be an NP-complete one [3], in the holonic scheme it can be efficiently handled at the level of each contractor agent: a holonic agent will not make a bid beyond its execution part capacity or without having an entirely specified plan.

The second advantage regards the way the plan actions can be ordered in an optimum manner at the appropriate level. If the holonic agent which is the manager in the CNP is using the plan (1) and according to its knowledge it can already decide on the most favourable order between the two actions, it will do so and accordingly use this order in announcing the corresponding goals and selecting the bids. On the contrary, more frequently the manager holon should leave this decision to be made later, based on the information provided by contractors. As an example from the considered experimental environment (see Section 4), if the two actions of the plan (1) refer to placing two parts of an assembled product, they will be done successively when the two parts are handled by a single robot (as the parts are in its storage device), or the actions are done simultaneously when the parts exist in the storage devices of two robots and can be handled by them. It means the decision is taken according to the bids proposed by contractors, the holonic agents of the two robots in this case. It is to point out that in HAPBA the concern is on the planning aspects, the scheduling problem being partially tackled: the resource allocation and the order of actions are decided, without giving a detailed time schedule.

The Petri nets provide the means to suitably model and integrate the above considered form of plans with the holonic agents' operation. For instance, taking into account the basic model of Fig. 1 and the plan (1), used by a product holon, the Petri net model of the execution phase for the product holon when it succeeded to finalize the planning phase is represented in Fig. 2. The places P_{Ra1} and P_{Ra2} represent the availability of the actors for the actions a_1 and a_2. When the plan is considered from the library of the holonic agent, its model appears as in Fig. 2, but without tokens in the places P_{Ra1} and P_{Ra2}. During the planning phase, the product holon acting as a manger sends goals to the potential contractors in order to find execution entities for the two actions of the plan. When contracting is successfully ended, the places P_{Ra1} and P_{Ra2} get a token, according to the holon chosen to execute the corresponding action. In this situation the model reflects the way the product holon should conduct and monitor the execution phase.

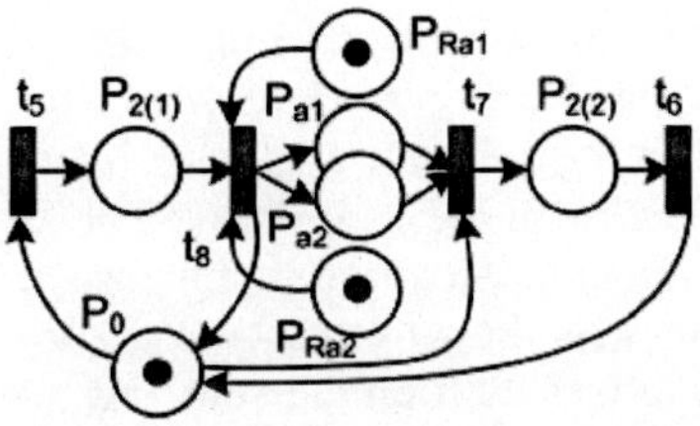

Fig. 2 The model for the execution phase of a product holon

As explained in Fig. 1, the transition t_5 starts a first execution stage marked as a distinct position, $P_{2(1)}$ in Fig. 2. t_5 happens as the result of contract awarding. It means the product holon is itself a contractor for the goal having as solution the plan (1). According to the validated plan, the execution continues with awarding contracts towards the chosen sub-contractors – the transition t_8 in Fig. 2. The places P_{a1} and P_{a2} abstract the activity of the two contractors. In the presented example the Petri net represents the case when the plan was settled so that the two actions are executed in the same time (in fact, they are allowed to start in the same time, and the end can be different). When the product holon receives the feedback concerning the finalization of both actions – this means the transition t_7 is fired, it passes through the last execution stage, marked in Fig. 2 as the distinct place $P_{2(2)}$. Complying with the model of Fig. 1, the entire execution phase is ended by the transition t_6, which provides the necessary feedback to the holon's manager. It is clear that this example illustrates the general case of a holon being both contractor and manager.

Because the monochrome Petri nets have limited possibilities with respect to the control scheme analysis and as a step towards facilitating the holonic design and implementation, the conceived models were transposed into Coloured Petri Nets (CPNs) [6]. By making possible information processing, this approach allowed the development of a complete model for the proposed architecture. In CPNs, a certain data structure for the tokens within a place has to be established; for the transition firing, besides the common condition (each input place must contain the needed tokens), a guard test can be set up. This is an optional condition attached to a transition, which enables it only if the information of tokens within the input places of the transition makes true a test predicate. By using guard tests it was possible to group some places of the original monochrome Petri net in one place of the CPN (this is an advantage of the respective approach); such a place output transitions are distinctively activated based on the token information and their test predicates. It is the case of the places P_{Ra1} and P_{Ra2} (see Fig. 2) that are one place (P_{Ra}) in the corresponding CPN. This has the colour structure of its tokens as a 4-tuple:

```
(plan-id, goal-type, goal-id, {(action-type, actor),…})
```

The *plan-id* attribute is a unique plan identifier, the *goal-type* attribute specifies the type of goal that the plan can be used for, the *goal-id* attribute identifies the specific goal for which the plan instance was created, and the last attribute is a set of pairs; these indicate the action type and its selected contractor / actor, for each action of the plan. The content of these tokens is similar to that of the items in the agents' plan library, except for the *goal-id* and *actor* components of the last attribute, which are unspecified before starting the planning process. Only when this process is ended, its results are tokens of the place P_{Ra} with all attributes being instantiated. These tokens are used during the execution phase and removed at its end. As a connection towards implementation, different data carried by tokens are materialized by the means of holonic agents' belief sets. Some other types of token data exist in HAPBA model, which maintain information on the sent and received goals, bids, contracts, and feedbacks.

A token in the places regarding the planning/execution processes carries always information about the goal identifier. If an operation is made by another entity than the holon itself, then the pair information on the action and its actor has to appear. Thus, when the model of Fig. 2 is transposed in its corresponding CPN, the tokens of the places $P_{2(1)}$ and $P_{2(2)}$ are not carrying the information on the *action type* and *actor* attributes, but these appear for the tokens of the places P_{a1} and P_{a2}. As an illustration for the guard tests, the information on the *goal-id* in the place $P_{2(1)}$ is compared with that of the place P_{Ra} so that the transition t_8 should determine the sending of the contracts to the right actors, the holons that are supposed to carry out the actions of the plan chosen and instantiated for the respective goal.

The way CPNs can guide the HMES design and implementation in a modular manner is an advantage of their use, too. The holonic scheme can be modelled in successive levels, by substituting an entire CPN corresponding to a level with a transition of the higher level. The model of the execution process (as the one of Fig. 2) and that of the planning component are substituted in the upper model, namely the one of the holonic agent, by two transitions. Such a transition determines the same changes in all the input/output places to which it is connected. The BDI reasoning mechanism can get its own model, which is after that included only in the model of the planning process, as the execution phase of the holonic agent has a single course; when a failure appears, the planning component has to be called. On the next level, the holonic agent model is abstracted by a transition connected to one input and one output place of the entire HMES model (see Fig. 3, where the model of the product holon of type 1 is connected with the places I_2 and O_2). These places are buffers of the messages produced during different events in the HMES.

The model of the HMES used for tests is shown in Fig. 3; it contains different entities that are coupled together through the input (I_i) and output (O_i) message places by a model of the communication network infrastructure. This transfers tokens (messages) from the output place of an entity to input places of other components. To make easier certain simulation experiments it may be necessary that some models are reduced; for instance, when making the HMES analysis, in order to diminish the state space, all the resource holons are further abstracted as a single transition, modelling their behavior (*SIM Resources* transition in Fig. 3). In this way, it is possible to simulate and analyze the behavior of the entire holonic system or only of certain parts of it. Two aspects are considered based on the reachability graph: the *boundedness* and *liveness* properties, and the system reaction to specific events. The last aspect, made possible by examining the token contents, permitted us to distinguish all the HMES outcomes.

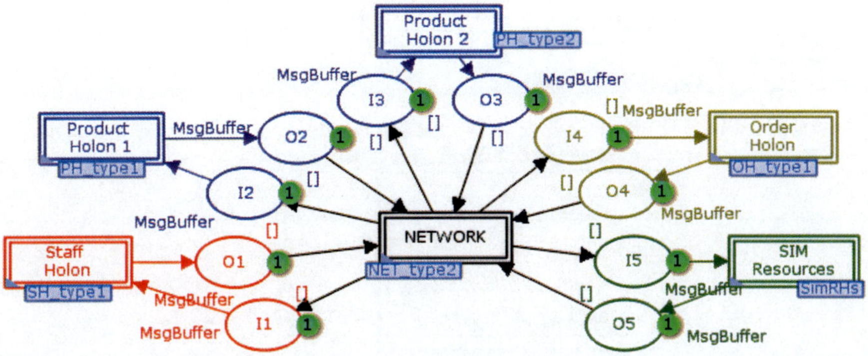

Fig. 3 The HMES model used for simulation experiments

3 The Need of Holonic Centralized Components – Staff Holons

After establishing the patterns for the operation of holons and the corresponding models, it is possible to analyse some drawbacks of the holonic scheme. These are not hard to explain considering how the plans are assembled from partial solutions proposed by holons having a limited view on the manufacturing environment. As an example, further commented in the next section, it may be that the HMES contains two product holons, each of them trying to solve a goal by using a plan of type (1), in the same time. If the available resource holons can achieve only two of the four actions within the two plans, it can happen that the bids are made so that neither of the two product holons is able to positively finalize its planning process. This stems from the way the CNP is a coordination scheme that is not able to always get a solution for a problem, even when this exists. The only possibility that the common CNP suggests is for the managers that failed to further announce their goals in a while.

A better remedy for an HMES is to introduce a centralized component, under the form of the *staff holon* according to the PROSA architecture. As in the respective methodology, a staff holon may not impose a certain result for a goal, but it can ensure a further coordination for the HMES, according to the global information it possesses. An enhanced form of the CNP is used in HAPBA.: it involves three types of entities – manager, contractor and staff holons. Because the operation of a contractor is not affected by the introduction of the staff holon, its cycle is the one of the normal CNP. For a holonic agent playing the role of manager, the cycle is added with initial and final communication with the staff holon. The following steps result (Fig. 4: the manager is a product holon and the contractors are resource holons):

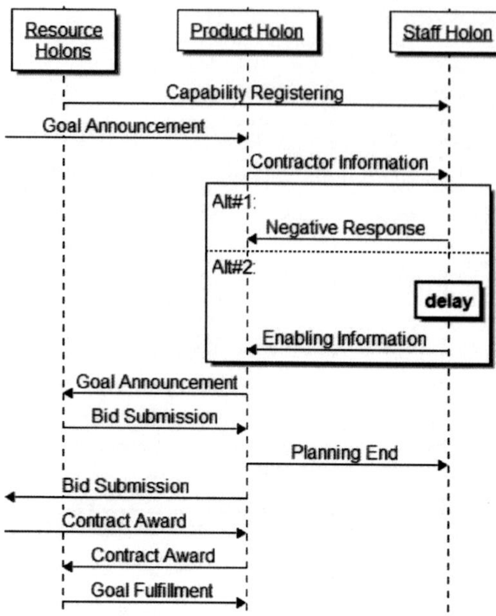

Fig. 4 Interaction diagram for the coordination protocol with a staff holon

1. According to the current goal (which was received from another holon – the first *Goal Announcement* event in Fig. 4) and the plan that the holonic agent of the product holon intends to use, it announces the components (actions) of the plan for which it has to find contractors (actors) to the staff holon; in Fig. 4 this step is represented by the *Contractor Information* arrow.
2. If the holonic agent receives a message from the staff holon about the impossibility of achieving the plan (due to the lack of potential contractors – the *Negative Response*), then it chooses another plan for the respective goal, according to the BDI mechanism, and restarts the cycle from the previous step.
3. If the holonic agent receives the enabling message from the staff holon, together with the set of potential contractors (*Enabling Information*), it sends the necessary goals to all the indicated contractors; this event is represented by the *Goal Announcement* arrow that points from the product holon toward the resource holons in Fig. 4.
4. The agent receives the bids for all actions of the used plan (*Bid Submission*).
5. If there is an action of the plan that received no bid, then the holonic agent aborts the plan and restarts the cycle from the first step, with another plan selected from its plan library. Otherwise it chooses the best bids and continues with the next step. In both cases the holonic agent announces the staff holon about plan ending (*Planning End* arrow).
6. According to the selected bids, if the holonic agent plays only the role of manager then it sends the message for the beginning of execution phase (the so called internal contract), else it sends the constructed bid towards its superior manager (*Bid Submission* event that follows the *Planning End* in Fig. 4).

It is to point out again that the above cycle regards the general case when the same holon is both contractor with respect to solving a goal received from a manager holon, and manager with respect to finding a solution by cooperation to those parts of the goal that is not able to achieve by itself. Only the top level holon of a holarchy will be just manager; in this specific case the sixth step of the presented cycle determines the internal contract issuing, and within the execution phase the holon will send messages of contract awarding to its contractors (see transition t_8 in Fig. 1). In the other case, when the product holon is also contractor, and successfully ends the planning phase, the CNP continues with its normal progress (see the final interactions in Fig. 4).

From the above cycle it results that a staff holon receives two types of messages. When it receives a message about the intention of a manager to use a plan, the respective message is placed in a queue. This data structure is preferred so that the goals to be considered in the order they appear in the HMES. The message contains the plan components, namely its actions. As an example, if the holonic agent selected the plan (1) to solve the current goal, then its message towards the staff holon contains the set formed of two actions, a_1 and a_2.

The second type of messages concerns the end of the planning process for a goal. This message is used to update the information for the filtering mechanism that the staff holon applies when it decides if a manager can proceed with a goal processing. It must be understood that the staff holon plays the role of enabling / blocking the processing of goals. This means that at a message of a manager regarding a planning process beginning, the staff holon replies with an enabling answer only when a filtering condition is satisfied. Otherwise, the respective request is included in a queue of delayed tasks. In this way the respective plan processing is postponed, as the staff holon does not provide the enabling answer to the requesting manager.

As already told, the staff holon should provide not only an enabling message, but also the set of potential contractors for each action (sub-goal). Thus in HAPBA by using this set, a manager announces a goal only to the indicated contractors. To be able to supply such information, it is further necessary that the staff receives and keeps the appropriate data from all holonic agents belonging to its domain. Any holon must begin its operation by announcing (registering) the types of goals it is capable to solve to the staff holon to whose area it is associated (this is the significance of the *Capability Registering* arrow in Fig. 4). Whenever a holon changes its capabilities (becoming able to solve more or less goals), it has to notify this modification to the staff holon. Consequently it is clear that a staff holon can detect the case when there is no holon able to execute an action comprised in the plan of a manager and it can provide the appropriate message (see the second step of the above cycle).

The information within the staff holon's knowledge base has a twofold utilization: it makes available the data needed for guiding the communication within the CNP, and it is involved in the staff holon blocking mechanism. The simplest approach is to stop a plan when among its potential contractors there are holons already involved in an ongoing planning process. This blocking condition for the plan π_i can be expressed as follows:

$$\exists \pi_j ,\ \pi_j \in \text{PIP, so that } H_{\pi i} \cap H_{\pi j} \neq \varnothing \tag{2}$$

where $H_{\pi k}$ is a set containing names of holons:

$$H_{\pi k} = \{H \mid H \text{ is a holon that can solve an action of the plan } \pi_k\} \tag{3}$$

and PIP (Plans In Progress) is the set of all already enabled plans (started and not validated / finalized, yet). The application of this enabling algorithm by the staff holon supposes that it keeps the information on the holons already involved in planning processes; this information is updated according to the messages received from the managers in the first and fifth step of the above presented cycle. The condition (2) is a rough one and its checking implies the complexity $O(n)$, where n is the cardinality of the union of sets $H_{\pi k}$, for all π_k belonging to PIP. It means the staff holon has to do more or less tests before deciding to enable a request, depending on the number of plans which are in progress at that moment.

A plan having its processing postponed is considered when the plans to which it can interfere are finalized. After the staff holon received a message from a manager about a planning process end, it must update the set PIP and re-apply the enabling test to all plans from the queue of delayed plans. Any such plan, if it does no more satisfy the blocking condition (2), is removed from the queue and the manager that originated the request receives the enabling message from the staff holon.

It is to notice that a holonic agent that plays the role of manager while not receiving the enabling message from the staff holon is not entirely blocked. This can be explained after completing the Petri net of Fig. 1 with the transitions regarding the holonic agent's relationship with its staff holon; they are shown with dashed lines in Fig. 1. The transition t_9 is fired when the holonic agent sends the message for requesting the staff holon's acceptance to use a plan. According to the output place of this transition, after the respective event the agent will be in the idle state, so that it can continue with planning/execution activities regarding other goals. The transition t_{10} represents the event when the agent receives an answer from the staff holon, which conducts to a further planning stage. The sending of the message at the planning end towards the staff holon is integrated within the transition t_4; when the holonic agent has finalized a planning process by composing a bid or issuing an internal contract, it also sends the message to inform the staff holon on the respective event.

About the number of staff holons, this depends on how a company is organized and the extension of HMES deployment. Different holonic levels can be separated: a multi-enterprise holonic level, an enterprise, a shop-floor, and finally the level of manufacturing cells [12, 14]. For each such level a distinct staff holon has to be created and the holons comprised in the respective domain must accordingly register their services, before beginning their operation, as well as whenever a change affecting their capabilities appears.

4 Experimental Results and Conclusions

The developed trials refer to a real manufacturing system (Fig. 5a) and to simulation tests. The experimental manufacturing system contains two industrial robots

(6 d.o.f., of ABB type) in charge with part handling, a machine tool for part processing, a conveyor for part transfer, a computer vision system allowing part identification and check, and several storage devices. In order to solve assembly tasks, a table is placed in the common working area of the two robots, so that they can both participate in the assembling process. The considered goals concern products as the one in Fig. 5b. Four parts are involved: part A is to be placed first, then the two parts of type B can be positioned simultaneously or in any order and the final product is obtained by screwing part C. The holonic agents associated to manufacturing resources, and to the product, order and staff holons are running on some computers or embedded devices, networked with device controllers. The holonic scheme deployment on the presented equipment is under progress.

Certain simulation experiments were carried out using the model of Fig. 3. The two product holons concern the manufacturing of two products supposing an assembling layout similar to that of Fig. 5b, with some particularities in the material- and processing flow. A product holon is in HAPBA an active entity (this is quite different from the PROSA scheme) managing the goals about a specific product processing. It should find a suitable plan, validate it according to the available resources, and then monitor the plan execution until the product is finalized. Though the plans of the considered products contain four actions corresponding to achieving the placement of the four parts, for simplicity there were considered only the two middle actions - referring to the arrangement of the parts of type B. Thus it results that the plans the two product holons are using are similar to the plan of expression (1) (the simultaneous execution of the two actions is possible when the two robots are involved), but with different actions: the plan of the first product holon (PH_1) contains the actions a_1 and a_2, and for the second product holon (PH_2) the actions of the plan are a_3 and a_4.

There is a simple scenario that certainly reveals the need of the staff holon. This is when the two robots, seen as the resource holons RH_1 and RH_2, are able to achieve the actions as stated in the first line of the Table 1; it happens that in the working area of the first robot there is a part that can be used to achieve any of the actions a_1 and a_3, while according to the available part for the second robot, it can achieve any of the actions a_2 and a_4. In the second and third columns, Table 1 presents a comparison between the operation of the HMES without and with the staff holon. The information regarding the reachability graphs obtained in these two experiments is given in the upper part of the Table 1.

The implication of the staff holon causes state space reduction: both the number of states and the number of transitions are diminished. The dead markings are the final states, i.e. the states from which no transition is possible. Based on the token information (e.g. the tokens of the place P_{Ra} of each product holon indicate the success or failure of its planning process), the dead markings can be classified in three categories. It is possible for one or the other of the product holons to finalize its planning process, or for none of them. The percentages in Table 1 represent the number of dead markings that belong to each of the three categories. As expected, the results of the Table 1 show that the staff holon removes the undesirable case.

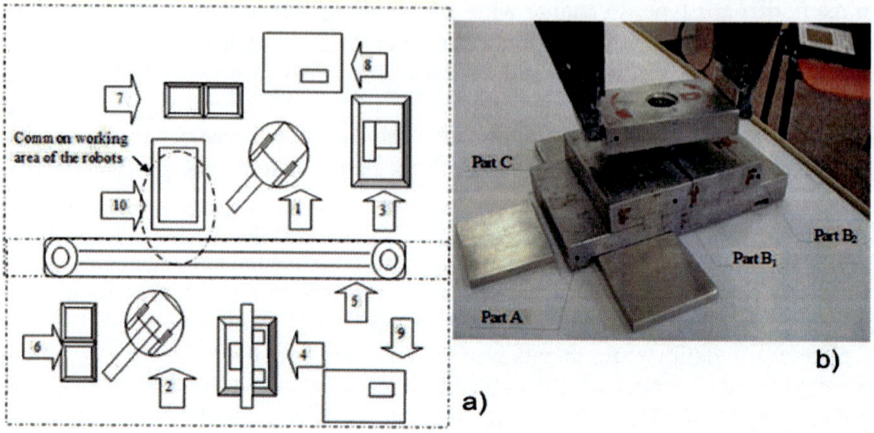

Fig. 5 a) Layout of the experimental manufacturing system: 1,8-IRB 1400 robots; 2,9-IRB 2400 robots; 3-Machine tool; 4-Computer vision system; 5-Conveyor; 6,7-Storage devices; 10-Assembly table. b) An example of assembled product

Another set of experiments is given in the 4th and 5th columns of Table 1 (the ones marked with *). They were obtained by assigning a higher priority for the transition that models the communication network (see Fig. 3). In this case the dimension of the reachability graph is reduced, which is due to the avoidance of messages from a transmitter interposing between messages of another transmitter. These experiments display the HMES results when the influence of the communication network is reduced to the maximum, and additionally prove the need of the staff holon in order to avoid a failure for both product holons; it is confirmed that the HMES malfunction is not caused by a peculiar progress of the holons' message exchange, but by a possible evolution of the planning process.

It can be noticed that the boundedness and liveness properties of the Petri net model are necessary but not sufficient conditions for a safe operation of HMESs.

Table 1 The results of the simulated experiments without/with the staff holon

RH Capabilities		$HR_1 \rightarrow \{a_1, a_3\}$, $HR_2 \rightarrow \{a_2, a_4\}$			
Method		Default	With staff holon	Default*	With staff holon*
Nodes		11148	5700	9643	4583
Arcs		24180	11018	16406	6626
Dead Markings		50	32	50	32
PH_1	PH_2	%	%	%	%
success	failure	48	50	48	50
failure	success	48	50	48	50
failure	failure	4	-	4	-

In the considered examples these conditions are met as shown by the reachability graph. The *boundedness* results because the reachability graph has a finite number of states, while the *liveness* is established through the analysis of dead markings. Namely, for any input goal (determining the initial marking) all the dead markings contain an answer provided by the HMES. It is not enough to ascertain the existence of a holon's answer, but the belief sets of each holon have to be checked in an additional analysis; this can be seen as a problem of knowledge base consistency. By comparing the token contents in the dead markings with that of the initial marking, it can be seen how the capabilities of holons are kept, namely at the end of goal processing the HMES has come back to the initial state. Except for the data needed in a learning procedure (this was not considered in our approach), the information regarding the history of solving a goal is not kept as it is not necessary. The performed experiences showed that in HAPBA the holonic agents' knowledge base consistency and efficiency are obtained. As an illustration for the usefulness of analyzing the holonic belief sets, a possible error that can be discovered refers to the case when a holon made a bid and it is not freed if it was not chosen by the manager; in HAPBA an explicit resource de-allocation mechanism is carried out during the planning and coordination processes.

Another merit of HAPBA is the way some cases of malfunction for the holonic planning component are discovered; namely, it may be that the HMES is not providing the right solution for certain goals or combination of goals, as illustrated in our examples when the staff holon is not present. The developed models allow the discovery of the possible conflict between the holonic processes. All these points support the fact that the formalism and the attached models are properly formed.

In conclusion, HAPBA is a holonic scheme that better clarifies some issues of the HMESs. It clearly separates the planning and execution phases for a holonic agent, based on patterns that can be modelled as Petri nets, and it highlights the three cases regarding the role of a holon during distributed planning. A holonic agent can be either only manager (e.g. the order holon that issues the goals to the product holons in our examples) or only contractor (as it happens for the considered resource holons) or both manager and contractor (the case of our product holons). There are certain weak points when applying the CNP, namely for the holonic agents acting as managers. This happens due to the lack of coordination between more managers using the same resources. The proposed solution is to include a staff holon in the coordination process. This one has to play an active role, enabling only the planning activity of those holons that cannot negatively interfere. Our analysis and experiments treated only three levels of the goal decomposition (the respective chain is manager– contractors / managers – contractors). This chain was considered without conflicts between the actions of different levels. As a future work, the staff holon operation may be improved to deal with the conflicting situations between different levels.

References

1. Bongaerts, L., Monostori, L., McFarlane, D., Kadar, B.: Hierarchy in distributed shop floor control. Computers in Industry 43, 123–137 (2000), doi:10.1016/S0166-3615(00)00062-2
2. Cheng, F.T., Chang, C.F., Wu, S.L.: Development of holonic manufacturing execution systems. J. of Intell. Manufact. 15, 253–267 (2004), doi:10.1023/B:JIMS.0000018037.63935.a1
3. Ghallab, M., Nau, D., Traverso, P.: Automated Planning - Theory and Practice. Morgan Kaufmann, Amsterdam (2004)
4. Hsieh, F.S.: Analysis of contract net in multi-agent systems. Automatica 42, 733–740 (2006), doi:10.1016/j.automatica.2005.12.002
5. Hsieh, F.S.: Holarchy formation and optimization in holonic manufacturing systems with contract net. Automatica 44, 959–970 (2008), doi:10.1016/j.automatica.2007.09.006
6. Jensen, K., Kristensen, L.: Coloured Petri Nets: Modeling and Validation of Concurrent Systems. Springer, New York (2009)
7. Leitao, P.: An Agile and Adaptive Holonic Architecture for Manufacturing Control. Ph. D. Thesis, University of Porto (2004)
8. Panescu, D., Sutu, M., Pascal, C.: On the Design and Implementation of Holonic Manufacturing Systems. In: Proc. of the WRI Congress, Los Alamitos, pp. 456–461 (2009), doi:10.1109/CSIE.2009.139
9. Panescu, D., Pascal, C.: Some Issues on Holonic Systems Analysis, Design and Implementation. In: Proceedings of the International Workshop RAAD 2010, Budapest, pp. 199–204 (2010), doi:10.1109/RAAD.2010.5524586
10. Panescu, D., Pascal, C.: On a holonic adaptive plan-based architecture: planning scheme and holons' life periods. International Journal of Advanced Manufacturing Technology (submitted for publication, 2011)
11. Smith, R.G.: The contract net protocol: High level communication and control in a distributed problem solver. IEEE Transactions on Computers C-29, 1104–1113 (1980), doi:10.1109/TC.1980.1675516
12. Ulieru, M., Cobzaru, M.: Building holonic supply chain management systems: An e-Logistics application for the telephone manufacturing industry. IEEE Transactions on Industrial Informatics 1, 18–30 (2005), doi:10.1109/TII.2005.843827
13. Valckenaers, P., Van Brussel, H.: Holonic manufacturing execution systems. CIRP Annals-Manufacturing Technology 54, 427–432 (2005), doi:10.1016/S0007-8506(07)60137-1
14. Van Brussel, H., Wyns, J., Valckenaers, P., Bongaerts, L., Peeters, P.: Reference architecture for holonic manufacturing systems: PROSA. Computers in Industry 37, 255–274 (1998), doi:10.1016/S0166-3615(98)00102-X
15. Wooldridge, M.: Intelligent Agents. In: Weiss, G. (ed.) Multiagent Systems. A Modern Approach to Distributed Artificial Intelligence, pp. 54–60. MIT Press, Cambridge (2001)

Integrating Intelligent Robot Services in Holonic Manufacturing

Florin Daniel Anton[1], Theodor Borangiu[1], Silvia Anton[1], Marco Ceccarelli[2], and Giuseppe Carbone[2]

[1] Department of Automation and Industrial Informatics,
University Politehnica of Bucharest,
Spl. Independentei, 313, 060042, Bucharest, Romania
{florin.anton,theodor.borangiu,
silvia.anton}@cimr.pub.ro
[2] Laboratory of Robotics and Mechatronics,
University of Cassino, Via Di Biasio, 43, 03043, Cassino, Italy
{ceccarelli,carbone}@unicas.it

Abstract. In the holonic manufacturing framework, production systems must process the orders, starting from the basic order submitted by the client, which gives the information only about the type of the products and the quantity (and maybe the delivery date). In order to obtain the final product(s) which has been ordered, the production system must decompose each order into a set of operations based on the capability of the production system. Each entity (in this case robots) can execute a set of operations – offer a service (object assembly, part inspection, etc.), based on its attributes (execution speed, working envelope, dexterity, etc.) and resources (tools magazine, raw materials, components for assembly, vision system, network connectivity, etc.) The chapter presents a case study for intelligent robot services in holonic manufacturing and is focused on how to create an automated system which is capable to decompose the initial order in sets of operations based on the services which the robots can offer, transform the operations into robot programs and offer high availability services.

Keywords: robot services, holonic production, high availability, fault tolerance.

1 Introduction

Nowadays in modern manufacturing environments, in order to optimize the production flow and to obtain high quality products, the automation process tends to be incorporated in each production entity. The human intervention must be reduced only to tasks which cannot be executed without human expertise, for monitoring or when the human knowledge cannot be integrated in an automatic process [1, 5, 16, 19, 23, 20, 24].

The holonic manufacturing concept is one of the most actual concepts which brings all these benefits together and offers a solution which can lead to an almost *autonomous production line*.

T. Borangiu, A. Thomas, and D. Trentesaux (Eds.): SOHOMA 2011, SCI 402, pp. 75–88.
springerlink.com

The following sections discuss such an approach where the online ordering system, the ERP system and the production system (PS - a robotized manufacturing cell) are linked together creating a self contained system (see Fig. 1) able to receive orders from customers, decompose orders into operations, and execute each operation in a predetermined way such that the products are obtained in a time- and energy-efficient way.

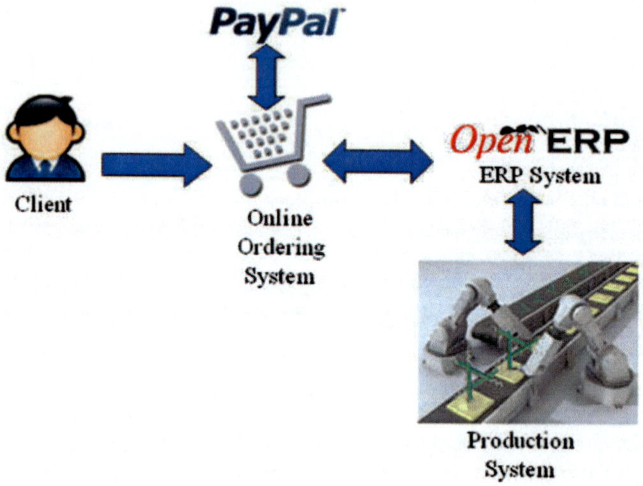

Fig. 1 The self contained production system

Upon receiving an order from a client, the system automatically validates the payment if PayPal is used or it waits human confirmation that the payment has been received; then the order (product type and quantity) is sent to the ERP system which locks the products which are already available in stock and then sends the modified order (the remaining number of products which must be manufactured) to the production system.

The production system is thus seen as a set of interconnected service suppliers (SS - PLCs which command the conveyor belts, robots for material handling and assembly, machine vision systems for robot guidance and quality inspection, and CNC milling machines) [21]. When the order is received from the ERP system the production system decomposes the order into sets of operations and raw materials lists needed for production. This is done by querying each service supplier about its raw material storage and operational capabilities (what are the operations that the SS is able to execute: for example assemble raw material R1 in position P1). After all SS have been queried the PS computes the number of each type of raw material and if there are needed more materials a request is sent to the ERP system which, using the MRP module (Material Requirements Planning) extracts from inventory the materials and sends them to the production system and /or orders more materials from suppliers.

When the materials are available and the operational capabilities are known, the production planning is ready to be executed to obtain a set of production plans which are then validated by simulation; finally, an optimal plan is selected and executed.

2 Decomposing Orders in Operations

As mentioned before, in order to obtain a final product, the customer order is decomposed first in sets of products (N products of type M, X products of type Y, etc.) and then each product is decomposed in sets of operations and raw materials.

Product decomposition is not an online operation but rather an offline operation which is done only once and is executed by a human expert. The product is associated with a set of operations which are executed quasi sequential in order to obtain a final product.

For example the product has the following structure:

- Product code: #IDprk;
- Associated operations set (operations array):
  - [o_k1, o_k2, ... , o_kf]
- Operations order (vector of predecessors/operation):
- [p(o_k1), p(o_k2), ... , p(o_kf)]

  where:
  - p(o_ki)∈ {Ø, ... , (f – 1)} predecessor operations
- Required materials (components) (array of materials [type; number] / operation):
  - [m (t, nm)] (o_k1), [m (t, nm)] (o_k2), ... , [m(t, nm)] (o_kf),

  where:
  - [m (t, nm)] (o_ki) = [t_1i | nm_1i, t_2i | nm_2i, ... , t_fi | nm_fi]
- Required tools (instruments) (tools array [type; number] / operation):
  - [s (t, ns)] (o_k1), [s (t, ns)] (o_k2), ... , [s (t, ns)] (o_kf),

  where:
  - [s [t, (ns)] (o_ki) = [t_1i | ns_1i, t_2i | ns_2i, ... , t_ui | ns_ui]
- Required programs (array of #ID programs / operation):
  - IDprg(o_k1), IDprg(o_k2), ... , IDprg(o_kf),

  where:
  - IDprg(o_ki) = IDprg_1i, IDprg_2i, ... , IDprg_vi

The operations are described in a similar manner, for example the operation of *processing a product* (PRP(pr, op, tool, rob)) using a robot "rob" (robotized processing of type tip.op) which has a tool "tool" in gripper requires one program specified by #IDprg() having one or more p_p (pick and place) routines and is described as follows:

- PRP(pr, op, tool, rob):
  - [pr(#IDpr, loc.pr); op(tip.op, tip.tool); tool(t, tool.hld, loc.tool);
  - robot(tip.rob, tip.grip, TOOL); message()]
  - pr(#IDpr, loc.pr): the code of the product located in loc.pr on a pallet which has been brought to the robot using the conveyor, and on which the robot of the type tip.rob must execute the operation of type tip.op, using the tool of type tip.tool.
  - op(tip.op, tip.tool): describes the operation executed on the product pr(#IDpr):
    tip.op: is the type of the operation which will be executed by the robot on the product (for example milling, marking, painting, welding);
    tip.tool: the type of the tool used by the robot rob(#IDrob) in order to execute the operation tip.op on the product #IDpr.
  - tool(t, tool.hld, loc.tool, TOOL): data which describe the processing tool of the operation tip.op using the robot tip.rob;
    tip.tool: is the type of the tool used for processing;
    tool.hld: the type of the robot tools storage which contains the tip.tool type tool;
    loc.tool: the location from which the tool will be retrieved from the storage tool.hld
  - TOOL: tip.tool type tool associated transform.
  - message(STATUS (PRP), time.r, time.p) where:
    STATUS (PRP):
    - Completed (ok / nok(cod.er)):
    - ok: normal completion of a PRP(pr, op, tool, rob);
    - nok: abnormal completion (abort), with error code cod.er transmission
    - Executing.

    time.r: execution time of the operation PRP( ) – counted.
    time.p: utilization time of the tool tip.tool during the operation PRP( ) – counted.

Each product is defined by a set of materials, tools, operations and programs. Based on the product specification, only a set of SS is capable to execute the operations required to create the final product: a set of operations can be executed only by a specific SS, and other operations can be executed by multiple SS entities. For the second case, each SS competes for each operation, the winner SS (chosen based on the global strategy: duration of operation, optimization of resource utilization, etc.) will execute the selected operation. Because the operation execution is quasi sequential, some operations can be executed in any order, but other operations can be executed only after a specific set of previous operations defined in the operations / predecessor array [7].

3 Transforming Operations in Programs

After each operation is associated with a service supplier (in our case a robot), the abstract form of the operation must be processed and transformed in executable programs. The program must be customized for the robot which must execute the operation.

Transforming the operations in programs can be done in two ways [22]:

- by using a human expert in an offline training session;
- automatically, using a parsing routine.

In the first case the human expert must train each location of the objects and write each program for each operation; this can be a time consuming operation, time which can be considered a downtime for the production because the manufacturing line must not be used in a degraded way, having one or more SS unavailable. The operation is time consuming because each operation must have a corresponding program on each SS (each program for each operation being customized for each SS); moreover the problem is more complicated when a new product must be defined to be produced and all operations / programs must be defined / created when a production is in progress.

This method is appropriate when the first definition of products is made, and the manufacturing line is using a small number of SS; the products are not changing dynamically and the line is producing a relatively small number of products [13, 14].

On the other hand if the production / products are changing dynamically and the number of SS / product types is high, the second approach is more effective. The only thing that must be done offline is the location training: location for each raw material storage, and the location for pallet frame when the pallet is stopped on the conveyor in the robot workspace.

The locations are stored on a *high available shared storage* which can be accessed by each entity on the manufacturing line [3, 10, 11, 17]. When an order is about to be executed each SS competes to execute the operations composing this order. After the competing process the operations are associated with the available SS; then for each SS a program builder entity parses the operations and creates the programs customized for each SS and the resulting programs are stored on the shared storage where each robot can load the programs and execute them [8, 15].

In order to use in an efficient way the previous product/operations definitions, they are transformed in a standard XML format which adds more obvious advantages, some of them being listed below:

- *flexible* (an operation or product specification described using XML is easy to be modified or changed);
- *easy to share* (the specification is saved in a plain text file which can be easily transferred from one system to another);
- *describes itself* (the product / operation can be easily interpreted by a computer but also by a human operator);
- *easy to extend* (new features and functionalities can be added);
- *vendor independent* (do not depend on the HW/SW infrastructure);
- *platform independent* (do not depend on the HW/SW infrastructure).

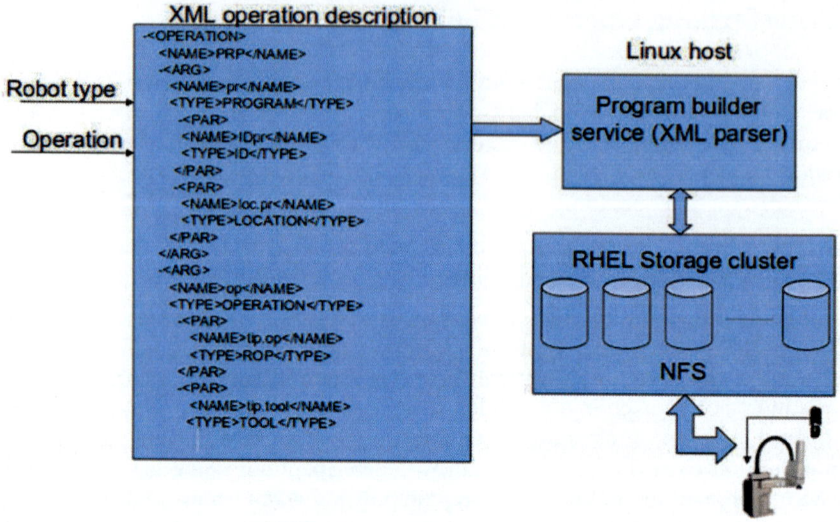

Fig. 2 The process of generating a robot program

For example the code below presents a sequence of the XML definition of the operation PRP:

```
-<OPERATION>
    <NAME>PRP</NAME>
    -<ARG>
          <NAME>pr</NAME>
          <TYPE>PROGRAM</TYPE>
                -<PAR>
                      <NAME>IDpr</NAME>
                      <TYPE>ID</TYPE>
                </PAR>
                -<PAR>
                      <NAME>loc.pr</NAME>
                      <TYPE>LOCATION</TYPE>
                </PAR>
    </ARG>
    -<ARG>
          <NAME>op</NAME>
          <TYPE>OPERATION</TYPE>
          -<PAR>
                <NAME>tip.op</NAME>
                <TYPE>ROP</TYPE>
          </PAR>
          -<PAR>
                <NAME>tip.tool</NAME>
                <TYPE>TOOL</TYPE>
          ...
```

4 High Availability Services

In order to implement a *High Availability* (HA) *solution* on the manufacturing line (a network of robot controllers), first a shared storage is needed, which must be reached by any controller from the manufacturing line [2, 6, 9, 13].

The file system from the storage is limited to NFS (network file system) by the operating system of the robot controllers (here Adept Windows). Five Adept robot manipulators were considered, each one having its own multitasking controller.

For the proposed architecture, there is no option to use a directly connected shared storage, because Adept robot controllers do not support a Fiber Channel Host Bus Adapter (HBA). Also the storage must be high available, because it is a single point of failure for the Manufacturing Structure, see Fig. 3.

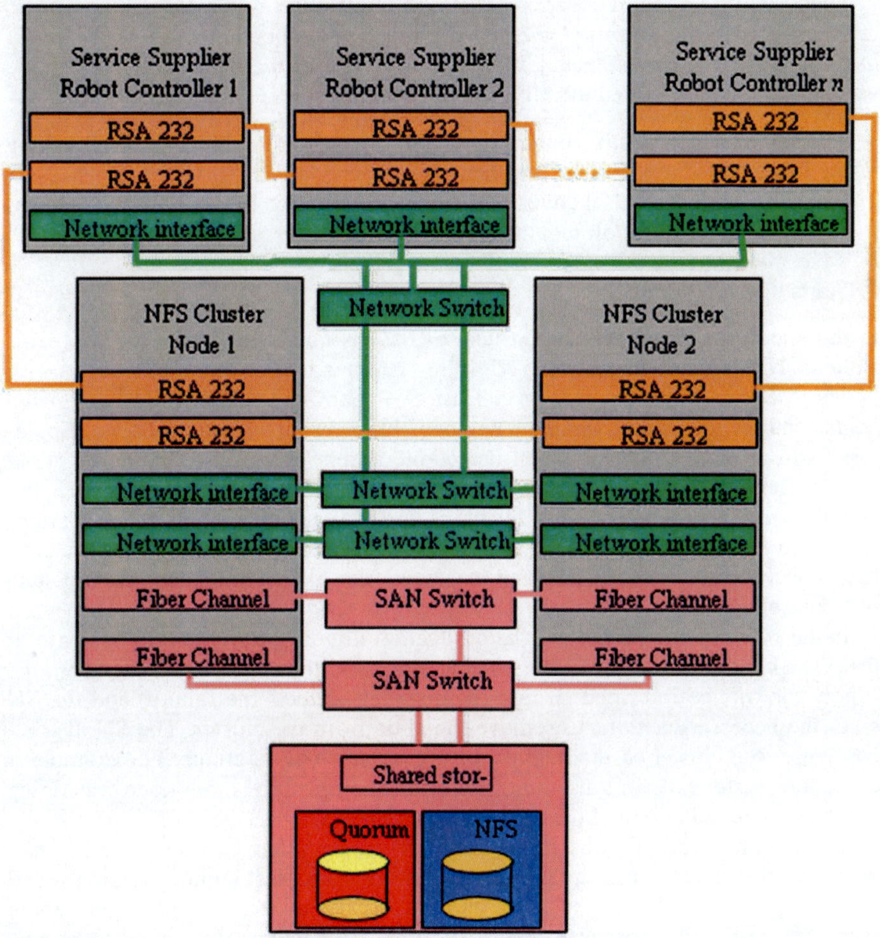

Fig. 3 Implementing the high availability solution for the networked robotic system

Due to these constraints, the solution is to use a High Availability cluster to provide the shared storage option (NFS Cluster), and another cluster composed by Adept Controllers which will use the NFS service provided by the NFS Cluster (see Fig. 3).

The NFS cluster is composed by two identical IBM xSeries 345 servers (2 processors at 2.4 GHz, 1GB RAM, and 75GB Disk space, two RSA 232 lines, two Network adapters, and two Fiber Channel HBA), and a DS4100 storage. The storage contains a Quorum volume used by the NFS cluster for communication between nodes, and a NFS volume which is exported by the NFS service which runs in the NFS cluster. The servers have each interface (network, serial, and HBA) duplicated to assure redundancy [4, 12, 18].

In order to detect the malfunctions of the NFS cluster, the servers communicate using heartbeat packets to ensure that the communication is established.

There are three communication routes: the first route is the *Ethernet network*, the second is the *quorum volume* and the last communication route is the *serial line*. If the NFS cluster detects a malfunction of one of the nodes and if this node was the node which served the NFS service the cluster is reconfiguring as follows:

1. The server which is still running writes in the quorum volume which is taking the functions of the NFS server, then
2. Executes the resource takeover operation (the resource being the NFS volume) and mounts the NFS volume, then
3. Takes the IP of the other server (IP takeover) and
4. Starts the NFS service.

In this mode the manufacturing structure (MS) is not aware about the problems from the NFS cluster, because the NFS file system is further available.

The MS can be composed by at least two robot controllers (nodes) – *group leader* and *group leader backup*. The nodes have resources like: robot manipulators (with attributes like: collision detection, current robot position, etc...), serial lines, Ethernet adapter, variables, programs, and NFS file system. The NFS file system is used to store programs, locations, log files and status files. The programs are stored on NFS to make them available to all controllers, the log files are used to discover the causes of failure and the status files are used to know the last state of a controller.

In the event of a node failure, the production flow is interrupted. In this case, if there is a connection between the affected node and the supervisor node (SN), the supervisor will be informed (if not, the heartbeat detects the failure) and the SN takes the necessary actions to remove the node from the cluster. The SN also reconfigures the cluster so the manufacturing process will continue. For example if one node cluster fails in a three-node cluster, the operations this node was doing will be reassigned to one of the remaining nodes.

The communication paths in the multiple-robot system are the *Ethernet network* and the *serial network*. The serial network is the last communication resort due to the low speed and also to the fact that it uses a set of Adept controllers to reach the destination (because it is a ring type connection). In this case the ring network will be down if more than one node fails.

Rebuilding the communication route in the network represents the last step required to restart the normal behave of the MS control system.

In case of malfunction of the communication network the following important cases can appear:

1. If the connection between the Switch and the Supervisor PC (the PC where the Server Application runs: ERP, Online ordering system) is down the remote control will be lost, but the MS will reconfigure as follows: the controller will use the Ethernet network for communication, and the controller with the first IP from the class will take the functions of the Supervisor PC. If the connexion is re-established the Supervisor PC makes a query, finds the replacing controller, transfers the databases and restarts the normal behaviour.
2. If the switch is not functioning, all the Ethernet connexions are lost, but the controllers will use the serial "network". The behaviour is like in the first case only that the web users can view the status from the Supervisor PC, including the images acquired by the observation cameras.
3. If a controller loses the Ethernet connexion, it will use one of the two serial lines to reach the Supervisor PC depending on the CPU time of the neighbours.

5 Conclusion

The proposed solution for product / operation description allows a minimal effort from the manufacturing cell operators to add / remove a product or modify existing products without any intervention on the robots programs. For example in some robot-vision applications (like quality control), in order to verify the quality of a product a pair of robot programs (running on robots) and vision sequences (running on the vision system) is needed in order to execute the product inspection

In many vision systems (Cognex, Adept Technology, Scorpion Vision, etc.) the vision sequences are defined and reside outside the robot controller (a PC, a smart camera, etc.) consisting of a series of vision tools (line finders, locators, calliper tools, etc.) applied on a specific location on the image (one vision tool can be applied more than once on the same object).

If the product is changed or another product is added, the verification sequences must be modified on the quality inspection station by the operator manually. If the proposed approach is used, the operator needs only to define a generic verification sequence (containing all the tools needed) and from the XML description of the inspection operation each tool is positioned by a robot program (communicating with the vision system) in a particular position, and executed. This is done in a loop, each tool being executed as many times as needed, until all inspection routines are executed and the product quality verification is completed. In Fig. 4 such a general vision sequence is presented which contains only three vision tools, but is able to verify a product which requires a set of vision tools containing a very large number of line finders.

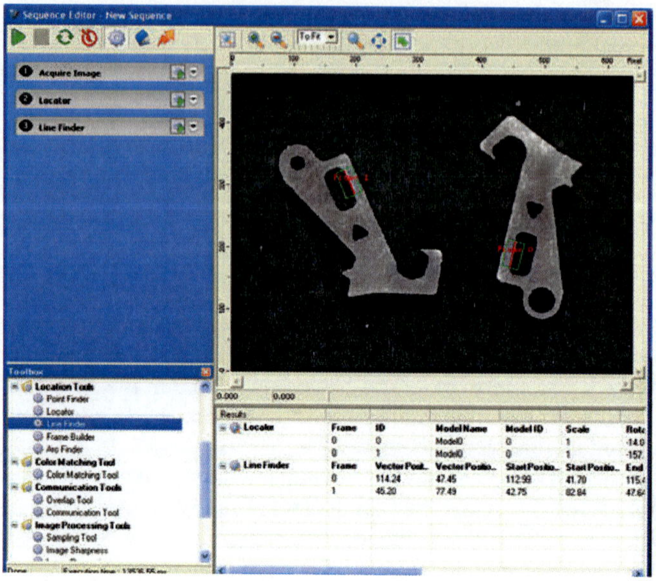

Fig. 4 A generic vision sequence used with XML operation definition

Fault-tolerance is provided to the cell communication system (Fig. 5), and therefore redundancy at both Station Controller level (a break down of a Robot Controller is detectable, the production tasks can be rescheduled to the remaining valid units for graceful degraded behaviour) and Station Computer level (replication of data bases for the IBM PC-type device terminals, reassignment of computers in case of break downs).

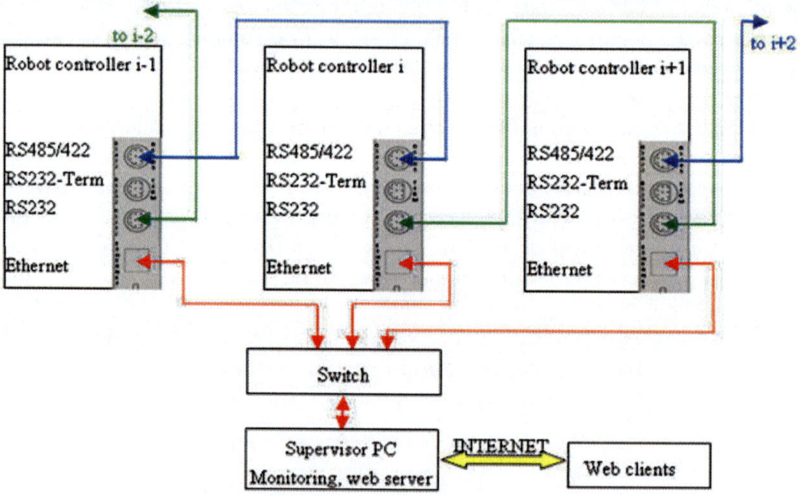

Fig. 5 The network connections between controllers

The fault tolerance solution presented in this chapter is worth to be considered in environments where the production structure has the possibility to reconfigure, and where the manufacturing must assure a continuous production flow at batch level (job shop flow).

There are also some drawbacks in this solution, like the need of an additional NFS cluster. The spatial layout and configuring of robots must be done such that one robot will be able to take the functions of another robot in case of failure. If this involves common workspaces, programming must be made with much care using robot synchronizations and monitoring continuously the current position of the manipulator.

The advantage of the proposed solution is that the structure provides a continuous production flow with an insignificant downtime (during reconfiguration).

The solution was tested on a five-robot assembly cell located in the Robotics and IA Laboratory of the University Politehnica of Bucharest. The cell also includes two CNC milling machines and one Automatic Storage and Retrieval System for raw material feeding and finite products storage.

During the tests, the robot network has detected a number of errors (end-effector collision with parts, communication errors, power failure, etc.) The robotic network has evaluated the particular situation, and the network was reconfigured and the abandoned applications were restarted in a time between 0.2 and 3 seconds. The network failure was also simulated during tests (Fig. 6). One robot (R2) was disconnected from the Ethernet network, the heartbeat packet sent by the robot to the other cluster members has detected the malfunction and the robot has switched the communication using the serial line; this was done in 0.3 seconds after the Ethernet cable was removed. The communication between the affected robot and its neighbours was done using the serial lines, and the communication with other robots was done by routing the communication using the Ethernet line of the neighbours (R1 and R3). In this way the communication latency was reduced.

After the communication was re-established, the serial lines of the robot were disconnected. The robot has detected the communication failure, stopped the manipulation program and retracted the manipulator in a home position in the exterior of the working area in 0.8s.

The neighbours have sent the heartbeat packets using the serial lines and detected that they do not have any connection with the robot; they announced the SN which has removed the robot from the cluster and reconfigured the cell (see Fig. 6). The neighbour R1 having the same working area as the affected robot R2 has loaded the values of the variables and the production program from the shared storage and started production, continuing from the point where R2 has been stopped. The cell reconfiguration from the point where the serial lines were disconnected has taken 2.2 seconds.

Another *communication test* consisted in disconnecting both the serial lines and the Ethernet line at the same time; in this case the cluster tested the communication sequentially and the cluster reconfiguration took 2.5 seconds. When configuring the cluster in order to test in parallel the communication lines, the reconfiguration took 2.3 seconds but the controllers processed a higher communication load.

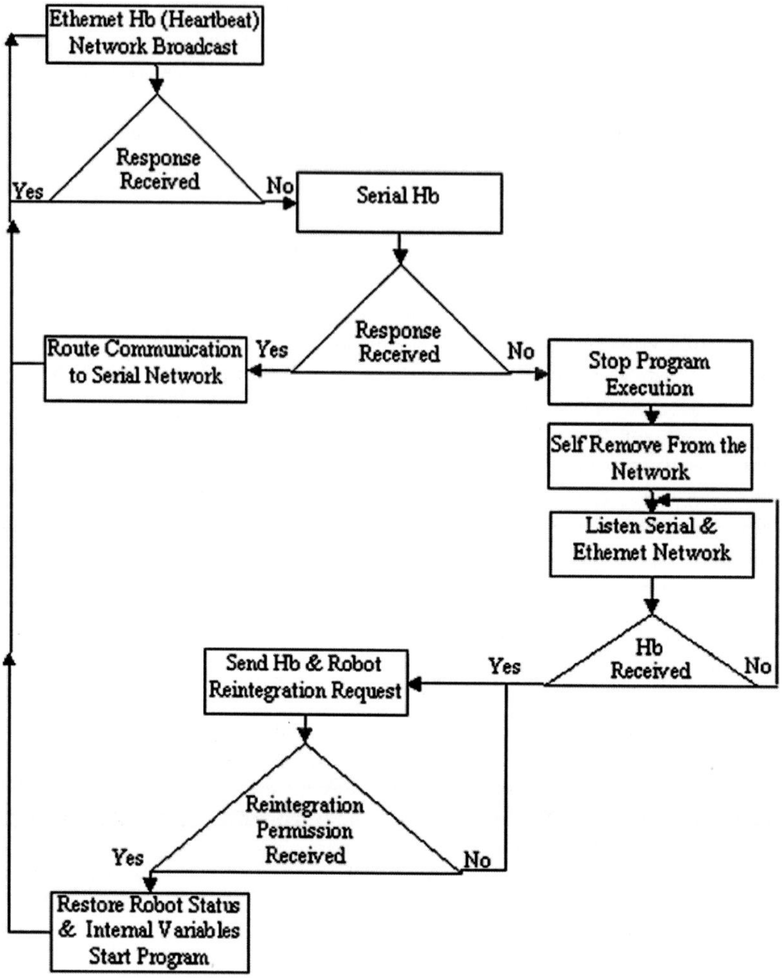

Fig. 6 Algorithm for network failure detection and reconfiguration

The most unfavourable situation occurred when a robot manipulator was down; in this case the down time was greater because the application which was executed on that controller had to be transferred, reconfigured and restarted on another controller. Also if the controller still runs properly it will become group leader to facilitate the job of the previous SN (that has also a manipulation task to do).

The proposed solution is not entirely fault tolerant; however, in some situations it could be considered as a fault tolerant system due to the fact that even if a robot controller failed, the production continued in normal conditions.

Acknowledgments. The work has been co-funded by the Sectoral Operational Programme Human Resources Development 2007-2013 of the Romanian Ministry of Labour, Family and Social Protection through the Financial Agreement POSDRU/89/1.5/S/62557.

References

1. Abdi, M.R., Labib, A.W.: Performance evaluation of reconfigurable manufacturing systems via holonic architecture and the analytic network process. International Journal of Production Research 49(5), 1319–1335 (2011)
2. Anton, F.D., Borangiu, T., Tunaru, S., Dogar, A., Gheorghiu, S.: Remote monitoring and control of a robotized fault tolerant workcell. In: Proceedings of the 12th IFAC Symposium on Information Control Problems in Manufacturing INCOM 2006. Elsevier (2006)
3. Bal, M., Hashemipour, M.: Implementation of holonic scheduling and control in flow-line manufacturing systems: die-casting case study. Production Planning & Control 22(2), 108–123 (2011)
4. Borangiu, T., Tomas Balibrea, L.M., Gonzales Contreras, L.A., Manu, M.: Object Oriented Model of Open Communication Architecture for Flexible Manufacturing Control. In: Moreno-Díaz, R., Pichler, F. (eds.) EUROCAST 1997. LNCS, vol. 1333, pp. 292–300. Springer, Heidelberg (1997)
5. Black, G., Vyatkin, V.: Intelligent component-based automation of baggage handling systems with IEC 61499. IEEE Transactions on Automation Science and Engineering 7(2), 337–351 (2010)
6. Borangiu, T., Anton, F.D., Tunaru, S., Dogar, A.: A holonic fault tolerant manufacturing platform with multiple robots. In: Proceedings of 15th Int. Workshop on Robotics in Alpe-Adria-Danube Region RAAD 2006 ((2006)
7. Brennan, R.W., Gruver, W.A., Hall, K.H.: Forward-Special Issue on Industrial Applications of Holonic Manufacturing Systems. IEEE Transactions on Systems Man and Cybernetics Part C-Applications and Reviews 41(1), 1–3 (2011)
8. Covanich, W., McFarlane, D.: Comparing the control structures of ISA S88-and holonic component-based architecture. IEEE Transactions on Systems Man and Cybernetics Part C-Applications and Reviews 1(1), 4–13 (2011)
9. Harris, N., Armingaud, F., Belardi, M., Hunt, C., Lima, M., Malchisky Jr., W., Ruibal, J.R., Taylor, J.: A guide to IBM Linux Solutions and Resources, 2nd edn. IBM Int. Technical Support Organization (2004)
10. Hsieh, F.S.: Design of reconfiguration mechanism for holonic manufacturing systems based on formal models. Engineering Applications of Artificial Intelligence 23(7), 1187–1199 (2010)
11. Hsieh, F.S., Chiang, C.Y.: Collaborative composition of processes in holonic manufacturing systems. Computers in Industry 62(1), 51–64 (2011)
12. Lascu, O., Sayed, R., Carroll, S., Coleman, T., Haehnel, M., Klabenes, P., Quintero, D., Reyes, R.: An Introduction to Security in a CSM 1.3 for AIX 5L Environment, 1st edn. IBM Int. Technical Support Organization (2002)
13. Lascu, O., et al.: Implementing High Availability Cluster Multi-Processing (HACMP) Cookbook, 1st edn. IBM Int. Technical Support Organization (2005)
14. Lee, H., Banerjee, A.: Executable design and control framework using reconfigurable manufacturing holons and its evolution algorithm. International Journal of Production Research 49(5), 1405–1423 (2005)
15. Lind, M., Roulet-Dubonnet, O.: Holonic shop-floor application for handling, feeding and transportation of workpieces. International Journal of Production Research 49(5), 1441–1454 (2011)

16. Madureira, A., Santos, J.: Inter-machine cooperation mechanism for dynamic scheduling. In: International Joint Conference on Computer, Information, Systems Sciences and Engineering, Bridgeport CT, December 05-13 (2008)
17. Manesh, H.F., Schaefer, D., Hashemipour, M.: Information requirements analysis for holonic manufacturing systems in a virtual environment. International Journal of Advanced Manufacturing Technology 53(1-4), 385–398 (2011)
18. Matsubara, K., Blanchard, B., Nutt, P., Tokuyama, M., Niijima, T.: A practical guide for Resource Monitoring and Control (RMC), 1st edn. IBM Int. Technical Support Organization (2002)
19. Paggi, H., Amo, F.A.: Beliefs, certainty and complex systems structure. In: 2nd International Conference on Computer Engineering and Applications, Bali Isl Indonesia, March 19-21 (2010)
20. Paggi, H., Amo, F.A.: Uncertainty and randomness: A holonic approach. In: 2nd International Conference on Computer Engineering and Applications, Bali Isl, Indonesia, March 19-21 (2010)
21. Salido, M.A., Peinado, M., Giret, A.: A holonic simulation environment for smart transportation systems. International Journal of Production Research 49(5), 1425–1439 (2011)
22. Vrba, P., Marik, V.: Capabilities of dynamic reconfiguration of multiagent-based industrial control systems. IEEE Transactions on Systems Man and Cybernetics Part A-Systems and Humans 40(2), 213–223 (2010)
23. Zhao, F.Q., Zou, J.H., Sheng, S.X.: A hybrid Pso and shuffled complex evolution algorithm for order assignment problem in holonic manufacturing system. In: International Conference on Manufacturing Science and Engineering ICMSE 2009, Zhuhai, P.R. China (2009)
24. Zhao, F.Q., Zou, J.H., Sheng, S.X.: A hybrid algorithm for task assignment problem in holonic manufacturing system. In: International Conference on Information Technology for Manufacturing Systems, Macao, P. R. China (2010)

Key Factors for Information Dissemination on Communicating Products and Fixed Databases

Sylvain Kubler, William Derigent, André Thomas, and Éric Rondeau

Research Centre for Automatic Control of Nancy, Nancy-University, CNRS, Boulevard des Aiguillettes, F-54506 Vandoeuvre-lès-Nancy, France
LastName.FirstName@cran.uhp-nancy.fr

Abstract. Integration of intelligent products carrying their own information is more and more present nowadays. In recent years, some authors argued on the usage of such products for the Supply Chain Management Industry. Indeed, a multitude of informational vectors take place in such environments like fixed databases or manufactured products on which one is able to embed significant proportion of data. By considering distributed database systems, one can allocate specific data fragments to the product in order to manage its own evolution. The chapter aims at analyzing the Supply Chain performance according to different strategies of information distribution between manufactured products and fixed databases. The purpose is to determine the key factors which lead to improve information distribution performance in term of time properties.

Keywords: Communicating Material, Data Dissemination, Distributed Database Systems, Product Driven Systems, Pervasive Systems.

1 Introduction

Intelligent products or products carrying their own information are more and more present nowadays. [9] quotes the example of clothes able to carry their own information and thus enabling the washing machine to automatically adapt its washing program. In one of our previous works [7], we highlighted several possible scenarios for intelligent products in different sectors: Supply Chain Management, healthcare [2], home automation. Such applications rely on ever more complex information systems using a multitude of vectors of information which allow product information to be available anywhere and at anytime. These vectors may be fixed (desktop computers) or mobile devices (PDA, laptops, RFID) or even invisibles (the concept of disappearing computer). Today, the concept of Internet of Things [5] based on the RFID usage enables to access to information disseminated on any kind of physical object and to develop new smart services and applications.

According to Meyer [11], in the context of supply chain management, few researches has been conducted on "intelligence at object", i.e. products carrying

T. Borangiu, A. Thomas, and D. Trentesaux (Eds.): SOHOMA 2011, SCI 402, pp. 89–102.
springerlink.com

their own information and intelligence. In fact, most of the time, products are only given an identifier (stored in a RFID tag) referring to a software agent or a database (approach used by [12]). This mode of information management is diametrically opposed to works initiated since 2003 by the PDMS (Product-Driven Manufacturing Systems) community, which advocates a physical information distribution on the product. In that case, a product carries physically a part, or even the totality of the information needed for its manufacturing or to manage its evolution all along its life cycle. Our previous work [8] aimed at prototyping a new type of materials, in which it is possible to write a significant quantity of information by inserting thousands of micro RFID tags. This new type of material is then referred to "communicating material". We developed an industrial process to produce a communicating textile with up to 1500tags/m2. Meyer concurs with the PDMS community by stressing the fact, in an increasingly interconnected and interdependent world involving many actors issued from different domains, supply chain information should not be stored in a single database but should be distributed all over the supply chain network. In fact, substantial information distribution improves data accessibility and availability, compared to centralized architectures. However, update mechanisms of the distributed information are needed in order to avoid problems related to data consistency and integrity. This type of architecture is thus more complex to design than centralized architectures. As a result, product information can be spread out on mobile or fixed devices or even directly on the product, via simple RFID tags or communicating materials. Centralized architectures or highly distributed architectures can be used. One might then wonder what the optimal information distribution is. The present chapter aims at studying the different ways to distribute information over a network composed of centralized, distributed databases and "communicating products", which may store information fragments as well. This study will determine the key factors which lead to improve information distribution performance. The performance is analyzed regarding the time required for accessing to the information system during the product life cycle. Based on this influent factors determination, in a further work, we will be able to implement an experimental design leading us to control the best way to disseminate information on the informational vectors.

This question is addressed in several steps. First, the data distribution is introduced, and then an overview on conducted researches on distributed databases over fixed and mobile devices is presented in section 2. Then, a case study extracted from this overview and adapted to our context is detailed in section 3. It only considers two types of informational vectors: fixed computers and communicating products. This case study is then used as a basis of comparison and evaluation between two different architectures of information distribution (one forbids data allocation on products while the other allows it). The evaluation process relies on several specific tools and a methodology using jointly two discrete-even simulators: *CPN tools* and *OPNET Modeler*. This piece of software is presented in the section 4 and assesses the manufacturing lead-time of a given number of communicating products all along the supply chain, by taking into account manufacturing run times, network delays, times to read/write information for both distributed databases and communicating products. Finally, the section 5 presents the results obtained with the case study and an analysis of the main factors impacting on the performance of the information distribution.

2 The System Characterized by a Guided Flow

2.1 General Data Distribution Framework

During the product lifecycle, users may access to product information for diverse reasons, either during the design phase, the usage phase or still the recycling phase.

As exposed before, information can be stored both on the product and or on fixed databases. Information are therefore bind to one or more relational data models, which have to be fragmented and distributed by the best way on these informational vectors.

One example retracing briefly a bobbin lifecycle is presented in the Fig. 1.We can see five data fragments [*F1..F5*] distributed between the product and the database (*F1*, *F4*, *F5* allocated to the database system and *F2*, *F3* to the product).

By reconsidering the example given by [9], the washing machine could access to data fragments located both on the product and on the database according to its queries. In our research, we are looking for assessing different distribution patterns of the data fragments between both informational vectors (manufactured products and fixed databases) by taking into account the access times for reaching information.

Work on distributed databases considering fixed and mobile environments is introduced in the next section.

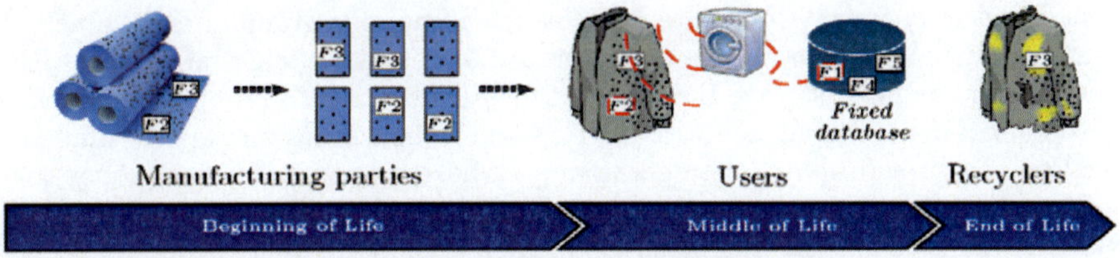

Fig. 1 Information distribution on products and fixed databases

2.2 Distributed Databases through Literature

The main constraint of the data dissemination in an information system is to make the dissemination process transparent for users: location, partitioning and replication transparency. Indeed, no matter why, where and how the data repartition is achieved from a user's point of view. Data distribution can generally be accomplished by two consecutive steps: the partitioning of the data model followed by the allocation phase of the resulting fragments. Many approaches and mechanisms exist for ensuring the best partitioning and allocation of the relational model regarding the environment and some applicative constraints.

Basically, the partitioning aims at subdividing the relational data model. Thus, the resulting fragments will be allocated to specific informational vectors in order to improve system performance. Three types of fragmentation exist: vertical [13], horizontal [1], mixed / hybrid [14]. The vertical fragmentation aims to break up a relation into a set of relations. It consists in dividing the attributes of a relation (i.e. the columns of a relational table). The horizontal fragmentation aims to break the large number of object instances into disjoint subsets. It consists in partitioning the tuples of a relation (i.e. the rows of a relational table). The hybrid fragmentation first divides the relation horizontally, and then splits each of the obtained fragments vertically or vice versa.

As stated previously, the allocation phase takes place subsequently to the fragmentation phase and its aim is to establish the optimal fragment assignation on the databases. Usually, methods tend to assign fragments to the clients requesting them mostly via objective functions to minimize or maximize [6]. Let us note that it is possible to perform data replications or in other words, to replicate a same fragment on several databases. This has the dual benefit of maintaining the system reliability and of increasing performance (e.g. reduction of the overload traffic, saving time for users) [15]. However, replication mechanisms are necessary for handling both the modification broadcast (updates) on replica and also the information access rights (to authorize one site or one group of sites to modify replica). The applicative expectations have an influence on the mechanism to implement and actually two parameters have to be characterized: *When* and *Where*? When do the updates have to be propagated? Two modes are available: Synchronous (*S*) and Asynchronous (*As*). The *As* mode makes it possible to carry out local modification without needing to inform its peers (contrarily to the *S* mode). Where do the updates have to be performed? Two principles exist: *Update everywhere* (*Ue*) and *Primary copy* (*Pc*). The *Pc* principle allows one site to perform modifications on a data fragment contrarily to the *Ue* mode which allows one group of sites. Finally, four types of replication may be considered: *Ue-S*, *Ue-As*, *Pc-S* and *Pc-As*. Also note that the memory storage limitation of mobile devices is a problem frequently encountered in the literature. Accordingly, some authors focus on the data summarization [3,10] (subclass of the data mining) whose primary aim is to reduce the information somehow. In [4] summarization methods used for distributed database systems are listed and the fragmentation method used in this study is described.

A multitude of interesting approaches are proposed in the literature; we therefore feel it is necessary to confront our proposition with them in order to compare and assess our distribution models. In this sense, works reported by Hababeh [6] seem interesting as basis of comparison. Indeed, a fragment distribution method is developed and then applied on a case study, which can be easily extended to our application. In what follows, two distribution architectures will be defined; the first one does not consider the presence of communicating products able to store data fragments, i.e. all information is located on databases. In fact, we rely on the distribution defined by Hababeh. The second one considers communicating products able to store data fragments, thus, diverse distribution patterns of fragments between the product and databases will be possible. The next section introduces this case study and then the adaptation realized in this chapter.

3 Case Study Presentation

3.1 Reference Distribution Pattern

Hababeh proposes a fragment distribution approach based on a two-step process: first, the sites (clients and databases) are clustered according to communication costs, and then data fragments are allocated to the different clusters via an optimization function. This approach is applied on a specific case study, including 3 databases, 3 clients which perform read and write accesses on a set of data fragments (8 in total: [*F1..F8*]). The resulting optimal allocation [6] is depicted in Fig. 2; the access pattern to the 8 fragments performed by each client is specified too (numbers in brackets indicate the number of bytes). The next section formulates the adaptation of this case study to our logistic scenario. In fact, we match parameters and data defined by Hababeh with the supply chain tasks, actors: number of databases and clients, query patterns, data fragments, etc.

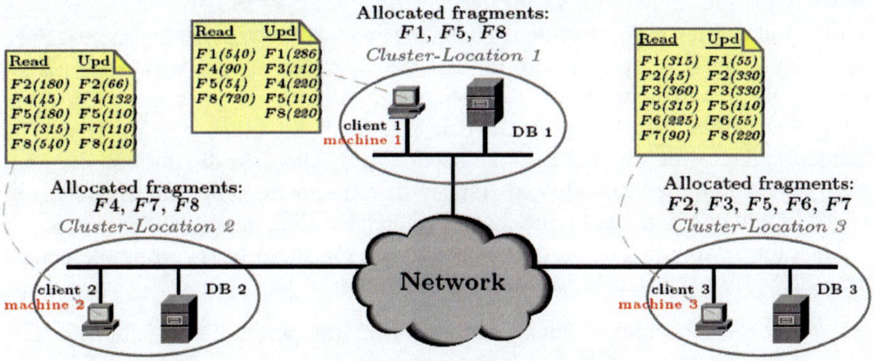

Fig. 2 Optimal distribution architecture established by [6]

3.2 Adaptation of the Logistic Process

A supply chain process consists of a set of tasks in a planned pattern or sequence (*rout sheet*). These tasks may correspond to manufacturing operations, transport phases... and can be performed by diverse suppliers. These suppliers may dispose of local databases where their own information system is implemented (related to their tasks), but they can also access remote databases if a collaboration between actors exists. As a matter of fact, databases are distributed (or federated) through one or more relational data models.

Inspired by our current researches, the applicative framework considered is related to a supply chain management process dedicated to the textile industry. Specifically, the scenario is related to the manufacturing of a simplified headrest composed of two textile parts. Each part is cut out of a different textile bobbin and then sewn together. As a result, this scenario is divided into three tasks carried out by three suppliers respectively. The operation *cutting 1* and *cutting 2* are performed in parallel by supplier 1 and 2, the resulting textile pieces are sewn thereafter by supplier 3. Each supplier disposes of one machine to achieve its own operation; this machine requires information after the arrival of products (range of product, production order…) and updates some of this information (notifications …). In order to adapt the case study of Hababeh to our logistic scenario, we assume that each supplier's location corresponds to the clusters 1, 2 and 3 introduced in Fig. 2 and by this fact applicative characteristics defined for each client in Hababeh are matched to each supplier's machine. In other words, the machine 2 has the same read/write access pattern on the set of data fragments as the client 2 defined in Hababeh and so on. Likewise, each location disposes of a local database and shares the same relational data model, distributed on the three databases. Taking into account the input parameters defined in Hababeh (query pattern, architecture…), the optimal distribution considered in our chapter is defined as shown in the Fig. 2: *F1*,*F5*,*F8* allocated to DB1, *F4*,*F7*,*F8* to DB2…

Fig. 3 illustrates in form of Petri Nets the synoptic of our logistic process. Each operation is defined by a transition, and the Petri Net model was designed by working on hierarchical views. Consequently, the distribution aspect - in other words the optimal distribution of the data fragments described previously will be detailed in the lower views (i.e. in the section 4.3). This first distribution does not take into account the possibility to allocate data fragments on products. However, we dispose of communicating products on which data fragments can be stored.

Therefore, two types of products through the logistic process are implemented and thereby, two types of architectures are feasible:

- *DiPA* (**Di**screte **P**roduct **A**rchitecture): no possibility to allocate data fragments on products; we consider only discrete products[1]
- CoPA (**Co**mmunicating **P**roduct **A**rchitecture): data fragments can be allocated to communicating products.

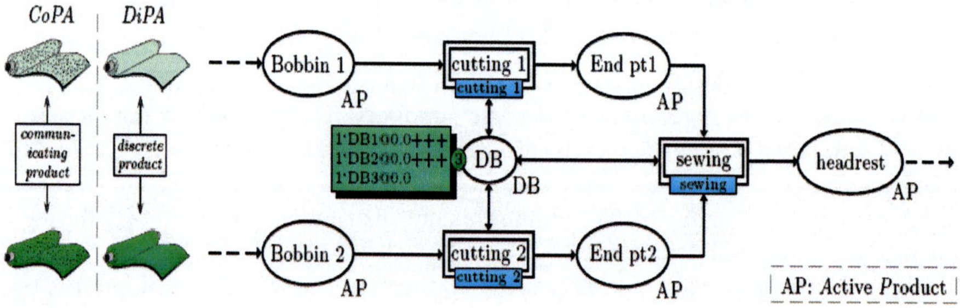

Fig. 3 Global view of the logistic process

[1] Products are only given an identifier (stored in a RFID tag) referring to a database.

Fig. 3 illustrates this point in which discrete products (discrete bobbins) are implemented on one hand, and communicating products (communicating bobbins) are implemented on the other hand through the logistic process. The idea is to highlight the benefits that can be achieved regarding one or the other of these architectures, bearing in mind that manufactured products act as mobile databases in the *CoPA* architecture as opposed to a classic one (*DiPA*).

4 *DiPA* and *CoPA* Architecture Modelling

4.1 Architecture

A description of how the assessment and the comparison are undertaken for both architectures (*DiPA* and *CoPA*) is proposed in this section. The evaluation architecture relies on two discrete events simulators and its usage process is depicted on the Fig. 4.

This architecture is composed by two sub-systems. The first one is a tool for editing, simulating, and analyzing Coloured Petri Nets (*CPN tools*). The logistic process sequence described in section 3.2 is simulated via this tool as shown in Fig. 4. It allows to deal with sharing of physical resources in the system (databases, machines, manufactured products...), operation times, queuing tasks, times for reading / writing information on databases or still on manufactured products (considering the *CoPA* architecture) and so on. The *DiPA* and *CoPA* distribution patterns are specified in this tool. Let us note that for the *CoPA* architecture, all the possible combinations of distribution between the product and the distributed fixed databases are realized (i.e. 2^k possibilities with k the total number of fragments). The second tool is the OPNET network simulator which is primarily aimed at developing and validating network protocols. However, it allows estimating various parameters on specific case studies, such as network times, overload traffic, equipment processing times, battery life, etc. In our study, the *OPNET* tool is used for assessing the *round trip time*[2] to achieve read/write queries on fixed databases. To do this, the physical architecture and the distribution adopted in the section 3 have to be specified in *OPNET*. The resulting times are then injected into *CPN Tools*. The following sections describe respectively each tool.

4.2 Estimated "Round Trip Times" via OPNET

First, the network interconnecting the client machines and the fixed databases is defined in OPNET (see Fig. 4). Thereafter, it is necessary to create system partitions on each server in order to allocate the data fragments as specified in the Fig. 2. Therefore, a replicating protocol has to be implemented owing to the replication of *F5*, *F7* and *F8*. In our application, we implement Synchronous and Primary copy mechanisms described in the section 2. Subsequently, it is necessary to specify the applicative exchanges between equipments, i.e. the query pattern (read / write) performed by each client machine on the databases. To do this, three models from OPNET are used: *Task*, *Application* and *Profile models*.

[2] Time between sending the first packet of the request and receiving the last packet of the response

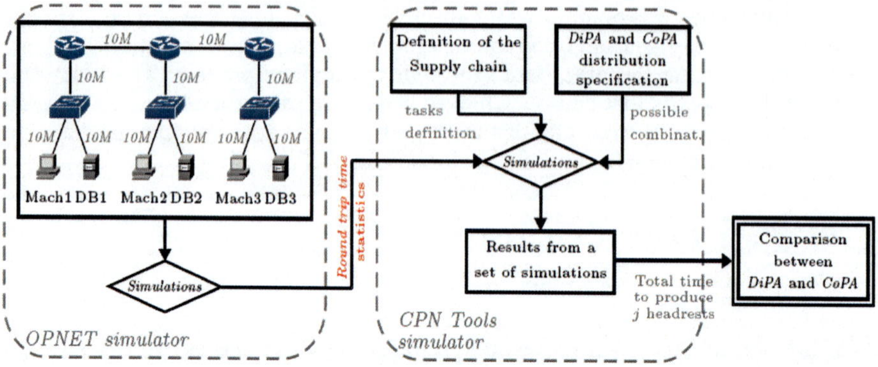

Fig. 4 Usage process of evaluation architecture

Finally, it is possible to estimate the *round trip time* for a specific query sent from a client to a database. Statistical tools are available in OPNET for computing averages, variances or still confidence intervals based on a set of simulations. In our study, both the *round trip time* average and the statistical variance have been extracted for each query and 50 simulations have been running for a same scenario. For instance, the Table 1 gives the *round trip time* induced by a read (R) or write (W) query on *F1* (fragment allocated to DB1) and *F6* (fragment allocated to DB3). The machine 1 requires 3.6ms on average with a variance of 9μs to access to this fragment and spends 7.6ms and 7μs respectively to write it. Likewise, the machine 3 requires 4.8ms on average with a variance of 23.24μs to access to *F6* and spends 10.02ms and 29.9μs respectively to write it.

Table 1 Evaluated times regarding access query patterns: S-Ue

		Machine 1			Machine 2			Machine 3		
		DB1	DB2	DB3	DB1	DB2	DB3	DB1	DB2	DB3
F1	R	3.6ms, 9μs	-	-	-	-	-	7.8ms, 12μs	-	-
	W	7.6ms, 7μs	-	-	-	-	-	11.3ms, 15μs	-	-
F6	R	-	-	-	-	-	-	-	-	4.8ms, 23.24μs
	W	-	-	-	-	-	-	-	-	10.02ms, 29.9μs

4.3 Petri Nets: DiPA and CoPA Architectures

As illustrated on the evaluation architecture (Fig. 4), the estimated *round trip times* are injected into *CPN Tools* and more exactly; they shall be set on timed transitions which reflect the read / write actions on databases. The second views of the Petri Net describe each operation. For instance, Fig. 5 shows the second view of the operation cutting 1. Both views of this operation, related to the *DiPA* and

CoPA architectures are shown in the same figure. The only difference between these architectures lies in reading and writing data fragments which will be discussed in more details below. Three places from these two Petri Nets are bound to the first view (Fig. 3), namely the place *Bobbin 1* (port: *In*), the place *DB* in which the three databases are defined (port: *I/O*) and the place *End pt1* in which cut pieces are stored (port: *Out*).

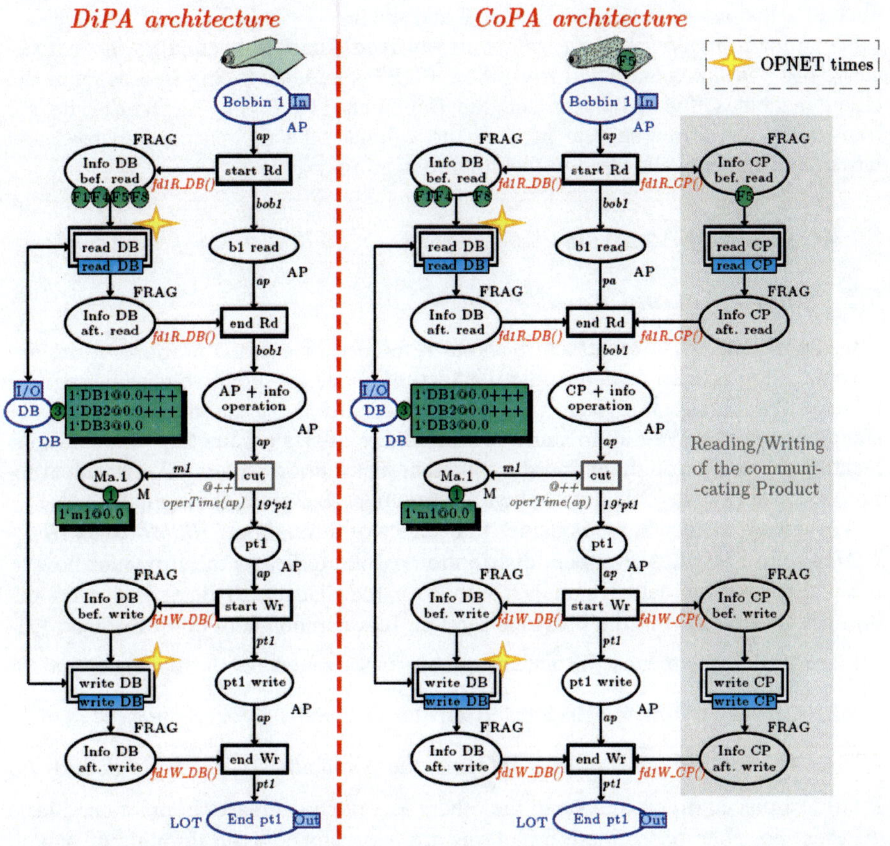

Fig. 5 Petri Net structure of the operation: *cutting 1*

Let us focus now on the Petri Net structure of the *cutting 1* operation (Fig. 5). When bobbin 1 arrives into the queue for being cut (i.e. in the places denoted *Bobbin 1*), we generate straight away the data fragments needed by the machine 1 for starting the operation (fabrication orders…). Machine 1 needs to read the following fragments: *F1*, *F4*, *F5* and *F8*. Considering the *DiPA* architecture, these fragments are initially all allocated to the fixed databases, i.e. into the place *Info DB bef. read.* Considering now the *CoPA* architecture, one part of these data fragments can be allocated on the product and another part on databases. To do this, we add to the *CoPA* view the right part (highlighted in gray in Fig. 5) to indicate that fragments should be read/write on the product rather than on databases. Thus, several distribution patterns between product and fixed databases can be defined. One possible combination might be to allocate *F5* to the product and *F1*, *F4*, *F8* to the databases, as shown in the Fig. 5. After having read the fragments, the cutting task (denoted "cut" transition) can start and then each resulting piece of textile is written by a method similar to the read phase. This principle is reproduced for the other operations: *cutting 2* and *sewing*.

The time to read/write data fragments on fixed database is equal to the statistical *round trip times* extracted from the OPNET simulator taking into account the client machines, the databases and the fragments (Table 1). For *read/write PA* transitions, we define several product throughputs (the time needed to read and write the fragments allocated to the communicating product).

5 Results and Analysis

5.1 Simulation and Results

Considering the *DiPA* architecture as our reference model, the purpose of the experimentation is then to determine the factors impacting positively or negatively on the CoPA architecture performance, and to identify configurations where *CoPA* should be benefit. We aim to study the influences of two parameters - the communicating product throughput and the fragment distribution pattern. Communicating products can exchange data with their environment at a given throughput.

For our experiments, we consider four levels of throughput: *100Mbps, 54Mbps, 11Mbps* and *1Mbps*. A fragment distribution pattern indicates the simulator how to place the different data fragments, either on the distributed database or on the product as explained in the previous section. It is composed of eight Boolean values [F1..F8]; $\overline{Fi}$ means the data fragment i is located on the database and Fi means the data fragment i is located on the product. For example [$\overline{F1}$, $\overline{F2}$, $\overline{F3}$, $\overline{F4}$, $\overline{F5}$, $\overline{F6}$, $\overline{F7}$, $\overline{F8}$] informs the simulator that only *F8* and *F6* should be placed on products and the others let on the database. In practice, all the different possible dissemination patterns are tested for a given throughput, which leads to 256 (2^8) experiments per throughput. Each experiment is simulated 10 times and the mean times needed to produce 85 headrests (number defined arbitrarily) using *DiPA* and *CoPA* architectures are recorded.

Table 2 summarizes the results obtained during the experimentation. Each line of the table corresponds to a given throughput value and each column to a specific data distribution pattern: *DiPA* (all fragments are allocated to the databases), full *CoPA* (all fragments are located on the product), and best hybrid *CoPA* configuration (some fragments are on the database, others on the product and we note down the solution given the smallest time). Time values obtained for a given throughput and configuration are then reported in the table. As can be seen, for our scenario, full *CoPA* and *DiPA* are quite similar in terms of performance when considering

100Mbps, *54Mbps* and *11Mbps* throughputs. Therefore, it might appear that disseminating information all over the different information vectors has no influence on the manufacturing time if the product throughput is high enough. With a correct throughput, it is then possible to imagine an information system completely distributed on a product network. When decreasing, the throughput yet acts as a very important constraint and full *CoPA* is clearly a bad solution. However, the best hybrid configuration always gives good results, which means some data can be stored on the product no matter what the throughput.

Table 2 Times obtained for 3 fragment distributions according to 4 throughputs

Product throughput	*DiPA* Architecture	*CoPA* Architecture	Best hybrid distribution
100 Mbps	2'27"	2'28"	2'27" (*F5, F7*)
54 Mbps	2'27"	2'28"	2'27" (*F5, F6, F7, F8*)
11 Mbps	2'27"	2'52"	2'27" (*F7*)
1 Mbps	2'27"	17'26"	2'27" (*F7*)

Two curves representing the *DiPA* and *CoPA* times (y-axis) for producing 85 headrests with a product throughput of *1Mbps* are plotted in Fig. 6. On the x-axis are represented the 2^8 possible combinations of distribution. Clearly, the distribution pattern has a very important effect in that case. In fact, the time needed to complete the production varies from 2'25" to 17'26". As a result, when weak throughputs are considered, it is then really important to know which pattern to use in order to prevent performance loss.

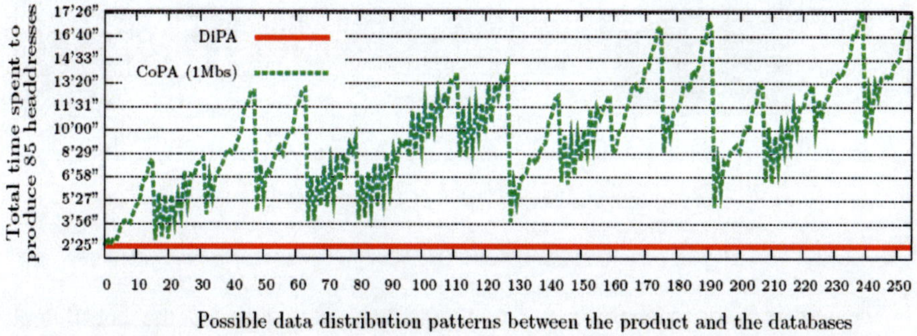

Fig. 6 Comparison between DiPA and CoPA (product throughput = 1Mbps)

5.2 *Key Factor Identification*

Based on these observations, it might be interesting to determine whether each data fragment has the same impact on the manufacturing time in order to identify the critical data fragments (impacting negatively on the manufacturing. time) and to identify the reasons. To do so, the impact of each fragment and of their interactions on the manufacturing time is first studied, based on a statistical analysis of the experiments done with a product throughput equal to *1Mbps*. The key factors are identifying in a second step.

A statistical analysis shows that all data fragments have a significant impact on the manufacturing time, but some of their interactions as well. An interaction between two fragments (Histogram denoted *Level 2*) means that the impact of these fragments all together on the product is different from the sum of the impacts of each fragment. For our scenario, there are up to 35 non-negligible interactions, as reported on the histogram x-axis in Fig. 7. The influence of each fragment and fragment interaction is then estimated thanks to a multiple regression analysis, represented by a coefficient value (y-axis) related to the linear regression equation. The value of this coefficient could be roughly considered as the effect of the data fragments on the manufacturing time when there are allocated to the communicating product. The higher the coefficient value, the more the manufacturing time increases. For instance, we can observe in Fig. 7 that when the fragment *F5* is allocated to the product, it impacts more on the manufacturing time than *F8*. Fig. 7 clearly shows that some fragments have a very important effect (*F1*, *F3*, *F5*) and others have a very moderate one, sometimes equal to interactions of *level 2* (e.g. the effect of *F4* is smaller than *F7*F8*). One can then wonder why some data fragments impact more on the manufacturing time than other ones? A study is carried out to identify the reasons of this behaviour and then the key factors.

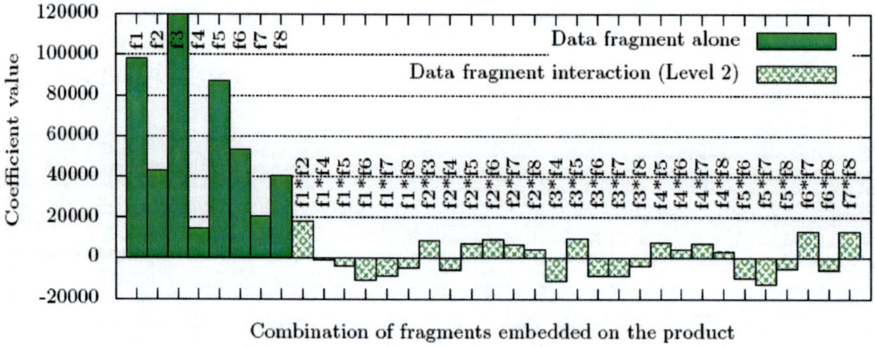

Fig. 7 Significant factors and their respective coefficient values

The aim is now to analyze and to identify the reasons why the coefficient values are more or less important. Each fragment size requested by each operation (*cutting 1*, *cutting 2* and *sewing*) is represented through the *Global process* histogram in the Fig. 8. Let us consider *F2*, the machine 1 does not read *F2*, the machine 2 reads and writes respectively 180 and 66 bytes of *F2*, and the machine 3 reads and writes respectively 45 and 330 bytes (see the access pattern in Fig. 2).

In total, 621 bytes of *F2* are requested. Then, we normalize the fragment size by dividing it by the higher fragment size, i.e. *F8* with regard to the Global process size (equal to 1810 bytes). On the second histogram, only the size of each fragment requested during the sewing operation (i.e. by the machine 3) is represented, since it is the bottleneck of the logistic process. Regarding *F2*, the number of bytes requested by machine 3 is equal to 375 bytes (45+330) as shown on the sewing operation histogram which is dividing by the higher fragment size, i.e. *F3* = 690 bytes. By focusing on the sewing operation, we can observe that *F1*, *F3* and *F5* are the biggest, which may partly explain the coefficient values of these fragments on the Fig. 7. However, it is not sufficient to explain it because *F2* has almost the same size than *F1* but the coefficient value of *F2* is really smaller than *F1*. But, if we look at the *Global process* histogram, we note that *F1* is bigger than *F2*, which may explain the significant difference about the coefficient values of *F1* and *F2* in Fig. 7. Conversely, one can see that the coefficient value of *F8* is not so high in spite of the size of *F8*, significantly bigger than the other data fragments (see the *Global process* histogram), but looking at the sewing operation, *F8* is not so big.

In conclusion, the sewing operation which is the bottleneck of the logistic process seems to impact significantly more the manufacturing time than the other operations. Thereby, it may be sensible to focus on bottleneck operations, tasks on the supply chain for the data dissemination issue, without forgetting to overlook the global process (as illustrated with *F1* and *F2*). Thus, the data fragment size and the operation characteristics influence strongly the manufacturing time. Consequently, it may be necessary to reconsider sometimes the fragmentation method, which generates the set of data fragments.

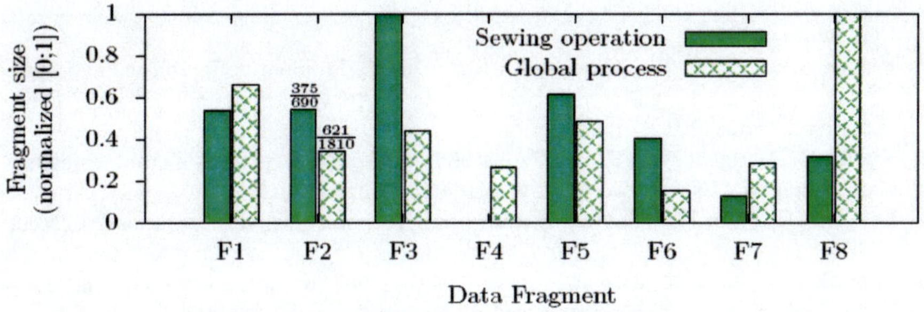

Fig. 8 Fragment size of each fragment with regard to specific operations

6 Conclusion

A multitude of informational vectors take place all along the Supply Chain environments as fixed databases or manufactured products on which one is able to embed significant proportion of data. By considering distributed database systems, specific data fragments can be embedded / allocated on these products (for example, data useful for their life cycle). The chapter analyzed various distribution patterns between manufactured products and databases in order to identify the parameters impacting the most on the manufacturing system performance and especially on the manufacturing time. This study shows that choosing a good pattern is not quite so simple. In a further work, we are willing to implement an experimental design leading us to control the best way to disseminate information on both vectors of information: communicating products and fixed databases.

Acknowledgments. We thank OPNET Technology Inc. for providing the software license to carry out the simulations of this research and, the financial support of the CPER 2007-2013 "Structuration du Pôle de Compétitivité Fibres Grand'Est", through local (Conseil Général des Vosges), regional (Région Lorraine), national (DRRT and FNADT) and European (FEDER) funds.

References

1. Apers, P.: Data allocation in distributed database systems. ACM Transactions on Database Systems (TODS) 13(3), 263–304 (1988)
2. Ausen, D.: Fobis: Foresight biomedical sensors. FOBIS-NICE Meeting (2006)
3. Chan, D., Roddick, J.: Context-sensitive mobile database summarisation. In: 26th Australasian Computer Science Conference, vol. 16, pp. 139–149 (2003)
4. Chan, D., Roddick, J.: Summarisation for Mobile Databases. Journal of Research and Practice in Information Technology 37(3), 267 (2005)
5. Gershenfeld, N., Krikorian, R., Cohen, D.: The Internet of things. Scientific American 291(4), 76–81 (2005)
6. Hababeh, I., Bowring, N., Ramachandran, M.: A method for fragment allocation design in the distributed database systems. In: The 6th Annual UAE University Research Conference (2005)
7. Kubler, S., Derigent, W., Thomas, A., Rondeau, É.: Problem definition methodology for the "Communicating Material" paradigm. In: IFAC Workshop on Intelligent Manufacturing Systems (2010)
8. Kubler, S., Derigent, W., Thomas, A., Rondeau, É.: Prototyping of a communicating textile. Industrial Engineering and Systems Management (2011)
9. Ley, D.: Ubiquitous Computing. In: Emerging Technologies, vol. 2, pp. 64–79. Becta (2007)
10. Lubinski, A.: Small database answers for small mobile resources. In: Intelligent Interactive Assistance and Mobile Multimedia Computing, pp. 9–10 (2000)
11. Meyer, G., Främling, K., Holmström, J.: Intelligent products: A survey. Computers in Industry 60(3), 137–148 (2009)
12. Morel, G., Valckenaers, P., Faure, J.M., Pereira, C.E., Diedrich, C.: Manufacturing plant control challenges and issues. Control Engineering Practice 15(11), 1321–1331 (2007)
13. Chirn, J.-L., McFarlane, D.: Application of the Holonic Component-Based Approach to the Control of a Robot Assembly Cell. In: Proc. of IEEE Conference on Robotics and Automation, San Francisco (2000)
14. Navathe, S., Karlapalem, K., Ra, M.: A mixed fragmentation methodology for initial distributed database design. Journal of Computer and Software Engineering 3(4), 395–426 (1995)
15. Padmanabhan, P., Gruenwald, L., Vallur, A., Atiquzzaman, M.: A survey of data replication techniques for mobile ad hoc network databases. The VLDB Journal 17(5), 1143–1164 (2008)

A Load Balancing Algorithm for Multi-agent Systems

Iulia Ştefan, George Moiş, Szilárd Enyedi, and Liviu Miclea

Automation Department, Technical University of Cluj-Napoca
{Iulia.Stefan,George.Mois,
Szilard.Enyedi,Liviu.Miclea}@aut.utcluj.ro

Abstract. Multi-agent societies are often used in manufacturing systems and other large-scale distributed systems. These systems often need an efficient task redistribution strategy in case of component faults or load variations. This chapter presents a simple, algorithmic approach for such a strategy, requiring low processing and communications resources.

Keywords: Load balancing, intelligent agent, distributed system.

1 Introduction

1.1 Generalities

Extremely dynamic customer requirements and global competition are shifting the production configuration of manufacturing organizations away from the traditional centralized, sequentially flowing planning, control and scheduling mechanisms. This approach is rendered too slow to adapt to evolving production styles and rapid variations in customer requirements, and limits the reconfiguration capability and the flexibility of the manufacturing system. The traditional, centralized organization may also easily lead to a large proportion of the system being shut down due to a single point of failure. The multi-agent society solution, being naturally distributed and decentralized, provides an easy method to overcome these disadvantages, and has been studied and employed in [9, 18] and [15].

However, reassigning tasks due to faults or load imbalances proves to be a difficult task even within such a society, and often results in wasted resources and poor system performance. This happens because the system architecture and the heuristics adopted by agents assigned to such production tasks are highly complex. Deen and Fletcher [7] have proposed a solution that attains load equilibrium in a multi-agent manufacturing system using concepts such as temperature, relative heat and latent heat. Employing the work accomplished in [4], the article proposes a solution to such a load-balancing problem, which makes use of a specific agent society structure and of simple computational algorithms to perform the load balancing.

T. Borangiu, A. Thomas, and D. Trentesaux (Eds.): SOHOMA 2011, SCI 402, pp. 103–114.
springerlink.com

Although primarily intended for flexible manufacturing systems, the viability of the approach extends to a wide spectrum of large-scale distributed, heterogeneous and/or geographically scattered systems, such as nationwide telecommunication or energy distribution networks. The article presents the concepts employed by the solution, the structure requirements an agent society must comply with so that the load balancing is applicable, and then details the various necessary algorithms. Note that aspects such as implementation alternatives or inter-agent communication protocol specifications are not detailed; the article merely remains at an abstract, algorithmic level for clarity.

A distributed security solution is also discussed, tackling the need for secure communication and the dynamic nature of the systems under scrutiny.

1.2 Load Balancing

Several articles describe solutions towards better production management using agent technology as optimal scheduling and cost evaluation, as it represents a priority for each business.

In [10], the design of the best structure for feasible manufacturing plans to obtain specific products is the main goal. The solution is based on the implementation of three types of agents: the *Process Planner Agent* (details about technological aspects and available resources), the *Manufacturing Planner Agent* (defines all the alternatives of the processes for manufacturing a product) and the *Production Planner Agent* (selects one of the available feasible plans). The Petri nets formalism is used to describe the manufacturing process structure and is implemented by the flexible structure of nonlinear process plan (NLPP). The tendency toward formalism, simulation and modelling representation of process structure and components is obvious.

Another solution for production-distribution planning, this time, is given by A. Kazemi et al. [11]. The authors use a multi-agent system to implement a genetic algorithm to solve production-distribution scheduling problems. Plans of the supply network for production and distribution are generated for the constraints and the global bill of materials. The authors present three different genetic algorithms, one for giving a solution from a centralized point of view, and two from distributed perspective. The parameters taken in consideration were those affecting time and costs, fixed production costs, unit holding costs, demand, fixed and variable transportation costs, processing time.

In [2], the load balancing algorithm is evaluated for geometric convex divided regions. These m regions provide m services situated at certain p_m distances. The idea is to divide the region in several sub-regions by a constant factor of the fatness of the original polygon. The chapter discusses the structure of an optimal subdivision, too.

For a conveyor distribution situation, for example, the best approach could be a decision differentiation by location awareness. The best suited machine to transport inside a plant the needed materials is the one located nearest. Division of region into sub-region and coast evaluation for minimal response time to a request could increase efficiency and avoid bottleneck situations.

In [11], the load balancing algorithm is presented in redundant services usage to speed up the requests evaluation. The presented model applies to Ethernet environment in an asynchronous communication. The redundant service is implemented as instantiation of the same class, it stays on a host and the host determines its load.

Load balancing approaches are often encountered in decongestion of wireless mesh networks, as well. In [6], the authors propose a novel solution for increasing bandwidth, not only by taking in consideration the signal strength, but also by the access point load. The solution regards the new access point association called Load-Balanced Access Point Association and could be considered to be an optimization result. The authors discuss both the centralized and the distributed approaches, because the load gathering information for the first option is too high, implying the possibility to break a current connection on a station inside a wireless network.

Load balancing approaches require information about the current situation and needs inside a system. If such information is not available, it must be obtained by sensor reading, communication, database interrogation.

A strategy for multi-robot system handling often used inside automated production lines is proposed in [12]. For coordinating two industrial robots, two methods for distributing unknown loads are discussed. One is based on force sensor reading, the other on leader/follower approach. The main objective of the author is to obtain a strategy that allows two robots to carry one heavy load.

For scheduling tasks, a load balancing algorithm could change those tasks' redistribution. Even if production planning involves evaluation and specification of every step, time interval, resource and equipment needed, in flexible manufacturing environments the adaptation to changes involves renewing the schedule. In these types of systems, scheduling is the main process affected by a load balancing approach.

The main objective is to avoid bottlenecks, meaning that the flow of the production lines stop because there are no resources available to continue. This also translates into the fact that one of the resources inside the system is overloaded and the production is delayed with Δ_i time units.

In [1], the authors define the load as the sum of the tasks given to a certain machine divided by its capacity. The load balancing algorithm intents to minimize the load on the system machines and to offer to the scheduler information in the form of a reassignment cost depending on the job. An on-line algorithm is described to solve the job assigning problem. The online algorithm must be able to satisfy and solve demands arriving sequentially, without knowing the final outcome.

1.3 Agent Society

Agents are combinational systems that can decide for themselves the actions they need to execute in order to satisfy their design objectives. The term "agent" does not have a universally accepted definition, because of the wide range of applications in which agents are encountered and because different systems benefit from different attributes they possess. The only characteristic common to all approaches is *autonomy* [21]. Thus, agents can perform their activities without human intervention or intervention from other external factors such as other agents.

A more general way of defining agents is considering them entities which can perceive the environment through sensors and which can act on it by using actuators. Fig.1 presents this simplified definition.

An agent has partial control over its environment and can influence it. Complete control is not possible and the same action might not have the same effects when it is performed at different moments in time. Each agent has its own effector capability and it must decide which of its available actions should be performed in order to best satisfy its goals, this leading to the conclusion that agent architectures are in fact decision-making systems.

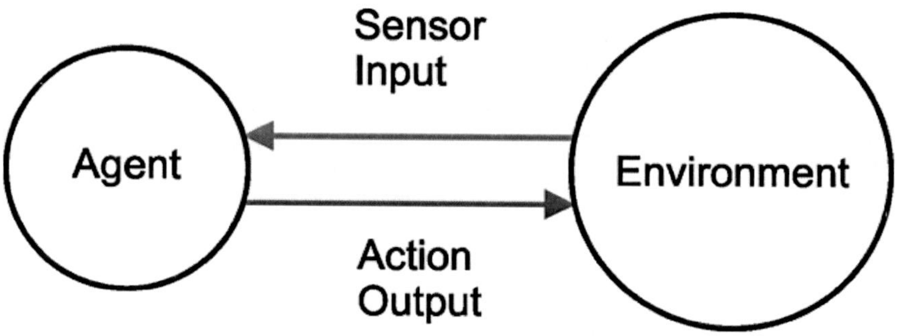

Fig. 1 General view of an agent

The problems that may be conveniently solved through the use of agents are the ones in which a classical centralized solution is not feasible and where the distribution of information and decision making is necessary. They are well suited in cases where a robust solution is required in a rapidly changing environment and where there is a continuous need for extensions and equipment reconfiguration [13].

The use of such computational entities provides great benefits in situations which involve the deployment and upgrade of distributed applications, services customization, support for disconnected operations, improved fault-tolerance, etc. [5]. It was shown that agent technology can provide efficient solutions in fields like: e-commerce [17], real-time monitoring and maintenance of distributed systems [17, 20, 19], communication networks [14] and information handling [3].

2 Proposed Solution

First, it is assumed that the distributed system where the multi-agent society is applied is geographically partitioned – i.e., smaller, spatially separated and network-connected subsystems are working together. Second, the objectives of the distributed system can only be reached by the composition of a large number of tasks. Each of these tasks is performed by a particular type of agent, which requires certain skills in order to function correctly. All these skills together form a set. Some of the skills in this set are related, logically leading to a type of agent that exhibits them. We name this subset of related skills a *skill class*. The bottom line is that an agent exhibiting a certain skill class may perform the duties of any other agent in that class. Such an agent is hereafter identified as a *task agent*, as opposed to the *service agents* employed by the solution.

A certain location *L*, corresponding to a specific geographical partition, requires a number of skill classes. A number of task agents of various types perform the system duties at that location. Let us consider now a single skill class at this location. This class contains a number of tasks:

$$C = \{t_1, t_2, ..., t_k\} \quad (1)$$

A numerical score is assigned to each task, s_i, $i = 1...k$, directly proportional to the complexity of the task (which includes, but is not limited to, required resources and task execution time). Obviously, more tasks of a given type can exist at a certain location. We denote the number of tasks of type t_i with n_i, $i = 1...k$. If the number of task agents residing at the considered location and exhibiting skill class *C* is *N*, we define the *load factor* of the skill class *C* at location *L* as:

$$LF_{C-L} = \frac{\sum_{i=1}^{k} n_i \cdot s_i}{N} \quad (2)$$

The purpose of the load-balancing now becomes clear: *to ensure a homogeneous load factor for every skill class over the locations requiring that class.* Of course, "homogeneous" is not an absolute term, and involves a tolerance interval. We assume that each task agent performs duties in a single skill class at a given time.

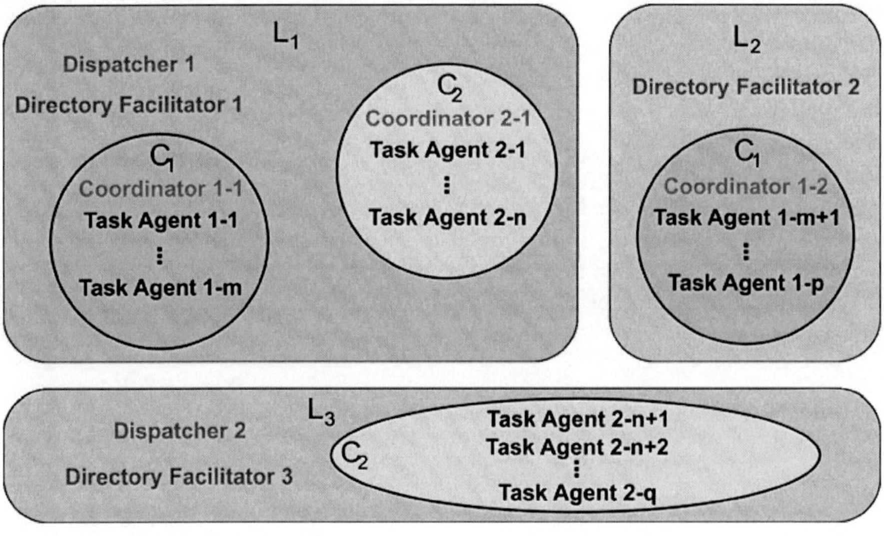

Fig. 2 Required Society Structure

The required components of the multi-agent society are presented in Fig.2. Locations are represented as thick-outlined rectangles, skill classes as ellipses and agents as labelled coloured rectangles. We therefore have three locations, one of them requiring two skill classes, and the second and third requiring each only one skill class common to the first location. The *Directory Facilitator* (DF) agents implement the directory service specified in [22]. All agents register the services they offer with the local DF, which may also be queried by any agent for addresses of agents advertising a certain service. Therefore, a DF must be present at each network (geographical) location. DFs will federate over the network. An arbitrary number of *task agents* may be present at each location for the skill classes represented there.

The key agent to our approach is the *Dispatcher*. The dispatcher performs the actual *load balancing*, using an algorithm we shall specify in the following sections. The purpose of the dispatchers is then to reach the following state of facts:

$$LF_{C_1-L_1} \cong LF_{C_1-L2} LF_{C2-L_1} \cong LF_{C_2-L3} \tag{3}$$

There is no need for dispatchers to be present at each location. In fact, a single dispatcher would suffice, but in order to maintain the advantages of distributed processing and minimize the communicational overhead, more dispatchers may work at the same time within the society, assigned to sets of skill classes specific to a group of locations (e.g. parts storage warehouses). The task agents will then maintain contact with the dispatcher closest to their location.

Note also that C_1 makes use of a *Coordinator* agent, while C_2 does not. The coordinator is a generic concept representing the agent which performs tasks distribution within a certain skill class at a given location. If such an agent exists, it naturally knows the load factor of its skill class. The difference between the two types of classes is that the dispatcher will poll coordinators for load factors each time the load balancing algorithm is activated, but will need to be informed actively by self-coordinating agents that determine the tasks they are currently performing in order to maintain a load factor for those agents' class.

All task agents will need to register themselves with the DF as providing the skill class they are part of, so that the Dispatcher may find them when necessary. The coordinators also have to register as suppliers of the coordination service for their skill class. We do not specify names for these services, leaving this for the implementation.

3 Skill Classes Awareness

In our solution, the context awareness characteristics are given by the dispatchers. They make use of a skill classes maintaining algorithm which ensures the flexibility and scalability of the distributed system. Therefore, they must become aware of new skill classes appearing within the system. The problem is in fact quite simple: for a skill class which does not employ coordinator agents, the dispatcher learns of its existence when the first agent performing the duties in that class begins functioning. The situation slightly changes if the class employs coordinator agents. The dispatcher will periodically poll the DFs about coordinator agents. The dispatcher becomes aware of a new class when it finds a coordinator agent representing that class.

The problem of skill classes appearing and disappearing at subsystem locations is similar. Since the classes' registry is dynamic, there is no trouble changing it in order to reflect the system's state at the given time.

4 The Load Balancing Algorithm

The dispatcher will run a *load balancing thread* for each skill class under its supervision. This thread consists of a balancing sequence repeating at variable intervals. During every such sequence the following actions are performed by the dispatcher:

- if the skill class makes use of coordinators, they are polled about the load factor at their locations. Otherwise, the load factor at every location is already known from the information actively supplied by the agents.
- an array of load factors is formed, with a distribution of the values similar to what is represented in Fig. 3. The vertical bars symbolize the positions of the load factors.
- the remainder of the sequence is executed only if the ratio of the greatest load factor to the least exceeds a *balancing activation threshold* specific to the skill class. Otherwise, the sequence is cancelled.
- the value $\delta = d \cdot \Delta$, where d is a ratio coefficient chosen between 0.1 and 0.4, gives the intervals at the end of the load factors span where task agents transfers will be attempted. The dispatcher chooses a number of *pairs*, each pair formed by a location with the load factor at the lower end of the Δ interval, and one at its upper end. The pairs are distinctively coloured in Fig. 3. The pairs forming stage ends either when one of the δ intervals has been exceeded or when a maximum number of pairs have been formed.
- for each pair, one task agent transfer is attempted from the less loaded location to the other. The agent transfer algorithm is detailed below.

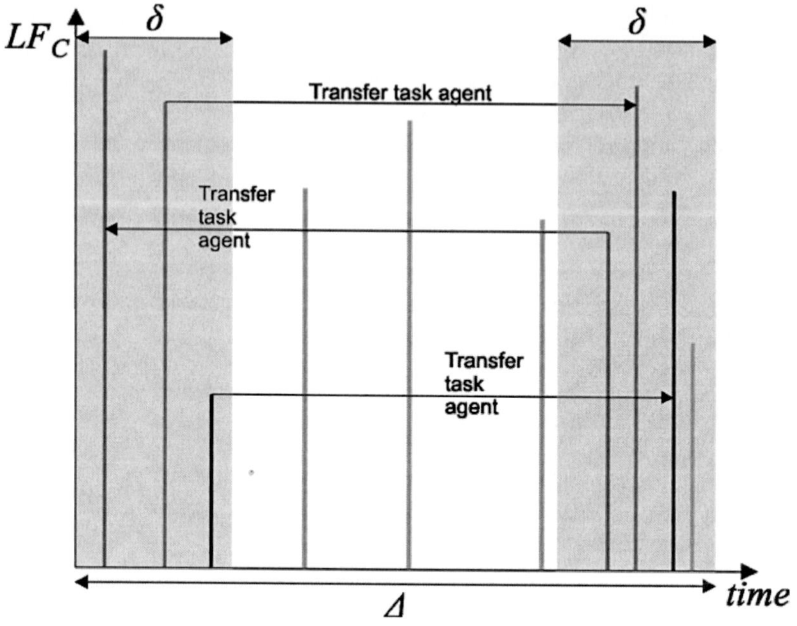

Fig. 3 Load factors distribution

The dispatcher will also ensure that the movement of the task agents does not become oscillatory, by never sending a task agent along a direction opposite to the direction a task agent of the same type was transferred at the last load balancing tick.

The dispatcher establishes the idle interval between two activations of a load balancing sequence (the *hyperperiod*) at every run, as follows: new load factors are estimated for the locations which exchanged task agents, and a new minimum and maximum are computed, together with their ratio *r*. The hyperperiod is then computed with the formula:

$$T = T_M - T \cdot \left(1 - e^{\frac{r-\sigma}{\sigma}}\right), T = T_M - T_m \tag{4}$$

where T_m and T_M are, respectively, the minimum and maximum hyperperiod values, and σ is the balancing activation threshold. The effect of this formula is the rapid (exponential) approach of the hyperperiod toward its minimum as the load imbalance rises above the threshold. Obviously, when the balancing is not activated (the threshold is not exceeded), the hyperperiod is maintained at its maximum value, as sketched in Fig.4. Of course, the load imbalance has a "bumpier" evolution in reality, than depicted in the figure.

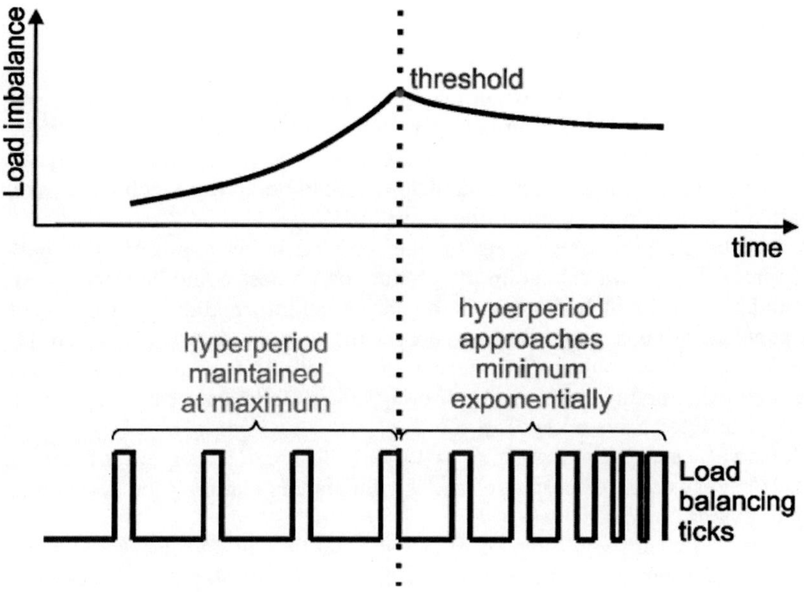

Fig. 4 Hyperperiod evolution

5 Task Agents Transfer

When the dispatcher desires to transfer a task agent exhibiting a certain skill class, it will query the DF for the presence of such agents at the less loaded location of the current pair (see the previous section). It will then randomly choose an agent and request it to transfer to the heavier loaded location.

A task agent is not required to comply unconditionally whenever a dispatcher requests it to move. If the agent knows that for some reason its remaining at the current location is critical, it will refuse the move. The dispatcher will mark such an agent in a registry it maintains and will not retry to transfer the agent for a certain amount of time, given in load balancing ticks.

If the agent is engaged in performing a task, it will respond with a "hold" message to the dispatcher, sending back an estimate of the time necessary to complete the task, together with its availability to transfer upon completion. If this estimate does not exceed a ratio of the minimum hyperperiod, the dispatcher will agree and consider the agent transferred. The ratio value will be constant and chosen somewhere between 0.4-0.75. If, however, the completion time is not satisfactory, the dispatcher will choose another agent from the set returned by the DF and try to transfer it. This repeats iteratively until either an agent is scheduled for transfer, there are no more agents or a timeout occurs. In the latter two cases, the transfer attempt is cancelled by the dispatcher.

6 Distributed Security

The issue of data security arises when taking in consideration the fact that representative information is transferred between agents, which may be connected by insecure communication lines.

One of the problems for agent communication could be, in high-tech industries, to maintain the secrecy of manufacturing procedures.

When scheduling and rescheduling for the load balancing approach, the message that imposes a certain change in the production process could be intercepted. What should be done in such situation is to encrypt the information. If every agent inside a population has a public key to encrypt the data and a private one to decrypt the message, this could offer a certain degree of security. If the population has a hierarchical structure, where inferior branches don't have access to all the information, distributed private keys is a solution.

The ElGamal decryption algorithm with k+1 differential degrees of access rights [8, 16] suggests a tree structure model - the entities claiming the decryption are located in nodes of a tree.

Thus, every post can be encrypted by X_i with public key and decrypted by Y_i with private key. There is also the possibility that a group of all posts or messages can be decrypted by other users, using a special private key. A grade *0* user can decrypt all the messages, a grade one user can decrypt a subset of messages that can be decrypted by grade *0* key and so on; a k user can decrypt a single message.

The information $\{m_i ... \mid i \in I\}$ is represented as a multitude of reduced size messages and is encrypted using a public key by entity X_i and decrypted by Y_i entities using private keys. The intention is that Y_{k-j} decrypts with a private key only the messages included in a partition of the set m_i. Y_0 will be the only one capable to decrypt the entire message.

The algorithm for messages is divided into several steps: generating the tree and creating the node indices, generating the keys, encrypting and decrypting the messages.

If necessary, an access level can be eliminated or added.

The tree structure includes the users, the access levels and the hierarchy: nodes, levels, arcs.

Every node and the user, associated to a leaf or node, will possess a private key; every leaf has a public and a private one. Decrypting a message by a private key is possible if and only if a chain exists of descendent nodes from the analyzed key to leaf z_i, where the message to be decrypted is situated. To generate the private keys, a cyclic group of order q will be chosen, q being a prime number, for which the discrete logarithm problem is difficult; g is its generator. In case of longer messages, the algorithm could be improved with a symmetric system.

7 Conclusions and Future Work

A simple load balancing strategy for large-scale multi-agent systems has been presented, together with the structure of the agent society it relies upon. The approach requires fewer resources in terms of communication bandwidth and processing time than the solution presented in [7]. Provided that the various coefficients, ratios and timeouts influencing the algorithms are well chosen, the stability of the strategy is guaranteed and the load factors for a skill class will converge into an interval situated around an average load factor after the various subsystem loads change or when faults occur. Therefore, in these stable states the system workload will be evenly distributed among task agents.

Also, a security solution with a distributed key approach was presented, trying to balance the dynamic nature of the agent society with the need for secure communications.

The first need to be addressed by future work is the implementation of a prototype and the study of its behaviour under various simulation conditions, for example in a system similar to the one discussed in [4]. Without any doubt, the load balancing threshold and the ratio coefficient *d* have the strongest bearing over the efficiency of the strategy. In the current form, they are chosen for each skill class and remain constant in time. Alternatives of adaptation algorithms for these coefficients will be studied, together with their effects on the strategy performance.

References

1. Andrews, M., Goemans, M., Zhang, L.: Improved Bounds for On-Line Load Balancing. Algorithmica 23, 278–301 (1999), doi:10.1007/PL00009263
2. Aronov, B., Carmi, P., Katz, M.: Minimum-cost load-balancing partitions. Algorithmica 54, 318–336 (2009), doi:10.1007/s00453-007-9125-3
3. Brewington, B., Gray, R., Moizumi, K., Kotz, D., Cybenko, C., Rus, D.: Mobile agents for distributed information retrieval. In: Klusch, M. (ed.) Intelligent Information Agents. Springer, New York (1999)
4. Buşoniu, L.: Multiagent systems in DBIST and distributed testing. Report of Technical University of Cluj-Napoca, Department of Automation (2003)
5. Carzaniga, A., Picco, G.P., Vigna, G.: Designing distributed applications with mobile code paradigms. In: Proceedings of the 19th International Conference on Software Engineering, pp. 22–32 (1997), doi:10.1109/ICSE.1997.610193
6. Cui, Y., Ma, T., Liu, J., Das, S.: Load-balanced AP association in multi-hop wireless mesh networks. The Journal of Supercomputing, 1–27 (2010), doi:10.1007-s11227-010-0519-7
7. Fletcher, M., Deen, S.M.: Task rescheduling in multi-agent manufacturing. In: DEXA Workshop, pp. 689–694 (1999), doi:10.1109/DEXA.1999.795268
8. Flonta, S.: Contributions to the development of models for accessibility and security of information systems. Report of the Technical University of Cluj-Napoca, Department of Automation (2010)
9. Handel, S., Levi, P.: A distributed task planning method for autonomous agents in a FMS. In: Proc. IEEE/RSJ/GI Int. Conf. on Intelligent Robots and Systems (IROS 1994), vol. 2, pp. 1285–1292 (1994), doi:10.1109/IROS.1994.407462
10. Imberti, L., Tolio, T.: Manufacturability models for manufacturing e-marketplaces. In: Perrone, G., et al. (eds.) Designing and Evaluating Value Added Services in Manufacturing E-market Places. Springer, Netherlands (2005)

11. Kazemi, A., Fazel Zarandi, M.H., Moattar Husseini, S.M.: A multi-agent system to solve the production–distribution planning problem for a supply chain: a genetic algorithm approach. The International Journal of Advanced Manufacturing Technology 44, 180–193 (2009), doi:10.1007/s00170-008-1826-5
12. Kim, K.: Two strategies for handling unknown loads of two coordinating robots. The Korean Society of Mechanical Engineers 13, 116–129 (1999), doi:10.1007/BF02943664
13. Marik, V., McFarlane, D.: Industrial adoption of agent-based technologies. IEEE Intelligent Systems 20, 27–35 (2005), doi:10.1109/MIS.2005.11
14. Pugazendi, R., Duraiswamy, K.: Mobile agents - a solution for network monitoring. In: RTCom 2009, pp. 579–584 (2009), doi:10.1109/ARTCom.2009.76
15. Shen, W., Norrie, N.: An agent-based approach for dynamic manufacturing scheduling. In: Working Notes of the Agent-Based Manufacturing Workshop, pp. 117–128 (1998)
16. Ştefan, I., Végh, L., Moiş, G., Flonta, S.: Multi-agent hierarchical system based on ElGamal decryption algorithm with K+1 access levels. In: ICAS 2011, pp. 51–56 (2011) ISBN: 978-1-61208-006-2
17. Tang, Q., Xie, F.: A multi-agent system for E-Commerce automation. In: IEEE CCECE, pp. 514–517 (2006), doi:10.1109/CCECE.2006.277722
18. Tzafestas, E.S.: Agentifying the process: task-based or robot-based decomposition? In: Proceedings of 1994 IEEE International Conference on Systems, Man and Cybernetics (SMC 1994), vol. 1, pp. 582–587 (1994), doi:10.1109/ICSMC.1994.399902
19. Vălean, H., Miclea, L., Enyedi, S.: Surveillance of a complex heterogeneous system using software agents – case study on a hydroelectric power plant chain. In: Proceedings of the 19th International Conference on Computer Applications in Industry and Engineering, pp. 126–131 (2006) ISBN: 978-1-880843-61-1
20. Weiss, G.: Multiagent Systems: A Modern Approach to Distributed Artificial Intelligence. The MIT Press, Cambridge (1999)
21. Wooldridge, M.: An Introduction to MultiAgent Systems. John Wiley & Sons, Chichester (2002)
22. Abstract architecture specification. FIPA (2002), `http://www.fipa.org/specs/fipa00001/SC00001L.html` (accessed May 15, 2011)

A Holonic Approach to Myopic Behavior Correction for the Allocation Process in Flexible-Job Shops Using Recursiveness

Gabriel Zambrano Rey[1], Nassima Aissani[2], Abdelghani Bekrar[2], and Damien Trentesaux[2]

[1] Univ. Lille Nord de France, F-59000 Lille, France - UVHC, TEMPO Lab., PSI Team, F-59313 Valenciennes, France - UVHC, TEMPO Lab., PSI Team, F-59313 Valenciennes, France - Pontificia Universidad Javeriana, Department of Industrial Engineering, Bogotá, Colombia
gabriel.zambranorey@meletu.univ-valenciennes.fr
[2] Univ. Lille Nord de France, F-59000 Lille, France - UVHC, TEMPO Lab., PSI Team, F-59313 Valenciennes, France - UVHC, TEMPO Lab., PSI Team, F-59313 Valenciennes, France
{nassima.aissani,abdelghani.bekrar, damien.trentesaux}@univ-valenciennes.fr

Abstract. This chapter's main interest is the myopic behaviour inherent to holonic control architectures. Myopic behaviour is the lack of coherence among local decision-making and system's global goals. So far, holonic architectures use mediator entities to overcome this issue, bringing the holonic paradigms more toward hierarchy than heterarchy. Instead, this chapter explores the recursiveness characteristic of holonic manufacturing systems (HMS) as a possible way to correct myopic behaviour, by distributing decision-making over adjunct entities. The chapter explains our approach and its agent-based implementation for solving the allocation problem in a flexible job-shop. Results from simulations are compared with a mixed-integer linear program to determine its efficiency in terms of makespan and execution time. Preliminary results encourage further research in this area.

Keywords: holonic manufacturing, myopia, resource allocation, recursiveness.

1 Introduction

Holonic manufacturing systems (HMS) claim to be the best-of-breed of hierarchical and heterarchical structures [8]. Hierarchies provide stability while heterarchical relationships introduce reactivity, flexibility and adaptability. Despite the effort to introduce HMS concepts in industry, holonic systems face important performance and development–related restrictions for their adoption [13]. This chapter targets on the HMS performance issues.

Performance issues arise because of the inherited heterarchical behaviour within holonic manufacturing systems. Heterarchical principles state that entities

T. Borangiu, A. Thomas, and D. Trentesaux (Eds.): SOHOMA 2011, SCI 402, pp. 115–128.
springerlink.com

should be autonomous and loosely coupled; therefore they should hold a minimum amount of global information. Theoretically speaking, global performance should emerge from the dynamic interaction among entities, but in reality it is hardly difficult to guarantee the system's outcomes [8]. Because entities do not have a global view of the system, they behave myopically [17]. Zambrano et al. [20] presents myopia and discriminates it into spatial and temporal myopia.

Spatial myopia is considered as the lack of information an entity has about other entities and its environment, while *temporal myopia* is the lack of knowledge and certitude an entity has over its future states [21]. Therefore, entities are not sufficiently informed to make the right decisions, not only in its best interest but also in the interest of the whole system. So far, two types of solutions can be found in literature, vertical and horizontal solutions.

Vertical solutions introduce mediators or supervisory entities into the architecture to impose (or advice) their point of view, creating hierarchical dependencies. On the other hand, horizontal approaches aim at integrating simulation functionalities to predict future states and find optimized solutions, without imposing any type of subordination, respecting heterarchical principles. A literature research is presented in [20], pointing out advantages and disadvantages of each type of solution.

Holonic systems are a kind of vertical solutions [20]. In general, these systems count on centralized direct supervision to cope with myopic behaviour and provide globally optimized performance [2]. These central deciders make the decision making complex and heavy to carry on. To avoid controlling myopic behaviour in a centralized way, our approach proposes to counteract spatial and temporal myopia using the recursiveness characteristic of holonic systems.

Recursiveness is the holonic concept that allows a holon to be a whole and a part at the same time. In this chapter this concept is assumed as the capability of a holon to create other holons and delegate them specific tasks, forming a semi-heterarchical holarchy. These delegates look for information and cooperate to propose solutions. The target is to count on more than one unit for decision-making and reacting against uncertain events.

For this particular case, a flexible job-shop problem in (FJSP) has been chosen to test the proposed holonic structure. The main idea is to study temporal and spatial myopia for the allocation process using HMS. The holonic shop-floor control is in charge of the short-term decision-making activities concerning the detailed assignment of tasks to production resources, at the proper times [6]. The complexity in the allocation process is mainly determined by the layout, the system's flexibility, the resources' capacity and the adopted control mechanisms.

Section 2 presents a discussion about myopic behaviour in HMS, as well as some insights of the recursiveness property. Section 3 explains the approach while sections 4 are 5 are dedicated to the agent-based implementation and the particular case study for the FJSP allocation problem. Section 6 discusses the results obtained from simulations and presents a comparison with some lower bounds drawn from a mixed-integer linear program. Some conclusions and future work will mark the path of this research.

2 Myopic Behaviour in Holonic Manufacturing Systems

Holonic Manufacturing Systems (HMS) are a class of distributed control mechanism aiming to provide effective decision-making processes by empowering autonomy, adaptability, agility and reactivity, among other characteristics [16]. HMS are composed of autonomous and cooperative entities called holons, that normally are assigned to manufacturing physical units, such as resources, products, handling systems, etc; or manufacturing operations, namely planning, scheduling, maintenance, etc. The allocation process makes part of the shop-floor control operation in HMS, and is carried out by a negotiation process among holons providing services and holons consuming services [10].

Holons commanding the shop floor are focused on accomplishing the manufacturing operations as optimal as possible to meet the objectives set up by the planning level. By auto-organizing (holarchy formation) and cooperating among each other, holons try to find mutually acceptable paths for achieving the expected performance. Despite this theoretical aspects, in reality HMS have two issues that inspire this work.

The first one is their overall efficiency. As HMS implementations are based on the multi-agent systems, there is an inherited tendency towards instability, unpredictability and even chaos [15]. Additionally, holonic systems do not guarantee a certain level of global performance, unless they use rigid hierarchical relationships.

The inability of holonic systems (and heterarchical systems in general) to achieve the desired level of global performance has been called myopic behaviour [17]. As each control unit inside these architectures is an autonomous entity with its own objectives, knowledge and skills, their decision-making process have a reduced global view. Decisions are based on local criteria making the entity spatially myopic. Moreover, as entities just consider the current state of the system, short-terms decisions are not necessarily lined up with system's long-term perspectives [13]. This latter type of myopia has been labelled as temporal myopia [21].

The second issue is related to the use of hierarchies to guarantee global performance. In general, HMS approaches incorporate direct supervision, coordination or mediation to tackle myopic behavior. By integrating these central deciders, holonic systems have had an inclination toward hierarchy more than heterarchy [13].

In HMS, direct supervisors or mediators are in charge of making the necessary adjustments for satisfying global optimality [1]. For instance, PROSA [18] and ADACOR [14] try to deal with myopia phenomenon by introducing a staff or a supervisor holon, respectively. Staff holons (PROSA) are coordinating entities who advise basic holons in their decision making. Although the basic holons are the ones making the final decision, they always try to follow their staff holon's advice. In the same manner, ADACOR's supervisor holons are coordinating entities that are in charge on triggering calculations for reaching optimized schedules. These recalculations are done periodically or in the occurrence of a disturbance.

In other architectures, mediator entities run on-line simulations. Temporal myopia should be diminished by the analysis of future states and the impact of local decisions into global goals [3]. In this case, the mediator holon is also the central observer and analyzer. Unfortunately, central deciders keep holonic architectures further away from industrial implementation [5], becoming a disadvantage of HMS.

These two issues motivated us to explore an interesting method to tackle spatial and temporal myopia without using central deciders. The objective is to consider distributed holarchies, combining a certain degree of coordination, but encouraging distribution and autonomy. The recursiveness characteristic of holonic systems is explored as a better approach for embedding hierarchy into heterarchy.

2.1 Recursiveness in HMS

Recursiveness is the holon condition that allows it to simultaneously be a whole and a part. This means that holons may include other holons and may be part of other holons. More precisely, this type of subordination might help to the effective operation of the whole system by mixing independence with hierarchical relationships. The difficulty of recursiveness is that agent systems per se do not count on this characteristic [11].

So far, certain definitions of abstract agents [9] and light-weight mobile agents [19] are the closest references to the implementation of recursiveness in agent-based holonic systems. In the first one, the AAgent (Abstract agent) can be a single agent or a multi-agent system. The idea is to forget what the structure of the AAgent is and just focus on the interactions of the current abstraction level. Nevertheless, the authors have not found any applications of abstract agents in allocation or other manufacturing applications.

On the other hand, light-weight agents are used for the exploration of alternative routings for vehicle traffic, so the principal agent in the vehicle helps the driver or make an informed decision [19]. Light-weight agents are created frequently by the vehicle agent depending on the amount of paths to explore. Once the vehicle agent gathers up the exploration agents with updated information, it makes a decision and creates intention agents to inform others. These mobile agents travel around for keeping the system updated and helping drivers to avoid traffic jams and arrive to their destination in the least amount of time.

It is possible to conclude that this method corrects the temporal and spatial myopia of the driver, since deciders collect updated information from the environment and inform others about the future states of the systems. The important issue is to consider scalability, mobility, behaviours and knowledge of these light-weight agents to be applied in a manufacturing allocation problem.

This chapter explores the idea of sending delegate holons into the production system's resources in order to reach for updated information and represent the interests of the product holons during the allocation process. In the following section, a recursive structure for task allocation is presented.

3 The Approach

This approach focuses on the allocation process at the shop floor. The proposed holonic architecture has two types of holons: *product holons* (PHs) and *adjunct product holons* (APHs). PHs and APHs are active components capable of carrying decision making processes. On the contrary, machines and shuttles are passive components and follow instructions from active components. Machines just update their state based on system dynamics. Passive components are not modelled by holons for now, but this will be taken into account for future work.

The approaches' objectives are described as follows:

- By decomposing themselves, PHs divide their problem into sub-problems (divide and conquer). The system tackles spatial myopia by sending delegates to each one of the resources involved in the task sequence. Delegates collect updated information and use it to make informed decisions.
- Each set of PH's delegates cooperate among them to propose solutions to PHs, considering not only the system's current states, but also possible future interferences with other products. Thus products can counteract temporal myopia.
- Additionally, APHs react locally under special situations where PH's goals are compromised. Corrections are done cooperatively with other APHs from the same holarchy, even before the PHs are affected.
- APHs of products relate indirectly and locally avoiding global information, excessive communication and negotiation. Through local intentions, they inform others about their plans. This mechanism aims at decreasing complexity.

3.1 Modelling Holons

Product Holons (PH) are composed of a processing unit and a physical product. The processing unit is in charge of decision making and it is implemented by an agent (Fig. 1). The following are its knowledge, behaviours and acquaintances:

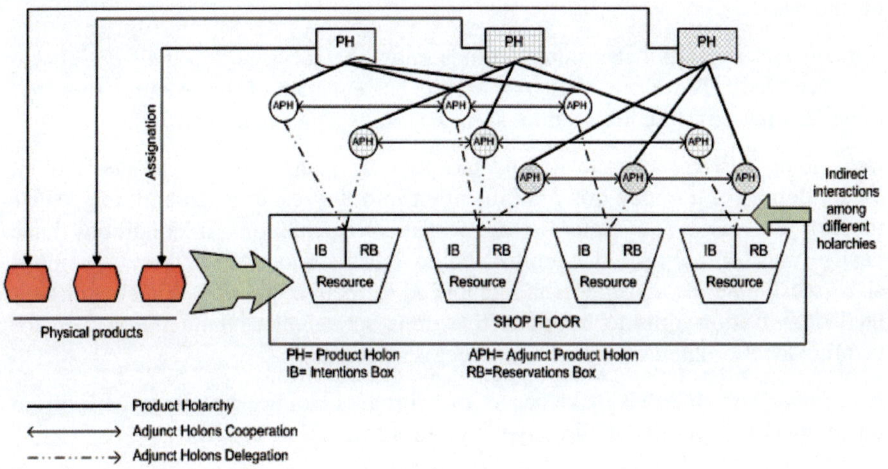

Fig. 1 Holon creation and delegation to the shop floor. Indirect holarchies interaction and holons cooperation

Knowledge: the Product Holon knows the task sequence it must follow to accomplish the product fabrication (global task knowledge). Its goal is to achieve the lowest fabrication time with the highest quality. PHs captures knowledge of the distributed system by means of its APHs. With this mechanism, PHs must not negotiate with each resource to allocate its tasks, thus decreasing communication complexity.

Behaviour: its main behavior is related to decision making. The first decision making process is related to the amount of holons it needs to create. This decision depends on resources current status, the quality of service the product is looking for and the number of tasks a resource can afford. The second decision is related to the allocation process. Its APHs propose various paths, but it depends on the PH to choose the best path among the proposals (see Eq. 3 in section 3.3) and ask APHs to reserve the resources for the winning path. If APHs find that conditions have changed from the moment they made the proposal, it is up to the PH to ask for recalculations.

Acquaintance: the Product Holon creates its own Adjunct Product Holons, and they are related only to them. Negotiation with other product agents is not considered for now since interactions are taken indirectly into account with intentions and reservations.

Adjunct Product Holons (APH) are holons created by PH holons, with a specific resource assignation. These holons are light-weight software holons with a local view of its PH's task sequence. Their main goal is to search for updated information into their designated resource and find the best choices for the allocation of their designated tasks. The following are their knowledge, behaviours and acquaintances:

Knowledge: the Adjunct Product Holons know the services that their designated resource offers. Therefore APHs capture the current state of the resource (reservations) and the future state (intentions) for making proposals to their PH.

Behaviour: APHs cooperate to find the possible paths for their product holon. While doing so, they lay down an intention into the resource. In addition, APHs monitor the resource's state to trigger path recalculations if conditions have change unfavourably for the product holon. In this way, APHs are autonomous since recalculations are done and announced to their acquaintances. This distributed organization aims to have reactive units spread throughout the system for tackling local problems before the product is affected.

Acquaintances: all APHs make part of their Product Holons' holarchy. APHs from one product do not interact directly with APHs from other products.

3.2 Modelling Resources

Resources are passive components. This means that the controllers associated to them do not make any decisions. Instead, they handle their state through their intention and reservation boxes and, their current status: available or busy. The intention and reservation boxes display the probability of the resource's workload. Resources have two types of relationships with active components. An indirect relationship is established with APHs through the intentions and reservation boxes. It is an indirect relationship because APHs do not have the power to change the resource status. On the contrary, the relationship with product holons is a direct one, because they alter the resource's status when they occupy the resource. A resource does not have any links with other resources, since allocation is not done by negotiation protocols and resources are not modelled as holons for now.

3.3 The Allocation Method

The allocation process is accomplished based on the assumption that the planning process issues a certain number of orders, which has to be done as quickly as possible, aiming at replenishing inventory (*make-to-stock* scenario). A middleware system is in charge of the interoperability between the planning system and the shop floor to guarantee a successful integration. The planning process and the middleware are out of the scope of this chapter. The recursiveness property builds a semi-heterarchical structure among PHs and APHs. APHs of the same PH have heterarchical relationships among each other, but hierarchical relationships with their PH. Nevertheless, the entire allocation process is done heterarchically because PHs do not have information about other PHs and just focus on their own chosen path.

The first recursive behaviour is the order holon's division into as many product holons as it has. For this chapter what matters is that each product holon knows its task sequence. Other behaviours and relationships among order holons and their product holons are not considered for now. The second recursive behaviour is the PHs division into as many APHs as it needs, which is in the interest of this work.

For product PH^n, let's consider s_j^n the cost of the *j*-path found by its APHs, and its relative cost RC_j^n is calculated as follows:

$$RC_j^n = \frac{s_j^n}{\sum_{i=1}^{j} s_i^n} \tag{1}$$

Now, let's consider t_j^n the expected occupancy (accumulated intentions) for the same j-path, so its relative expected occupancy RO_j^n is calculated as follow:

$$RO_j^n = \frac{t_j^n}{\sum_{i=1}^{j} t_i^n} \tag{2}$$

The chosen path is the one with the lowest relative cost and the least relative expected occupancy. These criteria aim at balancing the workload and evaluate not only current conditions, but also possible future states. Once the product calculates both criteria for each proposed path, the one with the minimum cost is chosen. The chosen path SS^n for product PH^n allows having a release time for the product and an estimative of the production time. The factor α is a term for balancing the importance of each parameter (α=[0,1]).

$$SS^n = \min_j \left(\alpha RC_j^n + (1-\alpha) RO_j^n \right) \tag{3}$$

Fig. 2 shows the theoretical graph and the UML diagram of a task sequence example. The allocation process starts by APHs building a graph in a distributed way. The last APHs for each one of the paths send the possible machine path, accumulated intentions and expected completion times to the Product Holon. Next, the PH evaluates the various paths, considering estimated work-balance and completion times.

3.4 Reactivity to Uncertain Conditions

Manufacturing systems experience uncertain situations that require them to be reactive. The system could be affected by external and internal perturbations. External perturbations are the consequence of changes in market conditions or clients' requirements, causing alterations in order quantities, orders' priority, product mixture and inventory status among others. Internal perturbations, on the contrary, are caused by the inherent stochastic nature of these systems. Machines failures, longer-than-expected processing times, longer transfer times and product interactions are some examples of internal disturbances. In both cases, entities within

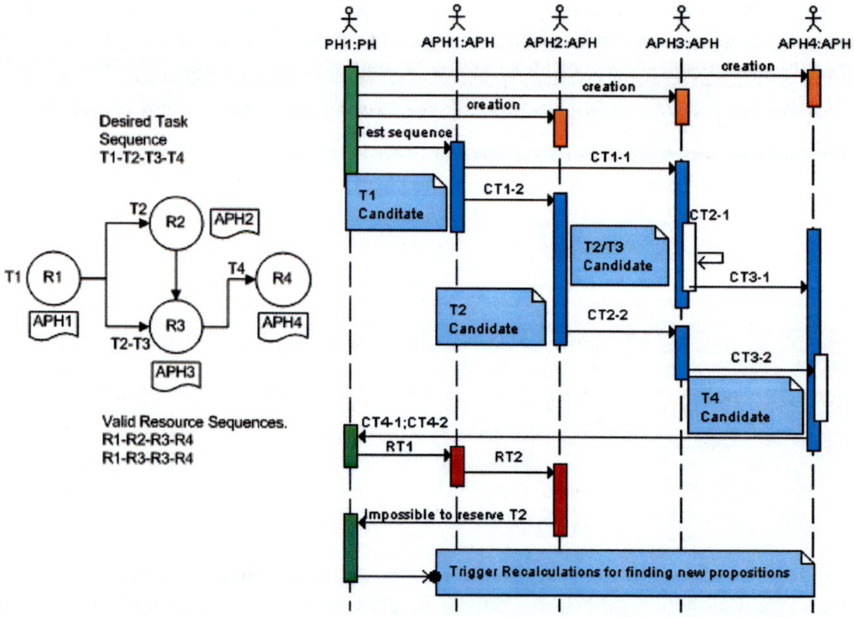

Fig. 2 UML for the task sequence example

the system need to react to avoid downgrading performance. For external perturbations the PHs will ask their APH to recalculate based on the new conditions. For internal perturbations, once APHs perceive them, they will trigger recalculations to estimate new processing times. Based on delegation, all concerned products will participate in recalculations to return to stability. Consequently, new products entering the system will count on updated information for their proper calculation.

4 Holons and Their Agent-Based Implementation

Holons are simulated using the Jade MAS development platform (http://jade.tilab.com/) in JDK 5 environment. The processing unit of the PHs holon is a class inherited from a basic Jade agent class. These PH holons create other software holons, called APHs, which are also an inherited class from a Jade agent class. Unlike PH holons, APHs do not have a physical part and they just have one class to make them light. This last characteristic is important because the system is susceptible of saturation, due to the number of APHs created. Resources' data are simulated using read/write files which are accessed by APHs in order to upload their intentions and reservations. For now, the simulation just considers APHs creation. Once the simulator start, each PH creates its APHs and they are alive until the product finishes all its tasks. In future work, recursivity will be applied from the planning to the operational activities.

5 FJSP Test Case and Lower Bound Calculation

The flexible job-shop system considered has *M* machines/resources strongly connected by a transportation system (Fig 3), must execute *N* products of four types ("b","E","L" and "T"). Each product has its task sequence as shown in Fig 3. Additionally, each machine is capable of servicing three different types of operations, one at a time (Fig. 3), except for M1 since it is the loading/unloading station. The allocation problem for this particular manufacturing configuration is thus assigning machines to specific operations and finding the paths that reach for the minimum completion times, otherwise known as makespan. This type of manufacturing scheduling problem is identified as a flexible job-shop problem (FJSP). The FJSP is known to be NP-hard in the sense that none solution in polynomial time is achievable [7]. For this test case, a mixed-integer linear program is used to calculate a lower bound and have a reference for comparing the proposed approach efficiency. The holonic implementation shares the same constraints as the MILP to allow a fair comparison. In the next section, the MILP constraints are presented.

5.1 Mixed-Integer Linear Program (MILP) for FMS

The MILP is inspired by the model proposed in [4]. Nevertheless, our model considers more than one transportation shuttle, contrary to the problem shown in [4] where just one automated guided vehicle is used for transportation. The objective of the MILP is to determine the schedule that minimizes the completion time of the last product including transportation times. Hereunder, the model's parameters, indexes and variables are presented first, and then constrains are detailed. The MILP is solved using the Concert Technology of IBM Cplex 12.2 [12].

Notations for parameters

- P: set of jobs to be executed, P={1,2,...*n*};
- R: set of resources, R={1,2,...*m*};
- I_j: set of operations of the job *j*, $I_j=\{1,2,\ldots,|I_j|\}, j \in P$;
- O_{ij}: operation number *i* of the job *j* ;
- S_e: type e of the operation O_{ij};
- p_{ij}: processing time of operation i ($I \in I_j$) for the product *j*;
- R_{ij}: set of possible machines for the operation O_{ij} ;
- $tt_{r1r2:}$ transportation time from resource r1 to r2.

Notation for indexes:

- $j \in \{1,2,\ldots n\}$: for jobs; $r \in R$: for resources and $i \in \{1,2,\ldots,|Ij|\}$: for operations,

Notations for variables:

- tij: completion time of operation O_{ij} (i Ij), $t_{ij} \in N$,
- μ_{ijr}: binary variable set to 1 if O_{ij} is performed on machine *r*, 0 otherwise,
- b_{ijkl}: a binary variable set to 1 if O_{ij} is performed before O_{kl}, 0 otherwise,
- tr_{ijr1r2}: a binary variable set to 1 if O_{ij} is performed before O_{kl}, 0 otherwise,

The objective of the MILP is to minimize the makespan $C_{\max}$, where

$$C_{\max} = max_{\forall i \in I_j \forall j \in P}\ t_{ij} \tag{4}$$

The constraints imposed in the MILP are:

- *Disjunctive constraints*: A machine can process one operation at time and an operation is performed by only one machine. Machines have unlimited queuing capacity.

$$t_{ij} + p_{kl} * \mu_{klr} + BM * b_{ijkl} \le t_{kl} + BM \ \ \forall \text{i,k} \in I, \forall j,l \in P, \forall \text{r} \in R_{ij} \tag{5}$$

where BM is a large number

$$b_{ijkl} + b_{klij} \le 1 \ \forall \text{i} \in I_j, \text{k} \in I_l \ \forall j,l \in P, \tag{6}$$

$$\sum_{r \in R_{ij}} \mu_{ijr} = 1 \ \forall \text{i} \in I_j \ \forall j \in P, \tag{7}$$

- *Precedence constraints*: ensure the task sequence of a product. The completion time of the next operation considers the completion time of the precedent one and the transportation time if the two operations are not performed in the same machine.

$$t_{(i+1)j} \ge t_{ij} + p_{(i+1)j} + \sum_{r1,r2 \in R} tt_{r1r2} * tr_{ijr1r2} \ \ \forall i \in I_j, \forall j \in P, \forall r1, r2 \in R_{ij} \tag{8}$$

$$\sum_{\substack{r1,r2\in R\\ r1\neq r2}} tr_{ijr1r2} \leq 1, \qquad \forall i\in I_j, \forall j\in P \tag{9}$$

- *Allocation and transportation relationship*: if successive operations of a product are performed on different machines, this implies that there is a transport operation between those two machines. Transportation delays are set to zero and the transportation system has unlimited capacity.

$$\mu_{ijr1} + \mu_{(i+1)jr2} - 1 \leq tr_{ijr1r2} \quad \forall i\in I_j\ \forall j\in P, \forall r1, r2\in R_{ij}, r1\neq r2 \tag{10}$$

$$\mu_{ijr1} + \mu_{(i+1)jr2} \geq (1+\varepsilon) tr_{ijr1r2} \quad \forall i\in I_j\ \forall j\in P, \forall r1, r2\in R_{ij},\ r1\neq r2 \tag{11}$$

where ε is a small number.

The remaining constraints of the MILP ensure the type of each variable.

$$t_{ij} \geq p_{ij}\ \forall i \in I_j, j \in P \tag{12}$$

$$b_{ijkl} \in \{0,1\} \forall i \in I_j, j \in P, \forall k \in I_l, l \in P \tag{13}$$

$$tr_{ijrr'} \in \{0,1\} \forall i \in I_j, j \in P, \forall r, r' \in R_{ij} \tag{14}$$

$$\mu_{ijr} \in \{0,1\} \forall i \in I_j, j \in J, \forall r \in R_{ij} \tag{15}$$

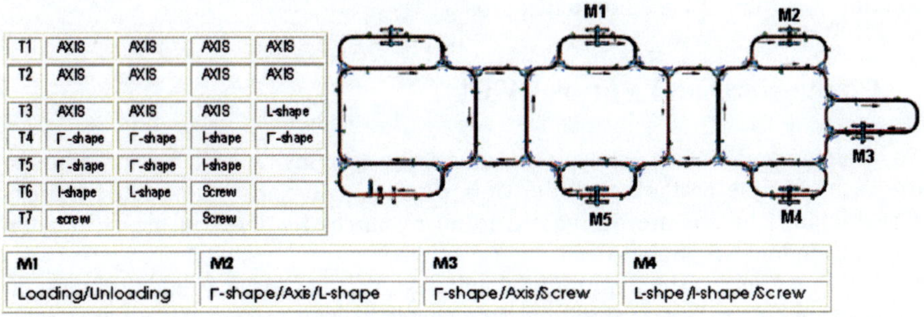

Fig. 3 Flexible manufacturing system. Services associated to resources

6 Results

Four mixed-orders were issued separately. A number of 4, 8, 12 and 16 products orders were launched into the Jade-based simulator as well as the MILP solver (Cplex). The results are presented in Table 1.

Although at first Cplex programs were launched with the whole set of constraints, Cplex just found an optimal solution for 4 products. For the other tests, Cplex is unable to find a solution. By relaxing the precedence and the transportation constraints, Cplex does not reach the optimum in one hour. The solution taken in these cases is considered as upper bounds for the relaxed problem and infeasible upper bounds for the original problem (not relaxed).

Table 1 Results from the proposed holonic system and the MILP (*optimal solution)

Order	Holonic System		MILP		Gap (%)
	Cmax (s)	Exec. Time (s)	Cmax (s)	Exec. Time (s)	
4	369	9.059	349*	165.63	5,73
8	583	24.937	549	3600	6,19
12	821	49.797	749	3600	9,61
16	1018	79.265	1083	3600	-6,00

From the Gantt diagram, an important behaviour needs to be highlighted. Even though a product could stay in one machine to continue its sequence, it goes to another one. This is for us a non-myopic behaviour because the product is not only aiming at its goals, but considering the system's goals. In other heterarchical systems already tested, such as Contract-Net or Potential fields systems, this feature had never been encountered because these models have instant-based decision making. These preliminary results let us think that creating delegates that interact with each other helps the system to find solutions that balance local goals with global goals. Even though delegates can be numerous (80 holons for a 16 product batch), the system does not reach saturation and finds similar results with very low execution times compared to an exact method. For the last instance an even better solution was found by the holonic approach.

7 Conclusions and Future Work

To diminish myopic behaviour in manufacturing systems, this chapter proposes to use recursiveness as the capability of a product to create task-oriented holons. These adjunct holons are designated to all resources involved in the fabrication sequence to look for information.

Two important aspects are highlighted: first, the fact that a product has an extended view of the system through its delegates helps it to consider a larger solution space and diminish its spatial myopia. Second, as products' decisions are based on direct information (current conditions) and indirect information (intentions), products achieve solutions that are not only better for them but for the whole system. Recursiveness not only helps to distribute problem solving, but also introduces semi-heterarchical relationships among entities, which is beneficial for tackling unexpected events. As delegates are autonomous entities, they can react and propose new solutions to cope with new conditions.

A comparison was made with a centralized system and results showed that the distributed approach reaches for good results in much less time. But as this is an exploratory work, several challenges come ahead.

First, it is necessary to formalize recursiveness for goal decomposition and goal assessment in order to define better the semi-heterarchical behavior among product holarchies.

Moreover, the recursiveness characteristic needs to be applied to order holons and resource holons also to achieve a coherent system. The decision making mechanisms based on intentions, and the scope of autonomy for APHs needs also to be studied further in depth.

Real experimentations are needed and the concept of mobile agents will be used to physically distribute delegates. Scalability, saturation, execution times and butterfly effects will be further studied.

References

1. Babiceanu, R., Chen, F.: Development and Applications of Holonic Manufacturing Systems: A Survey. Journal of Intelligent Manufacturing 17, 111–131 (2006)
2. Bongaerts, L., Monostori, L., McFarlane, D., Kadar, B.: Hierarchy in distributed shop floor control. Computers in Industry 43, 123–137 (2000)
3. Cardin, O., Castagna, P.: Using online simulation in holonic manufacturing systems. Eng. App. of Art. Intell. 22, 1025–1033 (2009)
4. Caumond, A., Lacomme, P., Moukrim, A., Tchernev, N.: An MILP for scheduling problems in an FMS with one vehicle. E. Journal of Operational Research 199, 706–722 (2009)
5. Chituc, C., Restivo, J.: Challenges and trends in distributed manufacturing systems: are wise engineering systems the ultimate answer? In: Int. Symp. on Eng. Syst. MIT, Cambridge (2009)
6. Chryssolouris, G.: An Approach for Allocating Manufacturing Resources to Production Tasks. Journal of Manufacturing Systems 10, 368–382 (1991)
7. Conway, R., Maxwell, W., Miller, L.: Theory of scheduling. Addison-Wesley (1967)
8. Dilts, D., Boyd, N., Whorms, H.: The Evolution of Control Architectures for Automated Manufacturing Systems. J. of Manuf. Sys. 10, 79–93 (1991)
9. Giret, A., Botti, V.: Towards an Abstract Recursive Agent. I. Comp-Aided Eng. 11, 165–177 (2004)
10. Giret, A., Botti, V.: Anemona: a multi-agent methodology for holonic manufacturing systems. Springer, Heidelberg (2008)
11. Giret, A., Botti, V.: Holons and agents. Journal of Intell. Manufacturing 15, 645–659 (2004)
12. IBM ILOG CPLEX Optimizer, High performance mathematical optimization engines, `http://www-01.ibm.com/software/integration/optimization/cplex-optimizer/` (visited June 15, 2011)
13. Leitão, P.: Agent-based distributed manufacturing control: a state-of-the-art survey. Engineering Applications of Artificial Intelligence 22, 979–991 (2009)
14. Leitão, P., Restivo, F.: ADACOR: A holonic architecture for agile and adaptive manufacturing control. Computers in Indistry 57, 121–130 (2006)
15. Shen, W., Hao, Q., Yoon, H., Norrie, D.: Applications of agent-based systems in intelligent manufacturing: an updated review. Ad. Eng. Inf. 20, 415–431 (2006)
16. Sousa, P., Ramos, C., Neves, J.: Scheduling in holonic manufacturing systems. In: Process Planning and Scheduling for Distributed Manufacturing. Springer, Heidelberg (2007)
17. Trentesaux, D.: Distributed control of production systems. Engineering Applications of AI 22, 971–978 (2009)

18. Van Brussel, H., Wyns, J., Valckenaers, P., Bongaerts, L., Peeters, P.: Reference architecture for holonic manufacturing systems: PROSA. Computers in Industry 37, 255–274 (1998)
19. Weyns, D., Holvoet, T., Helleboogh, A.: Anticipatory Vehicle Routing using Delegate Multi-Agent Systems. In: IEEE Intelligent Transport Systems Conference, Seattle, WA, pp. 87–93 (2007)
20. Zambrano, G., Pach, C., Adam, E., Berger, T., Trentesaux, D.: Myopic Behaviour in Heterarchical Control of FMS. In: Int. Conf. on Ind. Eng. and Syst. Management, IESM 2011, Metz (2011)
21. Zambrano, G., Pach, C., Aissani, N., Berger, T., Trentesaux, D.: An approach for temporal myopia reduction in heterarchical control architectures. In: Intl. IEEE Conf. on Industrial Electronics, Gdansk (2011)

Integrating e-IMS Platform via Interoperability within Collaborative Enterprises

Aurelian Mihai Stanescu, Mihnea Alexandru Moisescu,
Ioan Stefan Sacala, and George Burlacu

Faculty of Automatic Control and Computers, University Politehnica of Bucharest,
313, Spl. Independentei, Bucharest, 060042 Romania
ams@cpru.pub.ro, mamihnea@gmail.com,
{sacalaioan,burlacu.george}@yahoo.com

Abstract. Current R&D directions sustained by the European Commission and by FP7 and in the near future by FP8 research programs, are focusing on the development and standardization of new technologies to sustain the "Future Internet". In this context the development of new Internet-related concepts and technologies oriented towards providing positive benefits for economy has been included in a broad concept of "Future Internet Based Enterprise Systems". The Intelligent Manufacturing Systems (IMS) paradigms are leading our Information Society towards New Economy-driven Knowledge Society copying with Global e-Markets new list of System of Systems requirements. The Complex, Adaptive Non-monolithic Dynamic Systems (CADS) has been defined as oriented on Virtual Enterprise (VE) architecture, as well as involved in cross-enterprise methodology like Life Long Product/Service/Organization Management (LPM/PSO) and Business Intelligence (BI). Interoperating efficiently, both horizontally and vertically, is a basic requirement towards the aim of Systemic Integration within Collaborative Concurrent Competitive Enterprises.

Keywords: Future Internet Enterprise Systems, Web Service, Multi-agent Systems, Business Intelligence.

1 Introduction

"The world in which we live may be divided into the Natural World and the Man-mode World" [9].

In the famous book written by T.S. Kuhn (1970), the post war phenomenon that one could identify as "Scientific Crisis" is characterized by: ..."the best professional practices persistently prove inadequate for solving intrinsically important (complex) problems"... This phenomenon is calling for the definition of new concepts within a comprehensive methodological framework.

In this context we can consider T.S Kuhn's following important contributions regarding the progress of knowledge: "Science undergoes periodic paradigm shifts instead of progressing in a linear and continuous way. These paradigm shifts open

T. Borangiu, A. Thomas, and D. Trentesaux (Eds.): SOHOMA 2011, SCI 402, pp. 129–142.
springerlink.com

up new approaches to understanding that scientists would never have considered valid before".

T.S.Kuhn characterizes the paradigm as follows: "A paradigm is what members of a scientific community, and they alone, share". In this context a scientific paradigm can be considered not just a key concept but a comprehensive collection of mutually complementary ideas, some already well established and others still under development.

New manufacturing paradigms like: Concurrent Engineering, Virtual Organizations, Intelligent Manufacturing Systems and Networked Enterprises, have tried to make use of collaborative autonomous structures, and provided simple enough, versatile architectures based on elaborated communication infrastructure in order to ensure efficient behaviour patterns.

To manage the recent kind of complex systems, a new approach has to be developed, integrating Computer and Communications in order to reinforce the analytical power of control structures, giving way to new concepts, from Virtual Organizations to Intelligent Manufacturing Systems. We will further review these concepts as important steps towards the development of Future Enterprise Systems.

The added value of this chapter originates from two objectives related to the next generation of sustainable high competitive geographically dispersed Enterprise, coping with dual, vertical and horizontal interoperability requirements:

1. To learn lessons from the scientific "heritage" from isolated automation (1950's) towards Future Internet Enterprise Systems (2020);
2. To capture different contribution to define "the quality vector" of the System of Systems: involving availability, traceability, scalability autonomy, interoperability etc.
3. To discuss a new methodology oriented towards modelling, synthesizing and performance evaluating of Complex Adaptive System of Systems

2 Key Requirement towards the Development of a Methodology for Future Enterprise System of Systems

In this section the authors propose five steps towards the development of a Future Enterprise System of Systems, including the "scientific foundation" represented by the Digital Word Theory and Complex Adaptive Systems Theory, and the development of three fundamental tiers Collaborative Networks, Future Internet Enterprise Systems and Digital Business Ecosystems [12].

The Digital Word Theory proposes a unification of different physics fundamental concepts towards the foundation of digital "quantum" word including the **need** of *discrete models of finite type*, as the Feynman-Kontsevich graphs and our *Quantum Dot Resolution* (QDR). This theory has the advantage of being an algebraic approach, i.e. axiomatic, compatible with a top-down design methodology, in contrast with Newton- Leibnitz analysis (bottom-up methodology, constructive), or even Poincaré's topology ("in between", qualitative yet with a lot of "pathologies"). MacLane's category theory comes as the perfect object-and-relations oriented language for this purpose, where geometry and physics come with the needed "intuition" on top of its often mentioned "abstract non-sense" attribute.

The correlation between quantum physics (QP) and quantum computing (QC) as basis for the development of Systems of Systems (SoS) provides a basis for the development of CAS theories.

The present exploratory alternative to the *Standard Model* is motivated by the mandatory changes in our understanding of what space-time-matter really are. The "technical implementation", less viable, is only hinted, due to lack of time and expertise of the authors, and is intended first of all for exemplification purposes. An important point is that the technical tools are *already developed* (such as the Standard Model, String Theory, Loop Quantum Gravity, Lattice Gauge Theory, etc), yet they are "written" in the "old classical language": all we have to do is to re-write the code in Q++! This explains the style of the exposition i.e. that of a research report: second phase of the DWT-*project*. We "put the cards on the table", since the Linux open source development project showed that the WWW is the "perfect collaborator"!

The analogy goes even further; as in any top-down design of an informational system, we design the interface first, represented by names in italics, for which the implementation is just a matter of time or energy. And the Web is the interface to many more skilful mathematics- physics specialists capable of compiling the present authors' speculations from the high level language into solid mathematical-physics code. Then what remains is a "linkediting" with the current theories, to get a computational viable theory ("executable code").

The development of new concepts in the area of enterprise collaboration like: Collaborative Networks and Digital Business Ecosystem, the large scale utilization of the Service Oriented Architecture using the new infrastructure provided by Cloud Computing technologies and the ongoing demand for communication and collaboration is stressing to the limits the current internet technologies and infrastructure.

The Framework 6th-IST-project "Virtual Organization Road MAP" [4] provided an interesting assessment: "Every sustainable enterprise should reach, beyond 2012, the capability of networking in Virtual Organizations, within turbulent, global market".

Nevertheless, there is still a family of solid concepts around these topics like Collaborative Network Organization [4], Non-monolithic, Complex Adaptive System of Systems [12].

The research trend as stated by COIN FP7 IP research project is to develop pervasive Software as a Service – Utility that incorporates Collaborative Services and Interoperability Services as a basis for Enterprise Systems.

The concept of Digital Business EcoSystem (DBES) initiative responds ideally to this challenge of creating ICT instruments together with collaborative practices and paradigms that support economic growth and include all the societal and economic actors in the process. It has been commonly recognized as a new frontier for Research and Technology Development in the knowledge-based economy.

Indeed, Small and Medium Enterprises (SMEs) and local clusters are now competing in a global and dynamic market where they need more interrelations, more specialized resources, more research and innovation as well as access to global value chains and knowledge. The research driven within the DBE Initiative supports all these necessities by offering an open infrastructure that combines:

- human capital,
- knowledge and practices,
- technical infrastructure,
- business and financial conditions,

all modelled within the European industrial policy agenda.

The Future Internet networked enterprises [FInES] involve the following basic pillars: *Internet of Things* (IoT), *Internet of Services* (IoS), *Internet of People* (IoP), *Internet of Knowledge* (IoK).

In this context, the development of new Internet-related concepts and technologies oriented towards providing positive benefits for individuals, society, economy, culture and environment has been included in a broad concept o Future Internet Systems.

FInES have been oriented towards enabling enterprises and SMEs access to the full potential of Future Internet technologies through ICT.

As one of Internet's expanding directions is towards becoming a universal business environment, the business values associated may refer to:

- Revenue and profit,
- Reputation and level of trust,
- Resource planning towards efficient use,
- Green manufacturing,
- Enterprise social networking,
- Business partners collaboration,
- Customer relationships,
- Tacit and implicit Enterprise Knowledge Management,
- Business transparency and corporate governance.

In order to achieve these principles the characteristics that have to be taken into consideration when discussing future enterprise systems should include and not limit to: flexibility, adaptively, collaboration, innovation, knowledge manageability, openness a.o. The quality issues regarding the enterprise are another aspect of organization and management that has been greatly emphasized, in the last few years.

New concerns and constrains like: social and environmental impact, cultural diversity and ethical issues have to be taken into consideration and extend the Total Quality Management standards. As a consequence a new concept has been introduced: Quality of Being (QoB). QoB will have to incorporate along with the quality issues associated with the enterprise operation: products, strategies, staff and client relationships, new concepts like green manufacturing, social responsibility, and community vs. global impact.

In order to achieve such characteristics a new concept has been introduced: Internet Enterprise Resource (FInER) as describing a digitalization of enterprise entities. These entities will allow for identification, computational power, storage and communication. This concept will support the evolution of FInES as a network of different FInERs. Another important step in the evolution towards FInES is the ongoing research in the area of the Internet o Knowledge (Fig. 1)

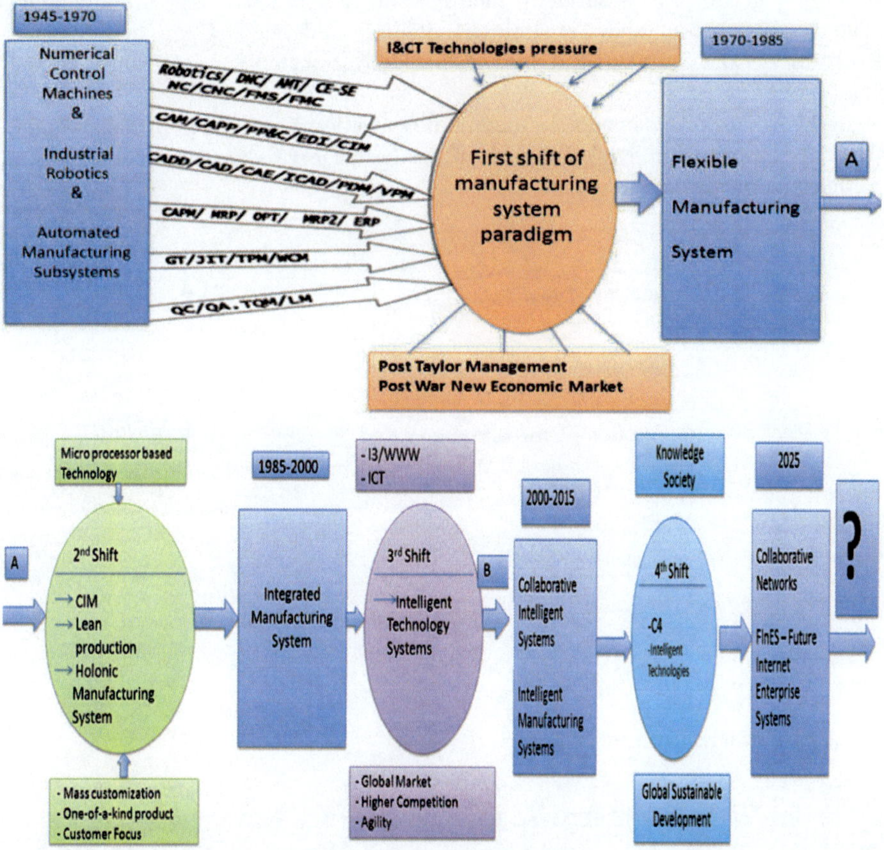

Fig. 1 The first 5 waves and 4 paradigm shifts to target FInES

The current knowledge management tools and knowledge repositories will have to be replaced by a flow of knowledge from distributed FInER collections accessible via internet infrastructure and with a meta-knowledge infrastructure represented with the help of ontologies and linked at semantic level.

The research objectives as stated by the FInES EU Cluster in relationship with Future Internet technologies and QoB can be identified as:

- Inventive Enterprise
- Cloud Enterprise

- Cognizant Enterprise
- Community-oriented Enterprise
- Green Enterprise
- Global Enterprise

A few technological paradigms that can sustain the future researches on FInES are presented below: Applications with proactive behaviour, IaaS or PaaS (Infrastructure / Platform as a Service), Interoperability Service Utility (ISU), Knowledge Representation and Semantic Modelling, Federated, Open and Trusted Platforms (FOT), Software as a Service (SaaS), Automated Service Discovery and Configuration.

In this context we can adopt a methodology as described in Fig 2, based on the five tiers described above, and founded on the Observe - Orient - Decide - Act paradigm principles.

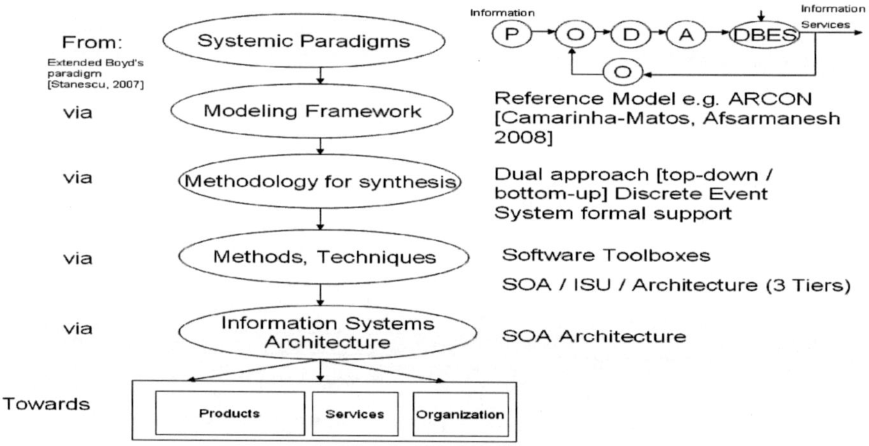

Fig. 2 Top-down holistic methodology

3 From Virtual Enterprise towards Future Enterprise

One could qualify the new type of distributed networked collaborative enterprise as a business integrator, having a specific life cycle of business duration. Each node has to perform a dedicated set of tasks, based on its core competence [4]. This new type of organization - Virtual Enterprise (VE), is faster reactive for facing the business process finite horizon opportunity than well-known traditional organizations (such as large enterprises or small- and medium sized ones (SMEs).

The Virtual Enterprise is the most appropriate concept in relation with its basic meta-system features:

- business integrator, allowing both the vertical and horizontally integration of SMEs [3];
- cyber space friendly user e-business approach for market place;
- multi-agent based internal coordinator, implementing Production Planning and Control;
- (PP&C) within wide-area-network of heterogeneous platforms; extended concurrent engineering oriented designer for product/ process / production system facilities.

The deployment of the virtual enterprise paradigm (VE) requires the definition of the reference architecture and the development of a supporting platform and appropriate protocols for open system architecture.

Clearly, there is not a unique definition of VE, but those usually used permit the underlining of its basic characteristics, as follows:

Definition 1: The Virtual Enterprise is a temporary alliance of enterprises that came together to share skills and resources in order to attend a business opportunity and whose co-operation is supported by computer networks and adequate IT (Information Technology) tools and protocols [3].

Definition 2: The Virtual Enterprise is a temporary alliance of partners focalized on their respective core competence that is capable to assure the co-operation as a support for the network innovative process and for a rapid response to business-market requirements.

Definition 3: The Virtual Enterprise is an open meta-system in a business universe where acts a set of Fabricators utilizing multi-model representations of products / processes and co-operating in the framework of temporary alliances for achieving a set of goals [12]:

VE = (U, F, Mv, G),

where:

- **U** is a universe of discrete business processes.
- **F** is the set of Fabricators [11].
- **Mv** is the set of representations of products / processes / facilities depending on the type of Fabricators utilizing them i.e. the set of multi-model representations [12].
- **G** is a structure of goal-type that is decomposable into a tree of objectives for every Fabricator implied in the VE meta-structure

Note: The CIM-OSA methodology was the first to formalize business as a specific type of processes triggered by market requirements.

The Fabricator is an autonomous agent having a set of core competence (i.e. a set of functions that it is able to accomplish). A Fabricator could co-operate with other Fabricators with similar / different core competences by communicating goals and data. By modifying its own performance evaluation measures according to a global goal and optimality criterion, a Fabricator defines itself as a fast-reactive component of a meta-system.

As underlined by the definitions above, the main features of a VE are:

- capacity of innovation;
- capacity of co-operation;
- capacity of internal reconfiguration;
- fast-reactivity;
- open system architecture compliant with well - known standards (STEP, COBRA, EDI-FACT)

Below we specify the basic model presented in the article [11] where was proposed this provoking concept as further research goal: "eEnterprise beyond the Concurrent Enterprising / Virtual Enterprise state of the art":

E = f(**Mp, Mo, Ma, MR, MK, MA, G**),

where:

- **Mp** = set of processes (manufacturing, design, business, a.o.);
- **Mo** = set of transformation objects encapsulating;
- **Ma** = set of intelligent agents (both natural and artificial);
- **MR** = set of resources (financial, capital, work-in-progress);
- **MK** = set of knowledge;
- **MA** = set of activities (flow-work activities);
- **G** = business goal based on decomposition tree (objectives / activities / tasks).

In the context of an enterprise or a network of enterprises, the agents can model human actors or intelligent machines, each one contributing to the execution of several different functions, sometimes sequentially and sometimes concurrently. Several decomposed functions must be allocated in such a way that the resulting workload does not exceed the capacity of each agent. A coordination layer to supervise the execution of the sub-functions as well as the interactions between people and machines should be designed. The functional decomposition is followed by the task allocation [11] - an optimization problem with a large number of constraints to be solved.

Instead of a coordination layer, we intend to propose a *negotiation protocol* according to which any of the involved agents can participate. The agents involved in the negotiation are associated with the Fabricators. The negotiation mechanism is of the type described in [12]. It can be initiated by any of the existent agents. The agent initiating the negotiation is the one in charge of the job execution and completion [11].

Some of the features of the negotiation mechanism are:

- there is a static game at each step of the negotiation; the stop conditions being: the accepting of an offer or the retiring of a candidate agent;
- an offer is submitted by a candidate agent only if it is acceptable as the best current offer; an offer is refused only if it has arrived too late;
- there are *n* parallel games - for each of the *n* candidate agents;

- through multi criterion optimization one obtains a change of the previously submitted offer by a move on the Pareto frontier determined in the bi-criterion optimization of each of the n static games.

The utility functions for both types of agents involved in the negotiation mechanism are the *profit functions* for each of them. In order to obtain the desired convergence we had to introduce a certain functionality of the candidate agent and two additional constraints:

- After each offer of the candidate agent refused, this agent must restart his evaluation mechanism in order to find a new solution different from the refused one and closer to the expectation of the agent in charge of the job;
- The candidate agent optimizes its criterion under constraints based on the requirements and on the value function of the agent in charge of the job. It is a bi-criterion optimization: the current offer must be better than the last value accepted of the utility function of the agent in charge, in order to make his offer acceptable;
- The second constraint is to have a certain minimal value for the own utility function of the candidate agent.

The necessity of the last constraint appears because in conformity with the conceived game the value, resulting from the evaluation mechanism, is on the Pareto frontier generated by the bi-criterion optimization. And, if the candidate agent wants to obtain a better offer than the last accepted one, it must perform a move of the solution on the Pareto frontier towards a better value of the criterion of the agent in charge of the job, which is, by definition, a better offer than the last one made by this candidate agent. But having in consideration that the two criteria are opposite, a better value for the utility function of the agent in charge of the job means a worse value for the utility function of the candidate agent. However, the benefit of the candidate agent must remain positive, so we need to impose a certain lower limit for his profit. If the candidate agent does not make its desired minimal benefit then it is not worth for him to win the negotiation.

The modelling effort described in this chapter aims at structuring a virtual enterprise in a robust and efficient way. The proposed model involves the eEnterprise objects, such as: machines, storage places, product structure, production orders, algorithms, and the enterprise agents which organize the objects in view of achieving particular goals. The eEnterprise agents may be human actors or software agents. They are in charge of executing the decisions of the strategic management board of the eEnterprise. They perform this role by activating the methods, implementing the workflow and organizing the information flows between the eEnterprise objects. Producer and provider agents are respectively associated with producer and provider.

In the literature, most of the studies on negotiation protocol between the VE agents have considered the non-hierarchical case. Typically, most of the lower-level agents are software agents, responsible for fully automated sub-process. On the contrary, most of the agents of the aggregated-level are actually human actors, responsible for complex and incompletely specified sub-process. The agents on the aggregated-level achieve particular roles of (internal) coordination and (external) negotiation. It appears here the necessity of an automated negotiation mechanism [11] at the lower-level of agents. This mechanism should be activity-oriented and it will have two sides, each agent should have covered in its negotiation mechanism:

- the capability of initialization of the negotiation mechanism, for the case when it is in charge of the considered job

AND

- the capability of response to another agent's initialization of a negotiation mechanism for the case that it can accomplish the required task.

4 Case Study Modules Operator for FInES Supply Chain

Internet of Future oriented Enterprise Systems will allow for the implementation of such new concepts and technologies aiming at improving enterprise efficiency. The research in the area of intelligent objects capable of unrestricted communication and interoperability with other similar objects as well as with any service within the entire enterprise system is focused on supporting that goal.

The "ubiquitous intelligent devices" concept will allow the possibility of information exchange between any intelligent objects. Another capability is the implementation of reactive behaviours according to a predetermined set of actions. Integrating devices and everyday objects in a smart environment is the first step towards the Internet of Things.

A modular design approach has been proposed to dynamically configure and reconfigure the FInES intelligent objects within the SoS environment based on market needs, on the experience and knowledge.

The operators proposed by our approach are:

a) α_{ad}: add intelligent object (I) in SoS FInES environment; it is a unit operator, which adds exactly one module Es,- from the Enterprise $Cloud_E$, to design model SoS:

$$\alpha_{ad}\ (I_m,\ldots,k,\ \mathrm{SoS}) : \mathrm{SoS} \rightarrow \mathrm{SoS}^*, \text{ where } \mathrm{SoS}^* = \mathrm{SoS}\ \mathrm{U}\ \{\mathrm{I}\}$$

b) α_{rem}: removes intelligent object (I) in SoS FInES environment; it is a unit operator which removes exactly one module I_i.., from the current design model SoS:

$$\alpha_{rem}\ (I_m,\ldots,k,\ \mathrm{SoS}) : \mathrm{SoS} \rightarrow \mathrm{SoS}\ ^*, \text{ where } \mathrm{SoS}\ ^* = \mathrm{SoS} \setminus \{\mathrm{I_m},.\}$$

It could be possible to begin the negotiation with an $I_{i,j} \in \{I_{;:.}\}$ agent, aiming at the integration in a SoS alliance based on a quality and quantity analyze of information existing in the distributed knowledge basis with Intelligent Objects modules. A new possible partner will be taken into account based on the presentation questionnaire or base on the recommendation of other business partners.

c) α_{rec}: reconfiguration of SoS FInES environment; it is a module operator that is applied to two design models, SoS_i, SoSj, which exchange module Intelligent object subsets at an arbitrary crossover point *j:*

$$\alpha_{rec}(j, SoS_1, SoS_2) : SoS_1 \rightarrow SoS^*_1,$$
$$SoS^*_1 = SoS_1 - \bigcup_{i=j}^{n} \alpha_{ren} (I_{i,.}, SoS_1) + \bigcup_{i=j}^{n} \alpha_{ad} (I_{i,.}, SoS_2, SoS_1).$$

The operator α_{rec}, applied to two SoS validated configurations, SoS_1, SoS_2, support the exchange of system set. This operator decreases the necessary time for a SoS configuration using the logistic chains previously validated.

d) α_{inl}: replace the intelligent object (I) in SoS FInES environment; it is a module operator that is applied on a design model SoS, to replace exactly one module Intelligent object at a certain point with another module from the set of validated partners $\{I_{i,.}\}$:

$$\alpha_{inl} (i, SoS) : SoS \rightarrow SoS^*,$$
$$SoS^* = SoS - \alpha_{ren} (I_{j,.}, SoS) + \alpha_{ad} (I_{j,.}, [I_{i,.}]_f, SoS).$$

e) α_{conf}: validate an object in the feasible intelligent object (I) set

$$\alpha_{conf} (I_{i,j}) : \{I_{i,.}\} = \{I_{i,.}\} \cup I_{i,j},\ [I_{i,j}]_f = [I_{i,j}]_f \cup I_{i,j}.$$

f) α_{inf}: invalidate an object from a feasible intelligent object (I) set

$$\alpha_{conf} (I_{i,j}) : \{I_{i,.}\} = \{I_{i,.}\} \setminus I_{i,j},\ [I_{i,j}]_f = [I_{i,j}]_f \setminus I_{i,j}.$$

Technical, extensible standards and protocols suited for "Internet of Things" are required in order to integrate all types of devices, including robots. Several standards have been proposed in order to identify and track objects based on RFID tags.

A framework for modelling the tag data, assigned to a real object has been provided in Fig 3.

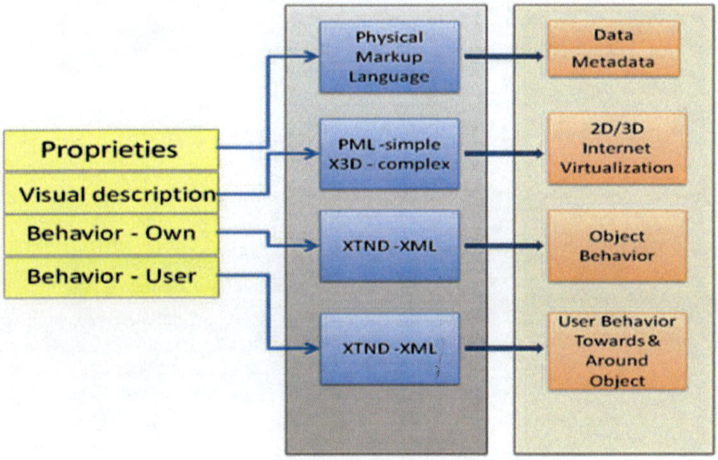

Fig. 3 Framework for local object identification and interaction

The framework adds to the data describing the object some data allowing for the virtualization of the object information regarding the object's own behaviour, or its functionality info about other object's behaviours related to this object.

As the object's properties and description have already been implemented in RFID tags, the focus of this research is to extend the object's description by defining and implementing object behaviours. All of these data and information can be stored in the tag's memory. A problem would be to find a wide used tool to describe this data and information. As discussed, a suitable and largely available way for data transfer is the Extended Markup Language. As PML (based on XML) has been widely adopted by the users of the EPC standard, we chose PML for the object proprieties description within the framework. PML also allows for the description of object shape and colour but we consider that a more appropriate way to describe the object's visual properties would be to use standard 3D web oriented language: X3D. X3D language is considered to be the next step after VRML (Virtual Reality Markup Language) aiming at communicating 3D on the web, between applications and across distributed networks and web services. In order to describe the behaviour of the object and the behaviour towards the object, XTND - XML Transition Network Definition Language has been chosen. We can model behaviours using transition networks. This type of networks is used to describe a set of states and the transitions that are possible between them.

In this context the development of methods used to transform common objects involved in the enterprise business processes into intelligent objects is regarded of high benefit. The first step for such an approach is to be able to identify the object and to distinguish it from the environment (Fig. 4).

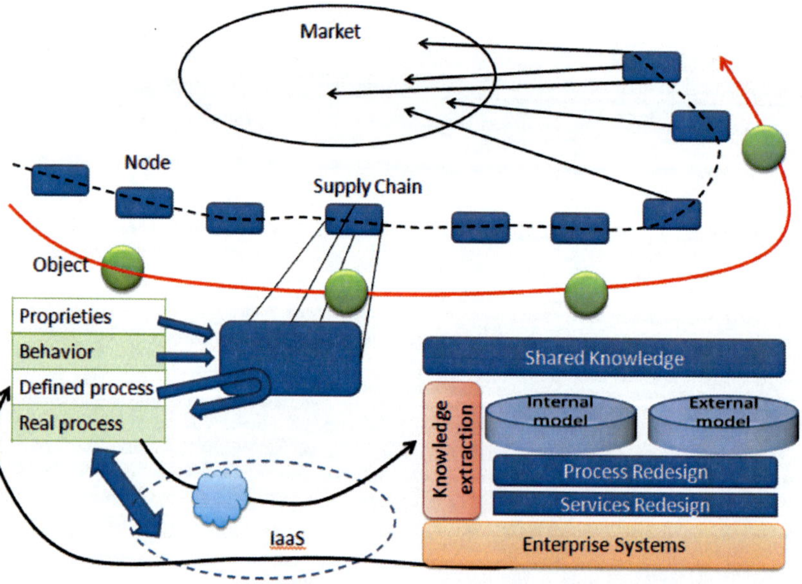

Fig. 4 Intelligent objects represented as FInER within the FInES Knowledge Management System

The recent progress of the RFID technology has boosted the capability of object identification and allowed for the development of new identification technologies as EPC code and *ucode* as discussed in the previous section. These technologies require that the information suitable for the identification of an object should be accessed from a remote database. With the development of new and more powerful data storage devices one can focus on a different approach: storing the data on the tag attached to the object and accessing that data when ether necessary.

The second step is to integrate such objects with the enterprise systems; the information attached to the tag should be not exclusively divided in the following categories: object proprieties, object functionality or in a broader meaning object behaviour, description of the planned processes (at task level) which involve the object, and the description of the actual processes as "experienced" by the object.

The defined process attached to the object is associated with the process models as described in the Process Modelling level of the Enterprise Architecture. This approach will allow for the creation of a map of the processes that the object will be subjected to and for a better integration with the services included in the Enterprise Systems. In such a case any enterprise system or production cell will be able to read the process map and proceed with the object accordingly. An example can be represented by a product that is going to be shipped. The shipping service will be able to read the process map, identify the shipping sub-process and act according to the predefined tasks.

As the predefined processes may differ from the actual processes completed by the object, a record to register the changes can be defined. An example in this area can be identified at the level of an adaptive supply chain. Such a record can provide valuable feedback related to the insight of the sub-processes and tasks. This information can be further processed in order to produce valuable knowledge within the Enterprise Management System.

5 Conclusions

The concepts, proposed within the Future Internet Enterprise Systems paradigm, are becoming a reality due to the research efforts leading towards the development of new services.

Enterprise Systems management in particular, owing to its evolution that parallels that of manufacturing paradigms, is expected to produce new methods allowing humans to both benefit from - and increase the value of - technological advances.

A hybrid Enterprise Systems can be foreseen, where the interaction between human and non-human knowledge stakeholders will became transparent and will allow creation and use of *metaknowledge* [12, 13].

The following conclusions should be drawn:

- The new Universe of Discourse (CE/CCE/C-Innovation) creates the great chance to "reload" the General Systems Theory & Applications.

- The digital Word Theory and the Internet of Things-based projects represent our choice to develop the scientific foundation of Internet of Future oriented Enterprise Systems.
- A new paradigm beyond the well known P-OODA, MAPE, SHOR, to sustain the Business Process Modeling, Monitor & Management so called OBPEIL is proposed.

References

1. ATHENA: Advanced Technologies for Interoperability of Heterogeneous Enterprise Networks and their Applications, FP6-2002-IST-1, Integrated Project – Annex I (2004)
2. Camarinha-Matos, L., Afsarmanesh, H., Novais, P., Analide, C.: Establishing the Foundation of Collaborative Networks. In: PROVE Virtual Enterprise 2007. Springer, Heidelberg (2007) ISBN 978-0-387-73797-3
3. Camarinha-Matos, L.M., Afsarmanesh, H., Ollus, M.: ECOLEAD and CNO Base Concepts. In: Camarinha-Matos, L.M., Afsarmanesh, H., Ollus, M. (eds.) Methods and Tools for Collaborative Networked Organizations. Springer, Heidelberg (2008)
4. FInES Internet of Things 2020, A Roadmap to the Future, European Technology Platform on Smart Systems Integration (2009), `http://ec.europa.eu/information/_society/policy/rfid/documents/iotprague2009.pdf`, Internet of Things: an early reality of the Future Internet (accessed August 10, 2009)
5. Kaposi, A., Myers, M.: Systems for all. Imperial College Press, Singapore (2001)
6. Kuhn, T.S.: The Structure of Scientific Revolutions. Chicago University Press, Chicago (1970) ISBN: 9780226458083
7. Lee, J.: Model-driven business transformation and the semantic Web. Communications of ACM 48(12), 75–80 (2005)
8. Missikoff, M.: Let's Semanticise the World!!... or not?? ICEIS 1, 17–24 (2009)
9. Molina, A., Chen, D., Panetto, H., Vernadat, F., Whitman, L.: Enterprise Integration and Networking: Issues, Trends and Vision. In: Bernus, P., Fox, M., Goossenaerts, J.B.M. (eds.) Knowledge Sharing in the Integrated Enterprise, pp. 303–313. Springer/Kluwer (2005) ISBN: 0-387- 26608-9
10. Panetto, H., Scannapieco, M., Zelm, M.: INTEROP NoE: Interoperability Research for Networked Enterprises Applications and Software. In: OTM Workshop, pp. 866–882 (2004)
11. Santucci, G.: From Internet of Data to Internet of Things (2009), `http://ec.europa.eu/information_society/policy/rfid/` (accessed January 4, 2010)
12. Stanescu, A.M., Ionescu, L.M., Georgescu, V., Badea, L., Moisescu, M.A., Sacala, I.S.: Toward Digital Business EcoSystem Analysis. In: Gunasekaran, A., Sandhu, M. (eds.) Handbook on Business Information Systems. World Scientific Books (2010)
13. Stanescu, A.M., Dumitrache, I., Pouly, M., Caramihai, S.I., Moisescu, M.A.: Towards a general systems theory approach to design the future of concurrent engineering science. In: Loureiro, G., Curran, R. (eds.) Complex Systems Concurrent Engineering Collaboration Technology Innovation and Sustainability, pp. 3–10. Springer, Heidelberg (2007)

Dynamic Bayesian Network for Decision Aided Disassembly Planning

Luminita Duta[1] and Sidali Ad Douche[2]

[1] University Valahia of Targoviste, 2 Carol Avenue, Romania
duta@valahia.ro
[2] IUT Montreuil, Université de Paris 8, 140 Rue de Nouvelle France, France
addouche@iut.univ-paris8.fr

Abstract. Disassembly processes of used manufactured products are subject to uncertainties. The optimal disassembly level that minimizes the costs of these processes and maximizes the end of life components values is hard to establish. In this work, we propose a method to find influences and causalities between the main disassembly performance indicators in order to decide the optimal disassembly policy. The proposed model highlights the temporal dependencies between variables of the system and is validated using the Bayesia Lab software. In the final part of the chapter, the results of method implementation on a reference case study are presented in order to demonstrate the performance of our approach.

Keywords: decision aid, disassembly planning, dynamic Bayesian network.

1 Introduction

In recent years, the interest for profitability of the disassembly process and in managing the end of life products has considerably grown. Remanufacturing, reusing and recycling are options for re-integrating the parts of the used product in manufacturing chains of new ones. Disassembly is the central stage of the three-R processes referred as "*product reconstruction*" by Pearce [1]. Through this process components and materials are obtained in view of their valorisation. Complex components are frequently considered as items to be disassembled.

Disassembly can be either a non-destructive process (addressing to parts rather than materials) or a destructive one (addressing on materials rather than items). After non-destructive disassembly, the items are sold, reused, recycled, stored for future use or disposed of. Similarly, after destructive disassembly (dismantling), materials are either recycled or disposed of.

The objective of disassembly planning is to identify the sequence of disassembly operations that will maximize the expected returns from the processed items. This is a decision making process. To maximize the end of life components revenue, an optimal disassembly policy must be applied. In this context, identifying the optimum sequence for disassembly operations is one of the most important objectives of disassembly planning. The optimal disassembly policy is

T. Borangiu, A. Thomas, and D. Trentesaux (Eds.): SOHOMA 2011, SCI 402, pp. 143–154.
springerlink.com

the one that minimizes the process costs and maximizes the final profit. However, a problem occurs: in the disassembly process of the used products, one could not know the state of the product component before the disassembly proceeds. Moreover, if the end-of-life revenues are deterministic and known, the operative costs as well as the operative disassembly times are stochastic and unknown at the beginning of the process. These values vary between certain values but have a probabilistic distribution.

In this chapter, an approach that integrates uncertainty in the model of components estate and stochastic values for disassembly operative costs and times is proposed. This approach takes into consideration the influences and the causalities between the main disassembly performance indicators in order to minimize the costs of the process and to maximize the end of life components values.

2 State of the Art

For disassembly planning and disassembly policy optimization a large variety of methods has been used. The literature is very rich from this point of view and in the book of Lambert and Gupta [2] a complete state of art is made in respect of the disassembly sequencing and planning optimization. In this chapter we will review only the works that dealt with the probabilistic disassembly planning or disassembly policies under uncertainties.

One of the first articles on the probabilistic disassembly planning is [3]. The authors used the probabilistic inference mechanism of a Bayesian network. They obtained a static and a dynamic mathematical model based on the product topology, deterioration processes, and alternative disassembly methods each of which is represented by means of a random variable in a Bayesian Network. However, the mathematical model proves to be very complicated for a complex product with more than 10 components. Dependences between Bayesian probabilities are hard to calculate. Grochowski [4] used hybrid Bayesian networks to estimate the optimal disassembly policy. He describes an expert system, consisting of a Disassembly Petri network and a Hybrid Bayesian network, for optimal disassembly planning. The expert system identifies a component defect and proposes a disassembly policy. It doesn't take into account the operative times or costs. Bayindir and al. [5] investigated the desired level of the recovery under various inventory control policies when the success of recovery is probabilistic. All used and returned items go into a recovery process that is modelled as a single stage operation. The recovery effort is represented by the expected time spent for it. The effect of increasing recovery effort on the success probability together with the unit cost of the operation is included by assuming general forms of dependencies.

Graphical models are used to represent the process of decision making in a process the variables of which are probabilistic. For example, adding specific nodes to a BN such as nodes of utility and nodes of decision, one can take into account the associated costs and uncertainties. This modelling technique, described by Jensen in [6], uses the concept of Influence Diagram (ID). These are graphs that take into account several scenarios (decisions) and make possible to evaluate the impact of decisions on the process costs.

The work of Clemen and al. [7] presents such influence diagrams in parallel with the corresponding decision trees. These two graphical methods resemble from the calculation point of view, but the representation using ID is more compact and more easily applicable in decision aid and large solution space problems.

Alternatively to recovered items, demand is satisfied by new items. In the field of dependability, recent studies based on modelling of industrial processes using the theory of BN (Bayesian Network) have been developed by Godichaud et al. [8]. His work demonstrates the relevance of the BN as a tool for modelling industrial processes functioning in the presence of uncertainties. He proposed an original use of DBN for decision aid in the dismantling processes subject to uncertainties. This uncertainty relates mainly to the requirements in parts resulting from dismantling process and to the arrivals of end of life products. He also proposed end of life destinations and certain dismantling levels through a forecast of the recycled spare parts supply. However, he didn't take into account any cost. Alami [9] treated the problem of uncertainty on the availability of recycled spare parts using an economical approach. His model of differential equation allows the evaluation of costs and revenues of the reconditioned spare parts obtained after the disassembly process. Unfortunately, the formalism of Alami has no long term effect. It is only a tool for static resolution that must be done periodically in time.

The aim of our work is to incorporate the dynamic approach into the disassembly planning process by using the Bayesian formalism so as to obtain minimum process costs in real time and to update the task assignment at each moment taking into account the type of operation. In the same time, we are looking to combine the dynamic inventory with Bayesian Network learning so as to obtain a dynamic model close to the real system. To accomplish this aim one has to use the Dynamic Bayesian Networks, a well suited modelling instrument for complex systems submitted to uncertainties.

3 Dynamic Bayesian Networks

A Bayesian Network (BN) is an appropriate graphical method for modelling of causal processes and probability-based knowledge representation under uncertainty. A BN is a directed acyclic graph whose nodes represent random variables and links define probabilistic dependences between variables. These relationships are quantified by associating a conditional probability table with each node, given any possible configuration of values for its parents.

Bayesian networks have the ability of capturing both qualitative knowledge (through their network structure), and quantitative knowledge (through their parameters). While expert knowledge from practitioners is mostly qualitative, it can be used directly for building the structure of a Bayesian Network. In addition, data mining algorithms can encode both qualitative and quantitative knowledge and encode both forms simultaneously in a Bayesian Network [10].

The static Bayesian Network can be extended to a Dynamic Bayesian Network (DBN) model by introducing relevant temporal dependencies that capture the dynamic behaviours of the system at different times. Two types of dependencies can be distinguished in a Dynamic Bayesian Network: contemporaneous dependencies and non-contemporaneous dependencies. Contemporaneous dependencies refer to arcs between nodes that represent variables within the same time period. Non-contemporaneous dependencies refer to arcs between nodes that represent variables at different times [11].

The advantage of DBN over Markov chains is that a DBN is a stochastic transition model factored over a number of random variables, over which a set of conditional dependency assumptions is defined.

Time invariance ensures that the dependency model of the variables is the same at any point in time. While a DBN can in general represent semi-Markovian stochastic processes of order k-1, providing the modelling for k time slices, the term DBN is usually adopted when only two time slices are considered in order to model the system temporal evolution. That's why such models are also called *Two time Bayesian Networks* (2-TBN) or 2-time-slice temporal Bayesian Networks [12]. Each time-slice contains a set of (time-indexed) random variables, some of which are typically not observable. When a first order Markov process assumption holds the future slice at time $t+1$, it is conditionally independent of the past ones given the present slice at time t. In this case, it is sufficient to represent two consecutive time slices called the anterior and the ulterior layer to represent the network. However, to specify the entire network and to correctly model the system, the next parameters have to be provided:

- the prior probabilities for root variables at time $t = 0$;
- the intra-slice conditional dependency model, together with the corresponding conditional probabilities;
- the inter-slice conditional dependency model and the transition model, which explicit the temporal probabilistic dependencies between variables;
- nodes from the anterior layer must contain only variables having influence on the same variable or on another variable at the ulterior level;
- the inter-slice edges connecting a variable in the anterior layer to the same variable in the ulterior layer are temporal arcs.

4 Proposed Model

4.1 Notations and Assumptions

The following notations are used in the text as well as in the graphical model:

Time	The time variable represented by 20 unities
State	State of the component at arrival
Rate	Percentage of reusable/reutilization
Utility	Result of decision in terms of profit
OpTime(*t*)	Operative time at the first slice of time *t*
OpTime(*t*+1)	Operative time at the second slice of time *t*
Cost(*t*)	Operative cost at the first slice of time *t*
Cost(*t*+1)	Operative cost at the second slice of time *t*
Revenue(*t*)	End of life revenue at the first slice of time *t*
Revenue(*t*+1)	End of life revenue at the second slice of time *t*

Constructing the model leads us to make the following assumptions:

- the model is conceived for *one type of product*;
- two type of operations are possible: *proper* disassembly and *destructive* disassembly (disassembly and dismantling);
- two types of end of life options are considered: *reuse* and *recycling*;
- the *initial values* (first slice of time *t*) for components end of life revenues, operative times and costs *are deterministic*, varying into known intervals;
- the *final values* (second slice of time *t*+1) for disassembly operative times and costs are *stochastic* / probabilistic since they depend on the state of the component which is not known in advance (at slice of time *t*);
- *probabilistic relations were provided for each node*, expressing the probabilities of that node taking on each of its values, conditioned on the values of its parent nodes;
- only *operative costs* are taken into consideration since the value of other costs is negligible;
- all costs and operative times follow a *Normal distribution*;
- the *state of the component* determines the *rate of utilization* and so the final revenue;
- two possible decisions are to be taken into consideration taking into account the state of the component : disassembly or dismantling;
- the decision is taken so as to *maximize the final profit*: the difference between end of life revenues and costs of the process.

A DBN model is designed to highlight the temporal dependencies and causalities between variables of the system (Fig. 1). The dynamic behaviour of the disassembly process is well described by modelling its stochastic variables: operative times, operative costs, type of operation, and components' end of life states.

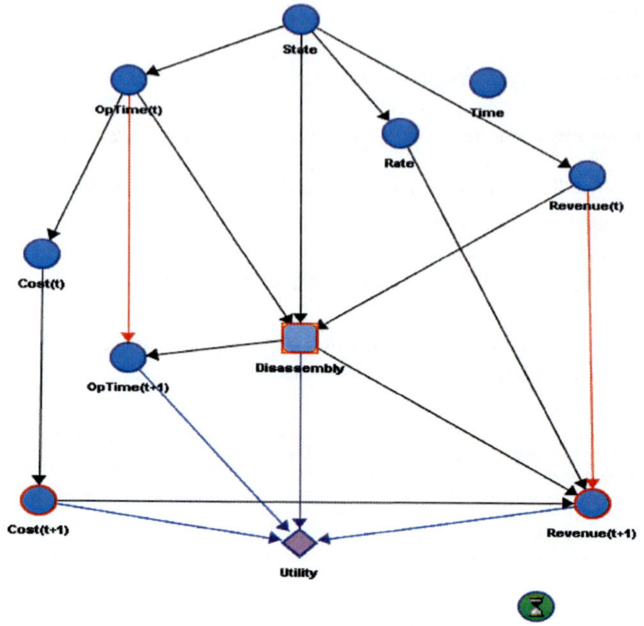

Fig. 1 Dynamic Bayesian Network of the model

This work aims at finding the best disassembly decision at slice time *t*+1 and the best combination between recyclable and reusable parts so as to minimize the costs and to maximize the final profit. Between two slices of time there are temporal arcs (Fig. 1) and the evolution of related temporal nodes is followed during simulation. The utility of the decision is calculated with a deterministic formula depending on the process cost, end of life revenue and rate of utilization at the slice time *t+1*. The proposed model is validated using the Bayesian Lab software.

5 Validation and Results

5.1 Case Study

We present the results of method implementation on a reference case study from literature, to successfully demonstrate the performance of our approach.

The case study refers to the disassembly of a notebook [13]. The case of three years old product is considered. Operative times are given in Table 1; the disassembly operative costs are given in Table 2, and the end-of-life values and the reusable/recycling percentage are specified in Table 3. Some data is taken from [14] whereas other data is calculated.

Table 1 Operative disassembly/dismantling times

Components	Disassembly Time [s]	Dismantling Time [s]
[A] LCD Monitor Type I	15	8
[B] LCD Monitor Type II	15	8
[C] Motherboard Type I	20	10
[D] Motherboard Type II	20	10
[E] Processor	5	3
[F] Memory	4	2
[G] Hard drive 20Gb	8	4
[H] Hard drive 30Gb	8	4
[I] CD Drive	6	3
[J] Combo Drive	6	3
[K] Network Card	4	2
[L] Modem	5	3
[M] Keyboard	8	4
[N] Battery	5	3
[O] Power Adaptor	4	2

Table 2 Operative costs

Components	Disassembly Cost [$]	Dismantling Cost [$]
[A] LCD Monitor Type I	0.25	0.15
[B] LCD Monitor Type II	0.25	0.15
[C] Motherboard Type I	0.15	0.10
[D] Motherboard Type II	0.15	0.10
[E] Processor	0.10	0.05
[F] Memory	0.10	0.05
[G] Hard drive 20Gb	0.15	0.10
[H] Hard drive 30Gb	0.15	0.10
[I] CD Drive	0.12	0.10
[J] Combo Drive	0.12	0.10
[K] Network Card	0.10	0.05
[L] Modem	0.10	0.05
[M] Keyboard	0.12	0.06
[N] Battery	0.10	0.05
[O] Power Adaptor	0.10	0.05

Components are named from A to O. The convention is that operative tasks are noted in the same way: **A** means the disassembly of the LCD monitor.

Table 3. End of life values and utilization rates

Components	Reusable value [$]	Recyclable value [$]
[A] LCD Monitor Type I	60[65%]	50[75%]
[B] LCD Monitor Type II	50[61%]	50[75%]
[C] Motherboard Type I	35[69%]	75[85%]
[D] Motherboard Type II	28[64%]	65[85%]
[E] Processor	30[61%]	180[90%]
[F] Memory	30[65%]	110[80%]
[G] Hard drive 20Gb	25[65%]	55[75%]
[H] Hard drive 30Gb	35[58%]	55[75%]
[I] CD Drive	16[70%]	25[70%]
[J] Combo Drive	32[69%]	50[70%]
[K] Network Card	16[69%]	50[80%]
[L] Modem	6[64%]	50[80%]
[M] Keyboard	6[73%]	15[65%]
[N] Battery	25[61%]	30[75%]
[O] Power Adaptor	15[65%]	25[70%]

With these values as initial data and taken into account the assumptions made in constructing the model, we performed the simulation.

5.2 *BayesiaLab*

BayesiaLab [15] is a tool for graphical manipulation of Bayesian networks. It allows defining, modifying, using and learning models based on Bayesian networks. It is also a decision aid instrument since it allows introducing decision and utility nodes. Therefore, the best decision can be found at the end of the inference process induced by simulation and the also the utility of this decision in terms of profit. The qualitative data in Bayesialab is represented by the structure of the BN graph (Nodes and Arcs), and the quantitative information by the Conditional Probability Table and Databases.

It is possible to learn a Bayesian Network considering an initial structure. To take into account this a priori knowledge, a virtual database with N samples (where N corresponds to the number of cases that have been used to set this a priori knowledge) must be provided. The distribution of these samples corresponds to the joint probability distribution represented by the initial Bayesian Network. This virtual database and the real database are then both taken into account by the learning algorithms to induce a new Bayesian Network. In addition to the unquestionable observations, BayesiaLab is also able to seize degrees of probability. Once validated, probabilities are used jointly with the probability distribution for giving a new Probability distribution. BayesiaLab allows the temporal dimension integration in a Bayesian Network.

A BN can be easily transformed into a DBN. Temporal nodes at instants *t* and *t*+1 can be represented and connected by temporal arcs. The parameters evolution of the DBN nodes can be so followed in time.

Concerning the way of reasoning of a DBN, different kinds of inference algorithms are available. Among those algorithms, particularly popular is the junction tree (JT) inference. It is based on the construction of a classical BN inference data structure called junction or join tree and belonging to the category of exact inference algorithms. Another inference algorithm is the Boyen–Koller (BK) algorithm, a parameterized procedure that, depending on the parameters provided, may return exact as well as approximate results [16].

5.3 Implementation

Launching the simulation one could observe the learning process of the DBN.

Using a learning algorithm the missing data is dynamically completed and the joint probabilities are updated at each iteration execution (Fig 2). One can observe that end of life options, costs and revenues influence the type of disassembly operation so as the decision is taken dynamically considering the state of the component. The main variables and their values are dynamically represented using histograms (Fig 3).

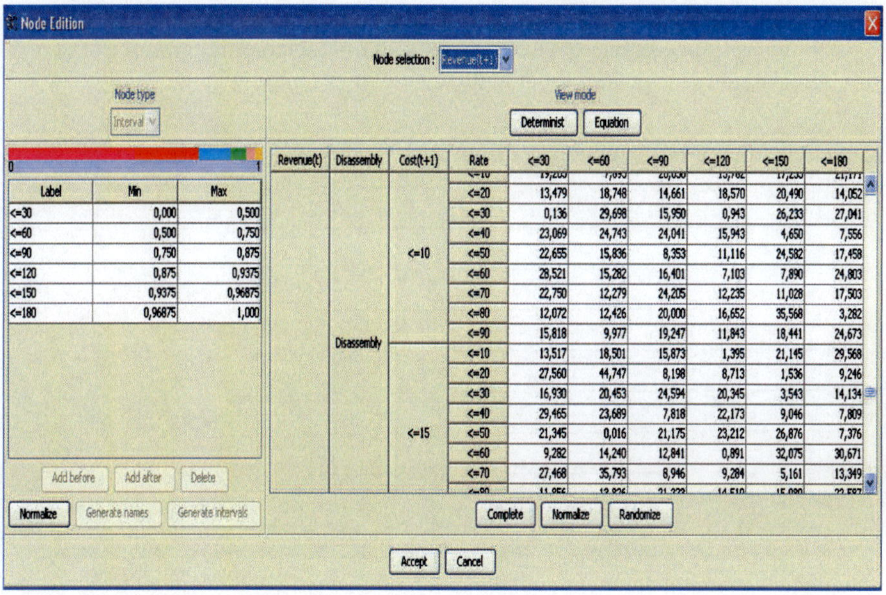

Fig. 2 Completion of variables values (revenue)

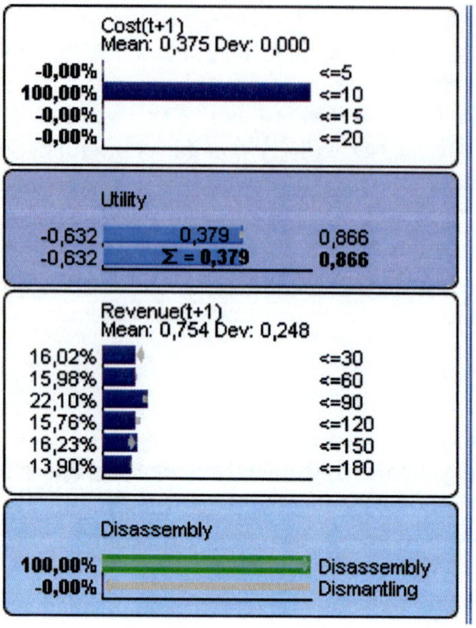

Fig. 3 Dynamical representation of variables and decision making during the simulation

Evolution of main indicators can be followed on the temporal graph. Here costs and revenues variability and decision utility are represented (Fig. 4).

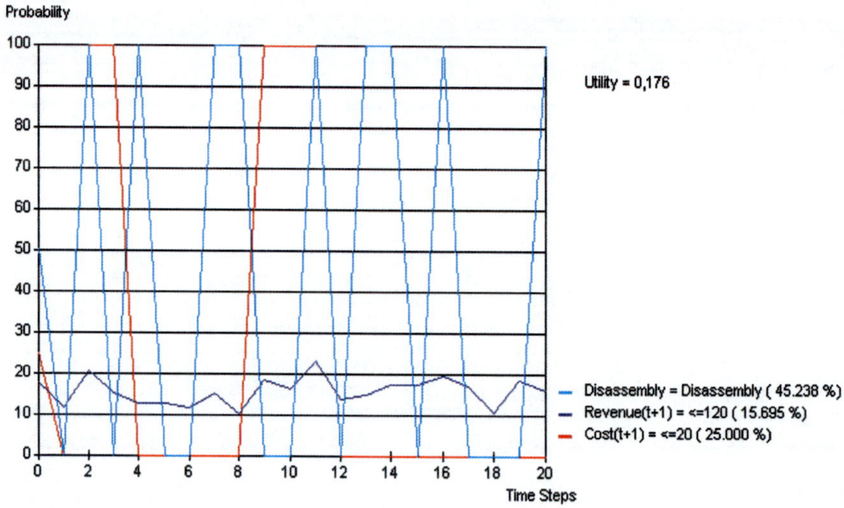

Fig. 4 Temporal graph for variables dynamic

Disassembly operations are fulfilled in 45% of cases, the rest of 55% being destructive operations. More than 75% of operative costs have high values while most of the revenues are situated under 120 unities. This result shows that the disassembly operation is expensive. Sometimes dismantling operations are preferred instead. In this case we deal with notebooks the components of which are themselves expensive, so one prefers to disassemble everything which is possible to obtain the maximal profit. The minimal value of the decision utility is 17% which is also the value of the minimal profit.

Decision values are given in the Fig 5. One can notice that components with high end of life values (such as monitors, network cards or keyboards) have to be disassembled. In the same time, most of the components suffer dismantling operations (such as motherboards, processors and batteries)

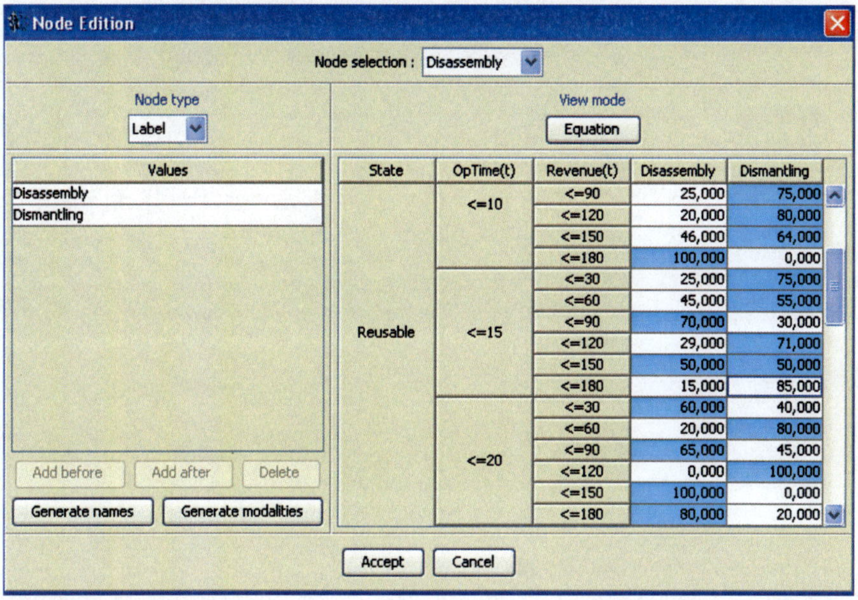

Fig. 5 Decision values

6 Conclusions

This chapter proposes a method to establish the influences and the causalities between the main disassembly performance indicators in order to decide the optimal disassembly policy. A Dynamic Bayesian Network model is developed to highlight the temporal dependencies and causalities between variables of the system. The dynamic behavior of the disassembly process is well described by its stochastic variables: operative times, operative costs, type of operation, components' end of life states. The proposed model is then validated using the BayesiaLab software. The method is implemented and validated on a reference case study from literature. The proposed model is useful in real time decision making process planning when one has to decide upon the type and the depth of the disassembly operations. However, the model does not take into account other end of life options (such as discarding, melting, grinding, etc) as well as different variants of the product.

References

1. Pearce, I.A.: The profit making allure of product reconstruction. MIT Sloan Management Review 50(3), 59–63 (2009)
2. Lambert, A.J.D., Gupta, S.M.: Disassembly Modeling for Assembly, Maintenance, Reuse, and Recycling. CRC Press, Boca Raton (2005)
3. Geiger, D., Zussmann, E.: Probabilistic Reactive Disasssembly Planning. CIRP Annals 45(1), 49–52 (1996)
4. Grochowski, D.E.: Parameter Estimation for Optimal Disassembly Planning, pp. 2490–2496. IEEE (2007)
5. Bayındıra, Z.P., Dekkerb, R., Porrasb, E.: Determination of recovery effort for a probabilistic recovery system under various inventory control policies. The International Journal of Management Science 34, 571–584 (2006)
6. Jensen, F.V.: Bayesien Networks and Decision Graphs. Springer, Heidelberg (2001)
7. Clemen, R., Reilly, T.: Making hard decisions with Decision Tools. Duxbury Thomson Learning (2001)
8. Godichaud, M.: Outils d'aide à la décision pour la sélection des filières de revalorisation des produits issus de la déconstruction des systèmes en fin de vie, thèse de doctorat, Université de Toulouse (2010)
9. Alami, M.: Lot économique de pièces de rechange à produire en tenant compte des différentes phases du cycle de vie du produit, Thèse de doctorat, Université Laval, Québec (2009)
10. Conrady, S.: Introduction to Bayesian Networks Conrady Applied Science, LLC - Bayesia's North American Partner for Sales and Consulting (2011)
11. Hu, J., Zhang, L., Ma, L., Liang, W.: An integrated safety prognosis model for complex system based on dynamic Bayesian network and ant colony algorithm. Expert Systems with Applications 38, 1431–1446 (2011)
12. Portinale, L., Raiteri, D.C., Montani, S.: Supporting reliability engineers in exploiting the power of Dynamic Bayesian Networks. International Journal of Approximate Reasoning 51, 79–195 (2010)
13. Duta, L., Caciula, I., Addouche, S.: On the profitability of the disassembly processes. In: Proceedings of the 18th IFAC World Congress, Milan (2011)
14. Imtanavanich, P., Gupta, S.M.: Generating a Disassembly-to-Order Plan. In: Proceedings of the 2007 Northeast Decision Science, on CD-ROM (2007)
15. `http://www.bayesia.com/en/products/bayesialab.php`
16. Murphy, K.P.: Dynamic Bayesian Networks: Representation, Inference and Learning, Thesis University of California (2002)

Service Oriented Architecture for Holonic Isoarchic and Multicriteria Control

Yves Dubromelle, Fouzia Ounnar, and Patrick Pujo

Aix Marseille University - Laboratoire des Sciences de l'Information et des Systèmes
Avenue Escadrille Normandie Niémen, 13397 Marseille Cedex 20, France
{yves.dubromelle,fouzia.ounnar,patrick.pujo}@lsis.org

Abstract. Faced with the limitations of methods fully or partially centralized for decision-making in workshop control, the PROSIS (Product, Resource, Order, Simulation, Isoarchic System) model is proposed; this model is based on the holonic paradigm and operates according to an isoarchic architecture that allows these holonic entities to use ambient services. All holonic entities contribute to the decision-making mechanism. Each holonic entity can be considered as the combination of an M_holon composed with the physical part of Holon, with an I_holon, adding to it intelligence and decision-making abilities. Each I_holon is synchronized with its M_holon via infotronic technology of RFID type. Furthermore, it supports the status information of the Holon and access to services allowing it to take decisions. The I_holon is hosted by an Ambient Control Entity (ACE) network, located, according to SOA, close to each resource. The ACEs provide to the Holons the *ad hoc* services they need.

Keywords: Holonic control, Isoarchic architecture, Ambient services, Multicriteria decision, RFID.

1 Introduction

The fully or partially centralized methods for decision-making in workshop control have limitations responding to the growth of productivity demand. Faced to that, an architecture without any hierarchical dimension in the mechanisms of decision-making is proposed.

The proposed approach, PROSIS (Product, Resource, Order, Simulation, Isoarchic System) [1, 2, 3], is based on the holonic paradigm [4]: all holonic entities are autonomous, cooperative and can be recursive.

The Holonic Manufacturing Systems (HMS) community [5, 6] recognizes that these holonic entities must belong to different heterogeneous types. The main research works [7, 8, and 9] converge towards the use of three types of basic Holons, namely Product, Resource and Order Holons, for which PROSA [7] is the archetype.

PROSIS also agrees with this consensus. However, in PROSIS we have deliberately chosen to focus on a decisional architecture approach outlawing any form

T. Borangiu, A. Thomas, and D. Trentesaux (Eds.): SOHOMA 2011, SCI 402, pp. 155–168.
springerlink.com

of hierarchy between Holons. Thus, only one architecture in 'flat holonic form' [10] can be considered and this requires giving unique features to these three types of Holons. PROSIS operates according to an *isoarchic architecture* allowing these holonic entities to use ambient services in the sense that the services are always available when the entities have to use them.

Isoarchy is a governance mode where only the holonic entities involved in a given decision contribute equally to the development of this decision, and they do so with a maximum of expertise and knowledge about their condition. The term "isoarchy" comes from the combination of the Greek prefix *iso* (equal) and the Greek suffix *archy* (power). In isoarchic architectures, decision making is as much related to self-organization as to the emergence, two interesting concepts [11].

In manufacturing, these concepts appear as ways to increase the production systems performance, either in terms of revenue or in terms of quality of service. Indeed, the successive evolutions of the industrial production context, such as greater variability in the markets or need for customer satisfaction, impose to the production systems to search a better productivity, particularly through the technological evolution of processes, the automation of their implementation, the flexibility of their operation and more recently their reactivity. The lifetime of products is becoming shorter as their variance grows. Such new production characteristics force one to rethink mass production strategies. In fact, the production is nowadays centred on manufacturing of small quantities in parts family with many variations, by reducing Work In Progress (WIP). For this, the production system must be more flexible, i.e. it must be able to easily handle a variety of products and batches. Such disturbances, external to the production system, are grouped under the terms of product flexibility and capacity flexibility [12, 13]. The objective of the PROSIS approach is to gain in terms of responsiveness and adaptability.

In fact, all holonic entities composing the overall system contribute in PROSIS to the decision-making mechanism. The decision-making architecture and information system are as close as possible to the organization of the material system.

For this, each Holon is composed of a physical part named M_holon, to which decision-making abilities named I_holon are associated, adding intelligence to the Holon (Fig.1). A M_holon can be constituted of an isoarchy of Holons (a Holon is at the same time a whole and part of a whole - recursion property).

In order to match the information and decision system with the physical system, each I_holon is synchronized with its M_holon *via* infotronic technology of RFID type. According to [14], this allows to completely control the distribution of the information system on all Holonic entities: each I_holon supports its state data and others information. Each I_holon can also access ambient services allowing it to take decisions and to interact with others Holons, having equivalent decision levels. Intelligence can thus be provided to the all Holons types. This proposal extends the work in [15]. According to the service oriented architecture (SOA) paradigm, each I_holon is hosted by an Ambient Control Entity (ACE), located close to each production resource [16]. The ACEs thus provide to the Holons all the various *ad hoc* services they need, such as, hosting services of the I_holons, synchronization services with the material system ({M_holons}), communication services between I_holons, and multicriteria decision aid services ...

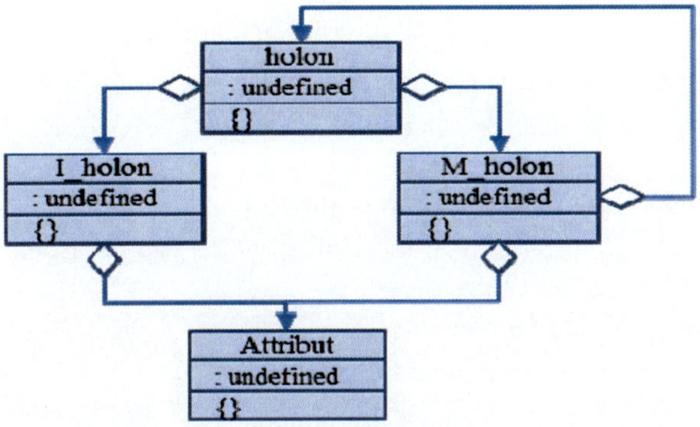

Fig. 1 UML model of Holon's composition

After a more detailed presentation of the PROSIS control model and its SOA, a description of the principal ambient services provided to holonic entities is given.

2 Main Characteristics of PROSIS Approach

2.1 Definition of Isoarchic System

A control system is characterized as isoarchic [2] when the I_holons that compose this isoarchy are able to organize the operation of their corresponding M_holons, in autonomous, coordinated and egalitarian manner, without the direction of any superior hierarchical decision centre. The ensemble of I_holons assumes jointly the overall operation of the system. This induces a number of properties:

- *Same decision-making capacity*: there is equality of power, even among Holons of different types; so there is no predominance of one type of Holons on another, but only different relative influences in thee decision taking, depending on the internal state of each Holon, taken into account in real time.
- *Decision-making mechanisms duplicated*: the processes and mechanisms to support decision-making are achievable by Holons through ambient services, according to the goal sought by the Holon calling service.
- *Interaction and coordination system*: the activities of each Holon being derived from the cooperation that it maintains with other Holons, each Holon must be able to communicate with any other Holon; in practice, this need of communication is related to decision problems, which limits the number of Holons belonging to identified isoarchies.
- *Self-organization of operations*: there is no attempt to forecasting optimization (which would be anyhow invalidated by the unexpected nature of the workflow and the appearance of random disturbances), but rather real-time search of local solutions (taking into account some overall constraints of performance).

We can give a formal representation of a Holon, at 'N' decision-making level:

$$\mathbf{H_N = <I\_H_N, (M\_H_N \vee \{H_{N-1}\}), \{(I_N)\}, T_H>}$$

where the quadruple is composed of:

- $\mathbf{I\_H_N}$: the informational and decisional part of the Holon;
- $(\mathbf{M\_H_N} \vee \{\mathbf{H_{N-1}}\})$: the material part of the Holon, which can be as appropriate:
  - □ $\mathbf{M\_H_N}$: an elementary material part;
  - □ $\{\mathbf{H_{N-1}}\}$: a material part composed of several Holons;
- $\{(\mathbf{I_N})_i\}$: the set of isoarchies $(I_N)_i$ related to this Holon;
- $\mathbf{T_H}$: the Holon type.

The holonic paradigm, which provides high opportunity in terms of decisional architectures use (theory of open hierarchical systems [4]), of defining interacting entities (possibility of considering different types of Holons) and of defining interactions' relations (bilateral, group, with or without bidding) ..., is explored through PROSIS in an 'extreme situation' with regard to the distributed architectures of the control study [17]. This choice in PROSIS facilitates the implementation of architecture oriented services.

2.2 Deployment of SOA in Isoarchic System

Currently, the integration of SOA approach in a company is usually achieved by using Web Services [18]. The Web Services approach doesn't characterize only the SOA concept [19]; SOA, as architecture, emphasizes different processes such as: 'craft' knowledge, support, management (different types of processes)) and separation of processes to allow reusability. SOA is often used to allow better use of available enterprise resources by identifying key processes of the information system. This vision of SOA can also be used within a software application, such as the implementation of PROSIS. Thus, the following services were identified: CRUD (Create Read Update Delete) services, support services (communication, RFID, Decision Aid, Time Management …) and 'craft' knowledge services (Holon Hosting, Call for Proposal ...).

To implement the PROSIS model, SOA is thus particularly well suited. Indeed, SOA allows having the overall vision of decision-making processes, which allows building the intelligence associated to the Holon, and highlights the interactions between services. This architecture facilitates the addition and composition of existing services to create new features. It can also characterize the performance of a control system, by establishing specialized and independent services from the chosen control mode.

However, in the context of PROSIS implementation as scientific demonstrator out of Web Services context, we have chosen to not use all the usual SOA environment tools, but rather equivalent best technical solutions. For example, in terms of communication, SOA uses XML as messages' support and SOAP (Simple Object Access Protocol) for exchanging messages and remote procedure control. We use Java internal serialization and the JGroup (Java Group) framework to communicate and avoid the overhead of XML.

The formal definition of a service λ is done by the following sextuple:

$$\mathbf{Serv_{\lambda} = < I\_H_i, T_{Hi}, (I_N), method_{\lambda}, param_{\lambda i}\ data_{Hi} >},$$

where:

- **I_H_i**: the I_holon which requests the service and waits for the response,
- **T_{Hi}**: the type of this I_holon (which can parameterize the service ...),
- **(I_N)**: the concerned isoarchy, defining the operation conditions of JGroup,
- **$method_{\lambda}$**: the invoked Java method to implement the service $Serv_{\lambda}$,
- **$param_{\lambda I}$**: the parameters transformed in arguments to the service $Serv_{\lambda}$, giving its context,
- **$data_{Hi}$**: the data provided by the ith I_holon.

Examples of services will be detailed later in this chapter. These services are accessible to different Holons *via* a network of ACEs (Ambient Control Entity) considered as support to the Holons.

2.3 ACE as Support System for Isoarchic SOA

A I_holon reduced only to its data structure cannot apply its autonomy: it needs a processing information system allowing it to interact with other Holons and build decisions. This is achieved through the provision to the Holons of a set of *ad hoc* ambient services, *via* network of ACEs. To cover the whole space occupied by the production system, an ACE is localized beside each resource (Fig.2).

The control of production system' holonic entities can be described as ambient due to the omnipresence of these computer units communicating with the Holons and providing them decision-making capacity which allows self-organization. Thus, each ACE gives its support for self-organization of an isoarchic set of Holons, where control decisions are made locally. PROSIS naturally reflects the physical organization of entities: beside each resource, a M_resource, *p* M_products and *k* M_orders can be founded physically. These latter orders concern the *p* corresponding products ($k \leq p$). If each of the *k* OH concerns q_i products, then $p=\sum_{i=1}^{k} q_i$. In the control, the self-organized decision-making requires the participation of all involved entities: resource, orders and products. For this, direct interactions between the I_holons of $(1+p+k)$ concerned Holons (*one* I_resource, *p* I_products and *k* I_orders) constitute, according to a 'flat holonic form', a local isoarchy ($I_{N\lambda}$), where the direct links between Holons (their interactions), are modelled by a complete graph.

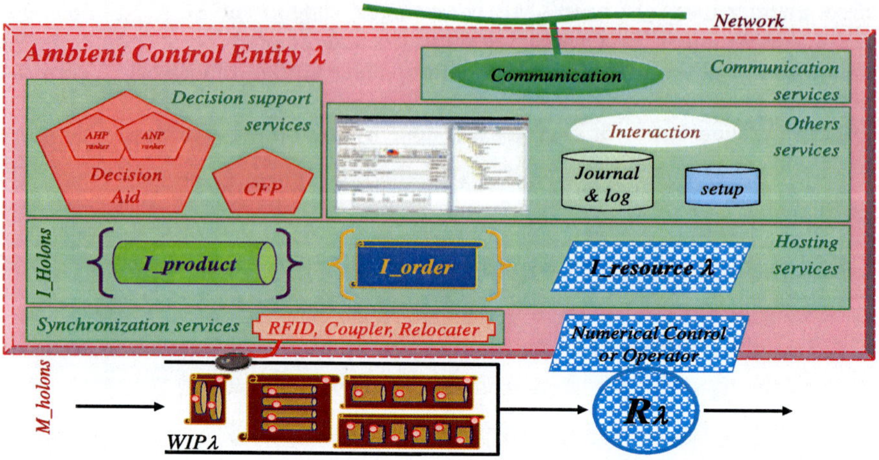

Fig. 2 Ambient Control Entity: Services provider

ACE is then a services provider for these I_holons. Each service works in the same manner for all Holons able to call it, whatever their type is. Fig. 3 shows the collaboration diagram between a H_i Holon, an ACE and a service $Serv_\lambda$ of 'Decision Aid' type, when invoking this service by the H_i Holon.

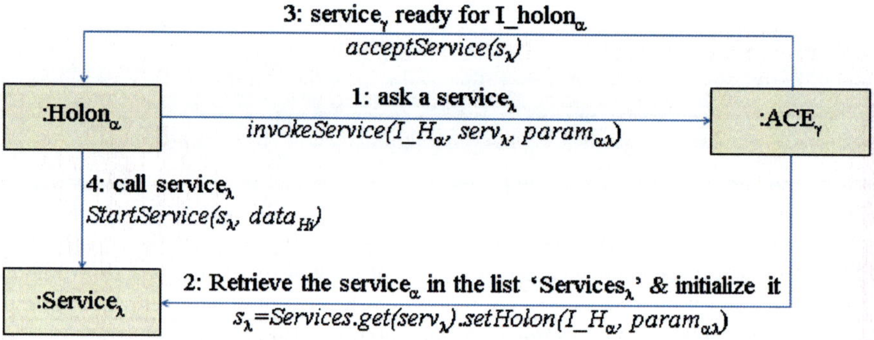

Fig. 3 Collaboration diagram illustrating the invocation of a service by a Holon

In fact, ACE prepares the service by configuring this latter one using '$param_{\alpha\lambda}$', and then the I_holon I_H_i can use the Java method '$method_\lambda$' having as arguments the data '$data_{Hi}$'. In most cases, $Serv_\lambda$ returns a result is based on data and behaviour related to the Holons' local isoarchy ($I_{N\lambda}$). In some cases, $Serv_\lambda$ requires a wider isoarchy, including Holons that can be located on different ACEs. This is particularly the case when $Serv_\lambda$ concerns a resources isoarchy.

Table 1 shows the ambient services classes proposed by each ACE to Holons.

Table 1 Typology of ambient services proposed by the ACEs to the Holons

Abstraction level

Service type	Service Name	Concerned Holon I_product	Concerned Holon I_resource	Concerned Holon I_order
Decision support services	Decision Aid	X	X	X
	Decision Aid Improver	X	X	X
	Call For Proposal	X	X	X
Evaluation services	Traceability	X	X	X
	Benchmark	X	X	X
Holon Hosting services	Holon Arrival	X		X
	Holon State Management	X	X	X
	Holon Departure	X		X
Synchronization services	Holon Relocator	X		X
	Integrity checker	X		X
	State Machine Sync.	X	X	X
Communication services	Communicator	X	X	X
	User Interface	X	X	X
CRUD services	Configuration Manager		X	
	Inventory Control	X		X

2.4 *Presentation of SOA in PROSIS*

Fig. 4 shows the general service oriented architecture of the holonic and isoarchic control system. On the right hand of the diagram, all the services that are being used in PROSIS are shown, along with the main utility classes that are used by the services. The left hand of the diagram contains the entities implied in the service

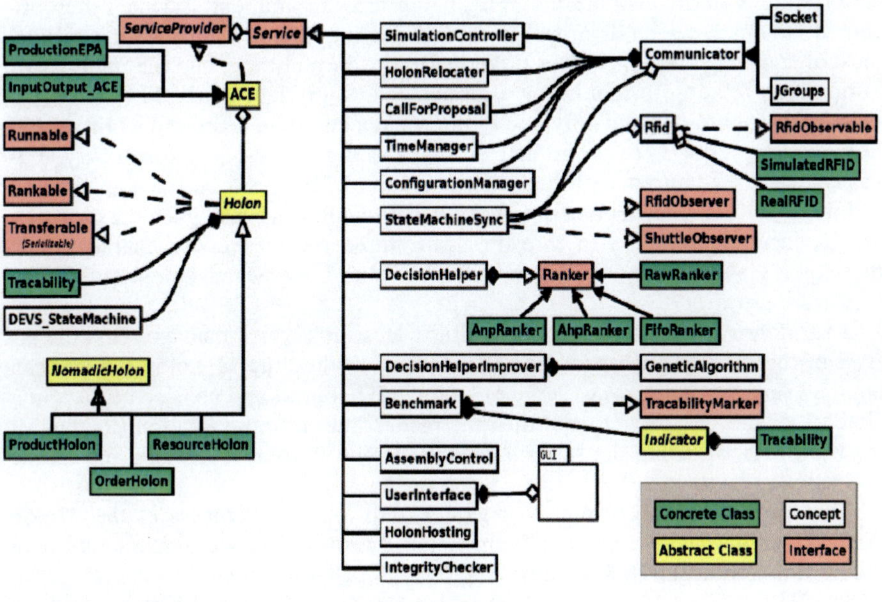

Fig. 4 SOA of the holonic and isoarchic control system

architecture, with some of their properties linked to the services. Two main entities compose the SOA in PROSIS: (a) I_holons (that are potentially nomadic entities) and (b) ACEs. Implementing all the functions (decision ...) as services outside the I_holons allow these latter ones to move more easily from a resource to another. The main service provided to the I_holons is their hosting by the ACE, in order to provide them with the necessary environment to their execution.

So, the I_holons share some properties, like their transferability or their execution through a state machine. Another advantage of this architecture is the modularity of the services. Thus, all the holonic entities are smart objects, with similar decision-making capabilities.

Finally, a large choice of services can be proposed to all Holons, to adapt to the different real manufacturing systems constraints in which PROSIS control is used.

3 Services Provided by ACE

These section details some services provided by the ACEs to the Holons.

3.1 Hosting Service

Each ACE supports the I_holons of the local isoarchy ($I_{N\lambda}$). In other words, each ACE hosts the I_holons and allows them to self-organize their operation. This is mainly to update the data during internal state changes and to activate decisional treatments by the invocation of *ad hoc* services. This host service concerns the I_resource, which is permanently hosted by the ACE. It also concerns the dynamic set of Holons that constitute the WIP, being thus an ambient service - the concerned Holons are seen as movable communicating objects. This hosting service must also manage input - output of these Holons in real time, and the integrity of information relating to them. For example, each I_product managed by an ACE must be related to one and only one I_order, but a common I_order may be related to several I_products (case of a batch). This hosting service is automatically activated upon the occurrence of the synchronization RFID event between an I_holon and its M_holon: a new Holon enters the WIP, and allows starting the creation of a self-management process (a thread). Thus, three services allow characterizing the Holon Hosting in ACE: Holon Arrival, Holon State Management and Holon Departure [16].

The 'Holon Arrival' service of the Holon Hosting service manages first the arrival of the I_holon (decapsulation of information) and the M_holon (RFID scan *via* the 'State Machine Sync' service), with an ID control and consistency (calling 'Integrity Checker' service) and then the reconstitution of the I_holon (loading of the right data, updating the states..., calling 'Holon Relocator' service and by the activation of its thread...).

The 'Holon Departure' service operates with a similar process as the 'Holon Arrival' service, but symmetrically opposite. It also manages the interaction with the M_holon transportation system.

The 'Holon State Management' service consists in giving full autonomy to the Holon. Self-management of I_holons is possible due to the activation of a thread managing the Holon state. For that, each Holon is associated to a state diagram described using the DEVS (Discrete EVent system Specification) formalism [20].

Fig. 5 shows the state diagram of an I_product. During the transportation of the M_product, the I_holon management thread is interrupted, which corresponds to a waiting state. When inserting the I_product in the WIP the thread is started, its internal states are initialized and the waiting start date of the product in the WIP is memorized. Then, the product is expected to be chosen by the Resource Holon; when this happens, start and end manufacturing dates are stored in order to archive traceability information and supply performance indicators. After treatment, when the M_product leaves the workstation, the traceability data are stored again.

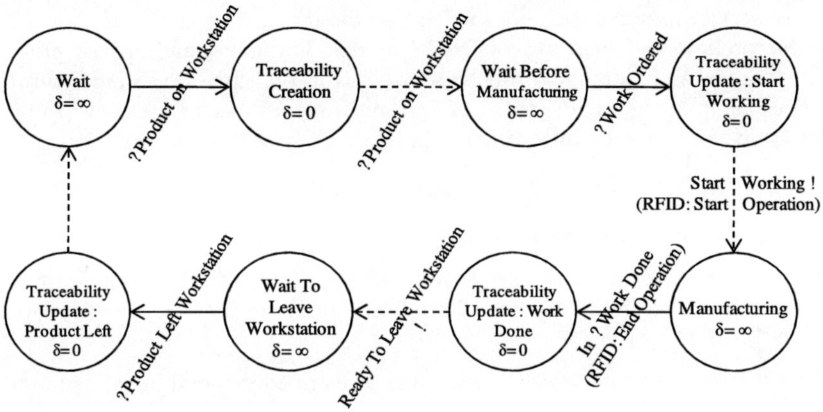

Fig. 5 DEVS model of I_product self-management

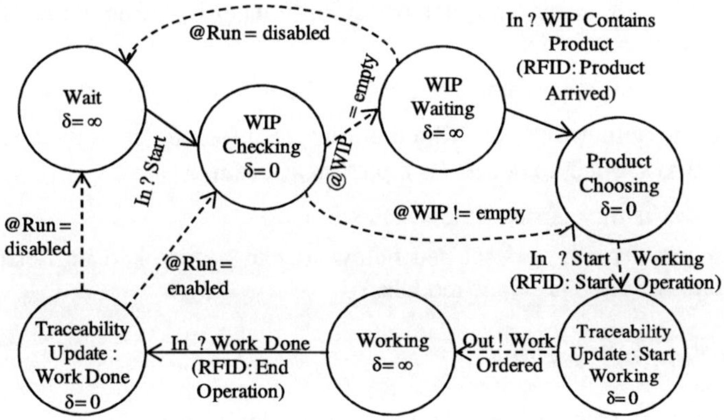

Fig. 6 DEVS model of I_resource self-management

Simultaneously with the *p* management thread of the *p* I_products composing the resource WIP, the corresponding I_resource is also managed by a thread. Fig. 6 shows the I_resource's state diagram. Once started, an I_resource seeks to begin manufacturing. So, when the WIP is not empty, its state changes. The new state aims at choosing the next product to be treated among all products located in the WIP. Otherwise, the I_resource waits the filling of the WIP. At the end of the treatment task, the I_resource checks the traceability information, and then seeks a new product to treat; this is done while it is in normal operation condition.

These state graphs provide the behaviour of the associated I_holons. Also note that some events change simultaneously the I_holons' states of different types. This is particularly true for events related to the external environment, and using specialized services, such as reading - writing on the ID tags.

The interpretation of such formal DEVS models is unique, and can be processed automatically, in simulation mode as well as in real-time operating control mode. For example, the DEVS model of an I_resource's self-management can be completely described by the following septuplet:

$$\mathbf{I\_R_{\lambda}\_SM} = <\mathbf{X, S, Y, \delta_{int}, \delta_{ext}, \lambda, D}>, \text{ where:}$$

- **X** is the set of external events types,
- **S** is the set of sequential states {Wait, WIP Checking, WIP Waiting, Product Choosing, Traceability Update: Start Working, Working, Traceability Update: Work Done},
- **Y** is the set of output events types, some of them correspond to calls of ambient services $\mathbf{Serv_{\lambda}}$,
- $\boldsymbol{\delta_{int}}$: $\mathbf{S} \rightarrow \mathbf{S}$ is the internal transition function defining the state changes due to internal events,
- $\boldsymbol{\delta_{ext}}$: $\mathbf{ST} \times \mathbf{X} \rightarrow \mathbf{S}$ is the external transition function defining the state changes due to external events,
- **ST** is the set of the system's states: ST = $\{(s, e)\ s \in S, o \leq e \leq D(s)\}$,
- $\boldsymbol{\lambda}$: $\mathbf{S} \rightarrow \mathbf{Y}$ is the output function, which allows, in the case where the output event is a call to service, to define $\mathbf{(I_N)}$, $\mathbf{param_{\alpha\lambda}}$, $\mathbf{data_{Hi}}$,
- $\mathbf{D{:}S} \rightarrow \mathbf{R^+} \cup \infty$, is the elapsed time's states function.

The Decision Support service is described below. It can be invoked indiscriminately by some states of the 2 previous models.

3.2 Decision Support Service

The Decision Support service enables a Holon to use a decision aid method which can be linked to a simple logic (for example, FIFO, priority rules or conventional heuristics for single criterion / single resource scheduling) or which takes into account interests sometimes conflicting. In the latter case, multicriteria decision aid methods can provide to I_holons substantial aid in their decision-making process faced to a complex decision problem. These methods propose to classify a set of potential solutions, called alternatives. For example, all the products in the queue (WIP) of a resource correspond to the set of alternatives. The ranking of all alternatives is done by examining the logic, the consistency of choices, then by aggregating preferences. The decision corresponds to the choice of the alternative having the highest ranking or could exploit the obtained ranking to derive information allowing decision taking. The Decision Support service must implement a method that is versatile, deterministic and simple to operate.

In the context of multicriteria problems treated in isoarchic control system, the solution is the result of interaction of all considered criteria and appears as being the best possible compromise. Therefore, no criterion related to different Holons involved in the decision-making process should be side-lined in the decision process. For this, the study focuses on complete aggregation methods. Complete aggregation is a strongly compensatory method.

Among existing methods, the AHP (Analytic Hierarchy Process) [21], [13] or ANP (Analytic Network Process) method can be chosen [21], [22]. In opposition to other multicriteria decision aid methods, AHP and ANP methods are the only ones allowing measuring the consistency of comparisons done by decision maker. They allow considering qualitative/quantitative criteria, decomposing complex problems and ranking them on several levels. Results are summarized by reducing complex decisions to comparisons. The algorithm doing this ranking is composed mainly of matrix calculations, which ensures efficiently achieving the solution.

The first example related to the use of this service concerns the local real time scheduling of a resource WIP. As previously discusses, the self-organization problem is based on how product flows are built in real time. In fact, the problem is to define which product will be processed by a given resource. The data needed for this implementation come from interactions between the concerned Holons.

Indeed, an ACE proposes a Decision Support service aiming to organise product flows between resources: control is done in real time *via* local scheduling. Through interactions between the I_Holons, an emerging mechanism organises the resource production according to criteria that can be parameterised. In fact, this service is invoked by the management thread of I_resource $_{\lambda}$, 'I_R$_{\lambda}$_SM', when this service is in 'Product Choosing' state. It generates the collaboration of all Holons of the local isoarchy ($I_{N\lambda}$). These various Holons do not all seek the same objective; it is thus necessary to find the best possible compromise. Faced with this problem multicriteria decision aid methods (especially the AHP method) were used in the 'AHP Ranker' implementation of the Decision Support service (Fig.4).

AHP is structured as a comparison tree (Fig.7). 'AHP Ranker' service can provide a solution whatever the problem to be addressed is: the AHP algorithm is implemented in a generic way, and is configured at the time of service invocation, depending on the expected result type. The tree root is the global objective to be achieved (Level 1). The second level is composed of the criteria that provide a complete view of the parameters of the problem.

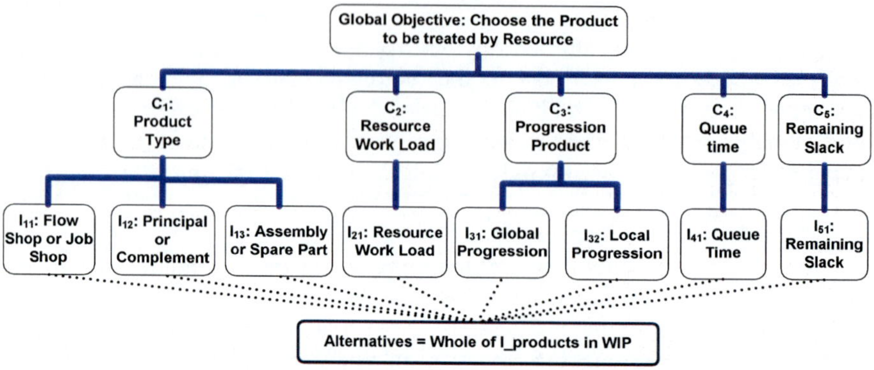

Fig. 7 Example of AHP structure

Then, each criterion has a number of indicators (Level 3) that are relevant to it. The indicators are calculated with some data from the local isoarchy ($I_{N\lambda}$) (Level 4). The alternatives' list (Level 4) corresponds to the set of the I_products contained in the WIP of a M_resource$_\lambda$. The list of alternatives is dynamic: products to be treated are added, and already processed products are deleted. It is therefore essential, for the robustness of the operation, to have a perfect match between the actual state of the WIP and the information system. This implies:

- the *necessary synchronization* between the I_holons and the M_holons: this is achieved by the use of RFID, with synchronization services ('Integrity Checker' and 'Holon Relocater' services - Fig.4) ;
- a *good management of the list*: there are the incoming and outgoing events of the RFID coupler allowing adding/removing items (I_products) from the list. This is generated by the Holon Hosting service in the previous section.

To achieve this objective, different criteria were considered representing three points of view (Product, Resource, Order points of view) on different Holon types. Fig.7 shows five criteria:

1. Criterion C1 [*Product type*], is related to Product Holon and allows prioritizing some types of products according to the classes they belong to.
2. Criterion C2 [*Product progression*], monitors the progress of the Product Holon through the production system.
3. Criterion C3 [*Remaining slack*], allows Order Holons deliver products in time.
4. Criterion C4 [*Queue time*], limits the presence of the PH in the WIP queue.
5. Criterion C5 [*Resource workload*], grants proper usage of the Resource Holon.

AHP algorithms implement in two phases: *configuration and exploitation* [3].

The *configuration* phase decides first on the relative importance of the criteria and within each criterion, on the relative importance of their associated indicators. This is done by filling pair wise comparison matrices for the root element and each criterion. Matrices are then converted into priority vectors giving the relative importance of each compared item. In this phase, the I_resource invokes the requested service and relays to it the parameter (**param**$_{\alpha\lambda}$), which indicates the tree comparison, the pair wise comparison matrices, the library used for the calculation of indicators and the access to the Holons' raw data of the local isoarchy ($I_{N\lambda}$).

The *exploitation* phase allows obtaining the vector of the relative importance of the alternatives compared to the global objective (level 4 related to level 1). In this phase, the real data (alternatives regarding the associated indicators) are taken into account: the data are directly read in real time in the I_holons of the ($I_{N\lambda}$) isoarchy.

This decision mechanism can be dynamically reconfigured according to obtained performances or to new productivity objectives. It is enough to change the service's configuration by using other values of (**param**$_{\alpha\lambda}$).

4 Conclusion

After presenting the control system based on holonic and isoarchic approach (the PROSIS model), the general SOA of PROSIS was described. The control is achieved through the support of Ambient Control Entities. The set of ACEs constitutes the implementation structure of PROSIS (ambient services to I_holons).

The chapter focuses on two main services provided by the ACEs to the Holons:

1. *Hosting Services*: supply the I_holons with the necessary environment to their execution. This service manages the arrival and departure of Holons (I_holons and their associated M_holons) as well as the management of the states of these Holons. The self-management of I_holons is possible by the activation of a thread, which manages the DEVS state graph representing the different states.
2. *Decision Support Services*: grant decision-making aid to I_holons. Through interactions between I_Holons, a mechanism organises production based on some criteria; this mechanism can be dynamically reconfigured according to obtained performances or to new productivity objectives.

More services can be added to PROSIS. One advantages of this service oriented architecture is the possibility of having a large choice of services without changing the I_holons. Thus, the integration of diagnosis service [23] is possible. Future research will be directed towards the system's global performance.

References

1. Pujo, P., Broissin, N., Ounnar, F.: PROSIS: An isoarchic structure for HMS control. EAAI 22(7), 1034–1045 (2009)
2. Pujo, P.: De l'Isoarchie pour le pilotage des systèmes de production. In: HDR, Aix-Marseille University, Marseilles (2009)
3. Ounnar, F., Pujo, P.: Pull control for Job Shop: Holonic Manufacturing System approach using multicriteria decision-making. JIM (2009), doi:10.1007/s10845-009-0288-4
4. Koestler, A.: The ghost in the machine. Hutchinson, London (1967)
5. Deen, S.M.: Agent-Based Manufacturing - Advances in the Holonic Approach. Springer, Heidelberg (2003) ISBN: 3-540-44069-0

6. Mathews, J.: Organizational foundations of intelligent manufacturing systems - the holonic viewpoint. CIMS 8(4), 237–243 (1995)
7. Van Brussel, H., Wyns, J., Valckenaers, P., Bongaerts, L., Peeters, P.: Reference architecture for holonic manufacturing systems: PROSA. CII 37, 255–274 (1998)
8. Leitão, P., Restivo, F.: ADACOR: A holonic approach to dynamic manufacturing scheduling. RCIM 24, 625–634 (2008)
9. Balasubramanian, S., Brennan, R., Norrie, D.: An architecture for metamorphic control of holonic manufacturing systems. CII 46, 13–31 (2001)
10. Bongaerts, L., Monostori, L., McFarlane, D., Kadar, B.: Hierarchy in distributed shop floor control. CII 43, 123–137 (2000)
11. Dalenoort, G.J.: The paradigm of self-organization: studies of autonomous systems. In: Dalenoort, G.J. (ed.) The Paradigm of Self-Organization. Gordon & Breach, London (1989)
12. Jubin, M.: Ateliers flexibles d'usinage. Cahiers de l'Ingénieur R7, 820 (1994)
13. Ounnar, F.: Prise en compte des aspects décision dans la modélisation par réseaux de Petri des systèmes flexibles de production. PhD thesis, INPG, Grenoble (December 9, 1999)
14. McFarlane, D.: RFID: Reducing the uncertainty. In: GS1 UK EPCglobal Conference and Exhibition, London (2005)
15. Wong, A., McFarlane, D.: RFID and product intelligence. In: Workshop on Ambient Intelligence Technologies to Enhance the Product Lifecycle, Brussels (2006)
16. Dubromelle, Y., Ounnar, F., Prunaret, J.P., Pujo, P.: Service Oriented Architecture in PROSIS holonic control. In: Proceedings of IESM 2011, Metz (2011)
17. Cardin, O., Castagna, P.: Using online simulation in Holonic Manufacturing Systems. EAAI (2009), doi:10.1016/j.engappai.2009.01.006
18. Fournier-Morel, X., Grojean, P., Ploin, G.: SOA: guide de l'architecte du SI, Dunod (2008)
19. SOA Manifesto (2009), `http://www.soa-manifesto.org/`
20. Zeigler, B.: Theory of Modelling and Simulation. John Wiley, New York (1976)
21. Saaty, T.L.: The Analytic Hierarchy Process. Mc Hill (1980)
22. Dubromelle, Y., Louati, T., Ounnar, F., Pujo, P.: AHP/ANP a Decision Making Service in PROSIS Model. In: IMS 2010, Lisbon (2010)
23. Legat, C., Thomas, M., Hubauer, C.-S.: Integrated Diagnosis for Adaptive Service-oriented Manufacturing Control with Autonomous Products. In: IESM 2011, Metz (2011)

Viable System Model Approach for Holonic Product Driven Manufacturing Systems

Carlos Herrera, Sana Belmokhtar Berraf, and André Thomas

Centre de Recherche en Automatique de Nancy (CRAN), Nancy Université,
CNRS UMR 7039 Faculté des sciences BP 239 54506 Vandœuvre-les-Nancy, France
{carlos.herrera, sana.belmokhtar,
andre.thomas}@cran.uhp-nancy.fr

Abstract. The chapter presents a generic framework for a Product Driven Control Systems (PDCS) dealing with production planning and control. The framework is based on Viable System Model (VSM) which is introduced for intelligent manufacturing systems. Based on suitable properties as autonomy, self-organization and adaptability, VSM allows modelling and considering these properties for PDCS. An application for a Manufacturing Planning and Control System (MPCS) is proposed. Different points of view are also presented based on this application which are shown and explained in the sense of VSM principles. Finally, a discussion is presented dealing with the main issues of the proposed approach.

Keywords: product-driven systems, intelligent manufacturing systems, viable system model, manufacturing planning and control.

1 Introduction

The economic context including globalization and free competition triggers quick evolutions of several technologies available for the industry and users. That pushes firms to be more reactive, more innovative and more agile.

In this context, Intelligent Manufacturing Systems (IMS) - arisen in the early 90's - aim at improving the conventional Manufacturing Planning and Control Systems (MPCS) in terms of flexibility, reconfiguration and scalability among others. In this class of systems, Product-Driven Control Systems (PDCS) [11], based on intelligent products [12], change the vision of products from a passive state to active actors in the decision process [14]. Nevertheless, one of the major limitations of these types of systems is the absence of specific frameworks to guide the process of analysis, design and implementation.

For that purpose, we propose in this chapter a general framework to model product-driven control systems applied to MPCS. The proposed framework is designed to be used in a wide range of product-driven fabrication or service systems. This leads to generic concepts which provide sufficient theoretical methods and implementation guidelines to manage the inherent complexity of such systems.

Our work has been motivated by the design of intelligent products as living beings, focusing our research in theories and models concerning the structure and

T. Borangiu, A. Thomas, and D. Trentesaux (Eds.): SOHOMA 2011, SCI 402, pp. 169–181.
springerlink.com

organization of these living machines [10]. The objective is to use such models to design manufacturing systems that can show properties like autonomy, self-organization and adaptability. The presence of these properties leads to the conceptual framework present in the Viable System Model [3, 4]. The choice of VSM has been made since this framework is suitable to model these properties.

The main property in VSM is recursion, which allows managing complexity and finding a compromise between control and autonomy to implement principles as heterarchy and cooperation [6]. These central elements of product-driven control systems are, in a more general way key objectives of IMS. Managing complexity allows considering a system as a whole preserving the emergent properties that can only be understood using a holistic vision, in contrast to reductionist ones.

Since its origin in the early 70's - a result of the research of Stafford Beer, VSM has been object of constant validation with several applications until nowadays. The pertinence of this model can be justified related to recent applications concerning: material flow management [8, 16], autonomous distributed systems [15], autonomous software [7, 9] or autonomous agent communities [20].

Our contribution deals with the use of VSM for PDCS as an integrated modelling framework. The most important paradigms arisen from IMS are bionic [21], fractal [23] and holonic manufacturing systems [22]. Properties like synthesis have been essential parts of bionic systems. Recursion is the basis property of fractal system, and the difficulty to separate parts from the whole is a main concept in holonic systems. Our intention is to show that VSM can be used like a consensual framework for all these research threads.

This chapter is organized as follows: section 2 presents a general description of the main concepts involved in VSM, and the approach focusing on the elementary object of the proposed structure that is the intelligent product. Section 3 describes an application of the proposed approach for a Manufacturing Planning and Control System. Section 4 discusses the main findings, and section 5 gives the conclusion.

2 VSM for Holonic Product-Driven Manufacturing Systems

The origins of VSM arise from the work of Beer applied to the steel industry in the fifties. This research can be placed in the line of works of Norbert Wiener, Warren McCulloch and Ross Ashby [5]. The main objective of the model was to identify and to explain how systems are viable, in other words, how it is possible that some systems can maintain an independent existence.

Although VSM is a general model for the study of any *viable system*, the most concerned application area has been that of human activity organizations, i.e., corporations, firms or governments [2]. In this domain, VSM changes the view of the traditional management model based on command and control, in which a control system is designed as a pyramid and such decisions are disaggregated in a top-down manner at different structural levels. The main difference, inspired by the biological organization, consists in mapping this hierarchy into a structural recursion.

The premise of this change of perspective was inspired from the living beings composition (cells, organs, systems, etc.). Indeed, they have properties of autonomy, self-organization and self-regulation, allowing them to have an independent existence. The differentiation of their functions and the relationships between these elementary components produce more complex systems; without those subsystems essential properties would be lost. However, one of the most important properties of a viable system is its intrinsic recursion. In fact, any viable system contains and is contained by another viable system. Every subsystem maintains its autonomy towards its environment, but it also contributes to generate the viable system in which it is included. In that way, a viable system and its different subsystems have the same structural requirements. A viable system supports its objectives thanks to an overall cohesion and adapts itself by the autonomy of its subsystems.

VSM was developed looking for invariances in organic systems. These invariances allow defining a homomorphism of their functions, organization and structure. Beer defines five elementary functions that any viable system must have: implementation, coordination, control, intelligence and policy. In the reminder of this chapter each one of these functions are defined more in details.

2.1 Implementation

This function refers to primary activities that materialize the identity of the system. These activities are identified after an analysis of what the system does, then directly related to the system transformation and environment interaction.

2.2 Coordination

This function corresponds to the coordination among primary activities. In a viable system, this coordination is not necessarily accomplished in a top-down manner like in a hierarchical management system. Primary activities can be coordinated in a centralized or distributed way thanks to cooperation and information exchanges.

2.3 Control

Control refers to the function which regulates and ensures the self- organization of the system. This is due to coordination and monitoring functions. A monitoring function (or sporadic audit function) allows at the same time to evaluate the actions of primary activities and to hold the coherence of the global activity (all primary activities). A control function is the function that defines the control / autonomy degree of primary activities.

2.4 Intelligence

The intelligence function is responsible for system adaptability. To achieve this adaptation, the intelligence function must be capable of treating the information which comes from the environment, with the objective to anticipate perturbations.

2.5 Policy

The policy function keeps the system objectives at its level.

As it was expressed by Ashby [1] in the law of requisite variety a controller will be effective only if it is capable of attaining at least the same number of states as those which it wants to control. The variety of a system is defined as all possible states of a system and it can be used as a complexity measure.

"A controller has requisite variety - that is, has the capacity to maintain the outcomes of a situation within a target set of desirable states - if and only if it has the capacity to produce responses to all those disturbances that are likely to take the outcomes out of the target set ".

To achieve that, the viable system tries to reduce the variety coming from the environment and to amplify the variety of its control function to reach a balance (homeostasis). Fig.1.a shows a classic cybernetic feedback composed by a system, a controller and their environment. Fig.1.b shows graphically the law of requisite variety. This figure shows that the controller must reduce the variety originated from the system and increase his own variety with respect to this one. The system realizes the same process with respect to the environment. In this way, the controller can be effective allowing the system to exist in the environment. Notice that there is not necessarily a physical separation between system and controller (for example, body and nervous system in living beings). Fig.1.c shows the basis structure of a viable system. This figure has the same components that in Fig.1.b.

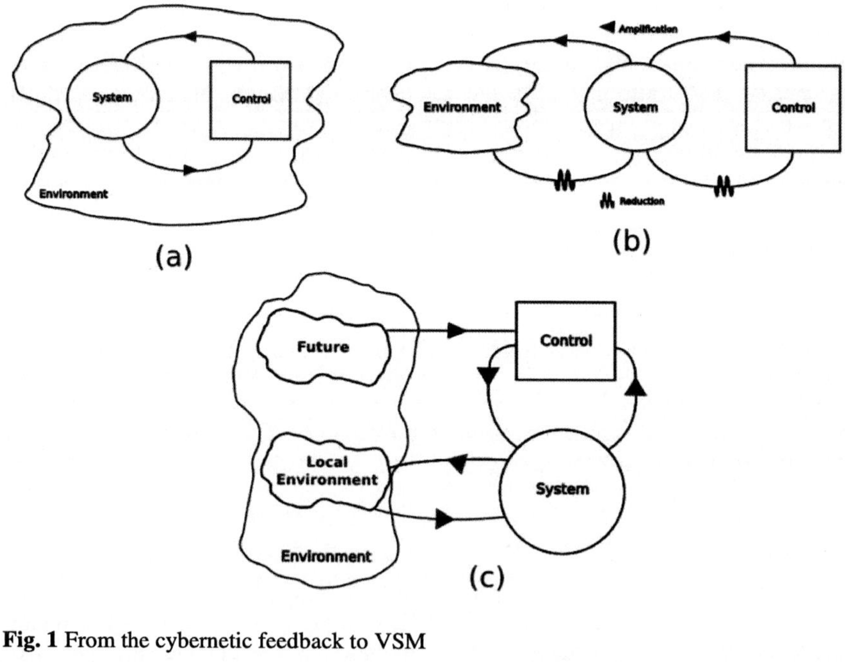

Fig. 1 From the cybernetic feedback to VSM

The main difference is that in the viable system model, the controller must develop its intelligence from information from the environment, with the objective to predict possible future states of this. It is important to notice that, in accordance with the concepts expressed previously, considering intelligent products allows the use of the viable system model.

With respect to structural recursion, Fig. 2 shows an example of a viable system which is decomposed into three subsystems. The recursion property of VSM allows continuing this process until the required level of detail. In the example presented in Fig. 2, it is possible to break down each of the three viable systems in new viable sub-systems. Also, the whole system can be considered as a subsystem of another viable system at an upper recursive level. Notice that all systems have the same structure.

A conventional partitioning of a production system will decompose the system, for example in the shop-floor, production cells, production lines, production modules and machines. In such case, machines would be the elementary (atomic) entities of the system at the lowest disaggregation level.

In contrast with this conventional interpretation, we consider the product to be the elementary decision-making entity responsible for driving work tasks. Therefore, it is considered as the basic element of the production process and it is designed as a viable system. Other system entities such as machines, conveyors, automated guided vehicles, team workers, etc., are then considered in the environment. These entities interact with the product only in terms of events leading to lack of availability, disruptions, constraints, etc. Fig. 3 shows the intelligent product structure which is proposed.

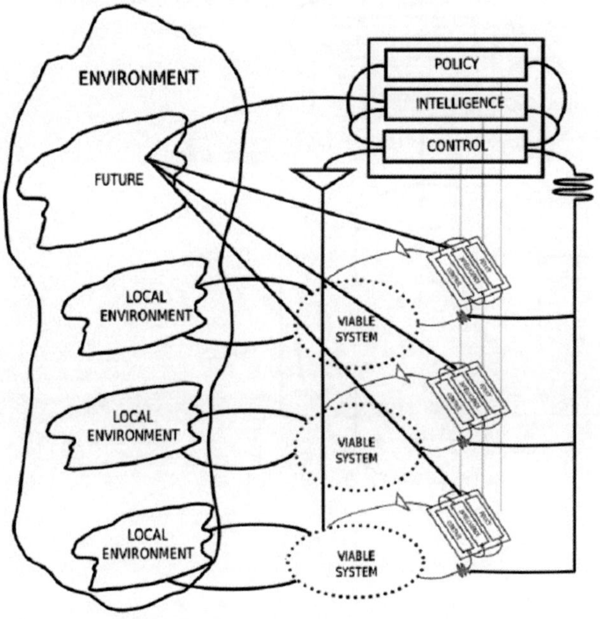

Fig. 2 An example of a Viable System Model

As has been said in the introduction, the *intelligent product* is defined as an entity able of interacting and communicating with its environment. When this physical entity transits on the shop-floor, it will be able to recover information concerning both its own state and the system's state. Then, focusing on the product, all the production planning system can be structured from the information recovered by this intelligent product. Also, decision-making problems are addressed to their corresponding recursion level.

The primary activities of the intelligent product are data collection and interaction functions. The data collection activity is designed as the action allowing the product to get information from other products and to give information to them. On the other hand, interaction functions allow products to interact physically with their environment.

An advantage of this approach is that the functional definition of every subsystem, provides technological specifications (requirements) needed for software and hardware implementation. This informational and physical nature of the system, associated to the concept of structural recursion is directly related to the holonic vision. Modelling agent communities as viable systems allows defining them as cognitive holarchies. In a recent work [13], author defines VSM like a holons organization. Then, a viable system is interpreted like a cognitive holarchy composed of sub-holons which are also viable systems.

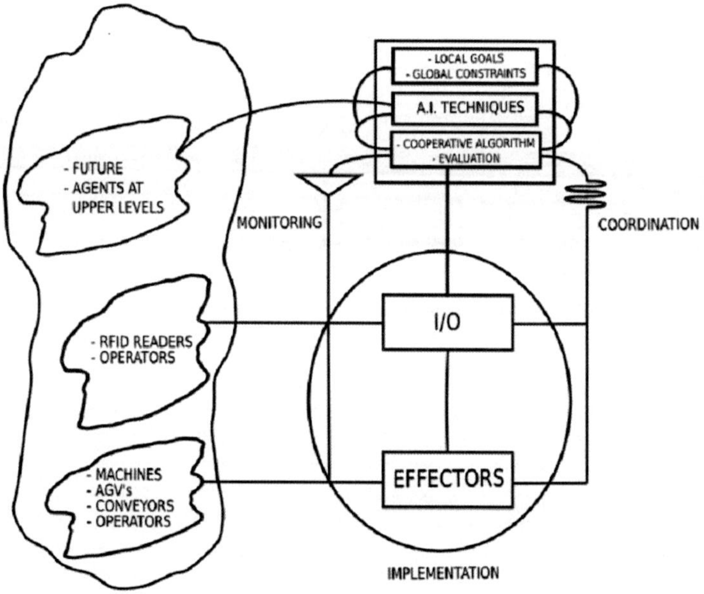

Fig. 3 An intelligent product modelled using VSM

In this way, the viable system model associated to the concept of product-driven control allows managing the complexity of a multi-level decision making processes. This is carried out considering a products holarchy such as the connection between the different decision levels. At every level it is possible to take into account entities and relations of the lower level without losing the richness of the interaction among levels. Indeed, it was proved that changes at one level can have big consequences at the global level. To take that into account allows preserving the emergent properties present in these complex systems.

As it was expressed before, intelligent products in their physical forms recover information about their states and the state of the system. This information will be shared at the concerned recursion level, which will assess the pertinence of a change of decision. This evaluation will have as effect information filtering, leading to a form of variety reduction towards the environment. In a more specific way, the reduction and amplification of variety will be managed by the cooperative algorithm implemented at each recursion level.

3 Application to MPC Systems

Manufacturing Planning and Control Systems contain five main functions which can be defined as: (1) Strategic Planning, (2) Sales and Operations Planning (S&OP), (3) Master Production Planning (MPS), (4) Scheduling and (5) Execution. Each of these functions corresponds to a level in the decision making process concerning different horizons going from a longer to shorter one. In practice, these decisions are taken using a rolling horizon to take into account the frequent changes that occur in the data (demand, capacity, etc.). Thus, the strategic planning is revised once a year, the S&OP is computed monthly, the MPS is get per week, and the schedules are performed daily or more frequently depending on disturbances.

Each function deals with a corresponding aggregation level of products respectively families, finished products and items (components). In this context, one of the major issues is to adapt decisions at each level when disturbances (internal or external) happen. Frequent modifications in the decision-making process lead to the so-called "nervousness" of the system which deteriorates the system performance (productivity and efficiency). One should notice that the shorter is the horizon the more frequent are the changes. Thus, the performance is more deteriorated at the lower level (scheduling level).

More precisely, our proposition is to provide a generic model based on VSM dealing with production planning considering both MPS level and scheduling (lot-streaming). The MPS allows defining quantities for finished products per periods on a specific horizon. This is usually performed with a capacitated lot-sizing problem which is classically formulated with a mixed integer program. This model is implemented at the higher level in the Fig. 3. Generally, once lot sizes are determined, sub-lots have to be sequenced and dimensioned to be launched on the shop-floor (lot-streaming problem) [17]. This model is implemented at the level two from the bottom in Fig.4.

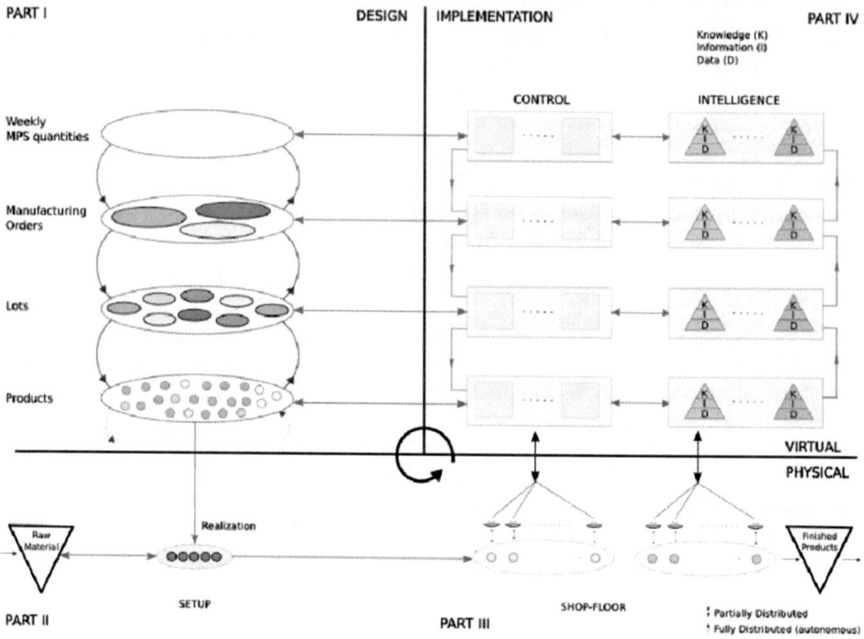

Fig. 4 VSM based product-driven manufacturing planning and control system

Thus, different objectives can be reached corresponding to each decision level (minimizing stocks and set up costs at the MPS and minimizing makespan at lot-streaming level). However, disturbances are frequent in this context; such an example is found in the demand changes or the capacity reduction (machine failure). As a consequence of these disturbances, the computed predictive planning would be inadequate and should be adjusted regarding to the new conditions (data). Nevertheless, the combinatorial nature and the variability of the different parameters of the considered problems do not allow an on-line computation. The characteristic of self-adaptation is then present with an adjustment of machine capacity when a failure happens in a certain limit. Obviously, when the capacity decreases to a certain down threshold, then it becomes impossible to find a solution. In this case the problem is forwarded to the upper level (a bottom-up process).

For external disturbances, the centralized system modifies the threshold and the global objectives to adapt itself in a top-down manner by defining new targets for the bottom level. In opposition with the bottom-up decision making process which has a cooperative characteristic, the adjustment process (top-down) has a coercive characteristic.

Fig. 4 shows an example of four recursion levels. This system has been proposed with the hypothesis of using intelligent products mainly concerned with RFID technology to embed and communicate information.

The objective of the proposed model is the decision making process adaptation towards external and internal disturbances. This system is based on a holarchy of products regarding to VSM modelling. Each decision level corresponds to a certain recursion level. The different levels of recursion are defined as agent communities which should adapt themselves to the decision making conditions. Moreover, they have the objective to collaborate in order to find a suitable solution. Fig. 4 is subdivided into four parts (I, II, III and IV) in order to simplify the explanations. The two axes allow representing different points of view where the horizontal axis distinguishes the physical world from the virtual one, whereas the vertical axis shows at the left side the design which is a representation of the system. At the right side, the implementation deals with both the knowledge management and decision making systems.

3.1 Part I: Design/Virtual

This part is related to the centralized or cooperative system representation. The representation is decomposed into four recursive levels of product entities. At each level, the concerned entities (weekly production, manufacturing orders, lots and products) are modelled thanks to the basic object proposed in Fig. 2. To take decision, a product or a group of products will have policies, intelligence and control functions, coordination and communication mechanisms.

Three alternatives are possible once the products defined as agents with a specific control / autonomy level. First, if the system is strictly distributed or heterarchical, agents communicate only in a cooperative way for decision-making processes. Second, if the system is strictly centralized, decision are taken with methods (models) that collect information thanks to agents and aggregate it to propose centralized planning orders. Third, between these two opposite solutions, many combinations can be implemented adapting different degrees of control / autonomy.

3.2 Part II: Design/Physical

This part of the figure is relative to the product instantiation at the physical level. Each of the four decision-making levels of the previous part could generate plans for the corresponding level. According to the physical/virtual nature of the entities, products will be built as holons in this part. They will have specific functions to allow the interaction with their environment. Each of the four recursive levels would be designed like holons in the same way. For example, the physical part of a manufacturing order holon would be the manufacturing order paper form or a kanban card would be the physical part corresponding to a lot.

3.3 Part III: Implementation/Physical

In this part product holons are able to make decisions according to events concerning their own evolution. Specific functions are embedded in these intelligent products leading them to react to events in an individual or collective way.

3.4 Part IV: Implementation/Virtual

This last part is relative to the treatment of information coming from the shop-floor and concerning disruptions and product actions, on one hand. On the other hand, decision making process is also handled in this part. In the same way, each recursive level corresponds to a decision making one which interacts with the corresponding knowledge level. This treatment could be defined as an information and knowledge model [25].

There are many implementation alternatives of manufacturing product-driven control systems. These implementation alternatives will depend on the characteristics and objectives of the manufacturing system. One of the technologies established as a good alternative in terms of costs/benefits is RFID technology [18]. This technology is part of the general class of Auto-ID technologies [24], which are usually constituted of two types of components: tags and readers, and other intermediate components necessary to establish a link with the computer system.

One of the advantages of Auto-ID systems, besides their reduced cost in comparison with other technologies based on WSN (Wireless Sensor Networks), is the possibility of taking into account the entire cycle of life of the product (even up to its stage of recycling). The cost reduction of this technology is one of the main factors which contributed to the increase of the use in industry and specifically in manufacturing. In opposition to these advantages, the Auto-ID technology reduces considerably the capabilities of intelligent products. This is due to the fact that low cost technologies do not allow today to implement information processing (passive tags only in reading mode). Nevertheless, the advances of these technologies make more and more feasible applications unthinkable some years ago.

4 Discussion

An important concern of VSM is the *adaptability* as it was described in the previous section; VSM was developed with the objective to understand the mechanisms which allow biological organisms to survive to changes in the environment. A viable system is designed for self-adaptation in a changing environment, by the implementation of the necessary mechanisms to support its viability. Indeed, it is not possible to identify a specific function which ensures adaptability, since the whole system is defined to fulfil this objective, thanks to its different elementary functions. This characteristic of adaptability is also one of the main objectives of the paradigms arising from IMS initiative.

One of the basic properties of the viable systems is *scalability* because these systems are structurally recursive. This notion of recursion has been mainly exploited by biological manufacturing systems, which make an analogy with biological organisms. Recursion has been also exploited by fractal manufacturing systems which show clear properties of *modularity* and *auto-similarity*. Auto-similarity in fractal systems is also included in the viable systems through its property of structural recursion. In that way, a viable system will keep the same structural requirements at any complexity level. These concepts have a direct relationship with *re-usability*, which is another objective of IMS systems. These concerns of *autonomy*, adaptability and scalability highlight the strong relationship between VSM and IMS objectives.

As it was previously expressed in the introduction, three main paradigms of manufacturing systems were developed from the IMS initiative: bionic, fractal and holonic manufacturing systems. These paradigms change a centralized vision to a distributed vision of the decision making process. In this context, the heterarchy concept is an example. It concerns the distribution of decision-making processes in entities near to the production flow, with the objective to improve reactivity, by granting autonomy to low decision levels for quick decisions. In this sense, an interesting question is: what is the desirable hierarchy/heterarchy degree which a system of this class must have? [19].

This trade-off between hierarchy/heterarchy has been widely studied in viable systems. Generally, VSM includes this relation in terms of a degree of control / autonomy regulation, implemented in the control function. The monitoring (sporadic audit) function is expressly defined with the intention of auditing the performance of primary activities (agents) and not in the way of coercive control as in conventional systems of command and control. This allows flexibility in the decision making process at each recursion level. These mechanisms can be found in the control of systems and their components in livings beings. Here, the objective is to maintain variables between certain limits rather than in a fixed value.

Considering the context of our study dealing with the Product Driven Control Systems which consider the product as an intelligent entity, the idea of the proposed VSM model takes the product as the elementary (primary) recursive component. The others recursive levels of our model are established according to the classical decomposition resulting from MPCS management which provides: products, lots, manufacturing orders and MPS weekly quantities. Note that the product is a generic term and could be extended to component or raw material. In such case, corresponding lower levels should be added. This organization is composed of agents which are conceived as autonomous and co-operative organisms which are organized with respect to the relation between hierarchy and heterarchy (control / autonomy), to assure a compromise between flexibility and global objectives cohesion.

On the other hand, another important objective of IMS paradigms is to improve the scalability of systems. This allows having structures which include a growing number of functions, while always keeping a global coherence. This property of scalability is also included in the viable systems through its property of functional and structural recursion. In that way, a viable system will keep the same structural requirements to all levels of complexity. A viable system will be able to extend itself with the same functional and structural requirements. This facilitates scalability and allows reusing components. The reuse of components has been another objective of IMS paradigms.

So, it seems that the use of VSM as conceptual and theoretical modelling framework gives answers to some of important question originated from the IMS initiative.

Finally, we would highlight that designing intelligent devices like living beings, and specifically like viable systems, can improve considerably the relationship between humans and these devices in terms of communication, coordination and interoperability.

5 Conclusion

In this chapter we have presented the main concepts involved in the VSM and how they can be applied to model product-driven control systems. It was highlighted that VSM has good properties to model intelligent manufacturing systems.

We have presented an application of the proposed modelling framework to model a MPCS product-driven system. The main advantage of this application is reaching coherence among different decision levels giving a high degree of flexibility to the system. The proposed modelling approach applied to product-driven systems can be implemented in different industrial context.

Acknowledgments. The authors gratefully acknowledge the financial support of the CPER 2007-2013 Competitiveness Fibre Cluster (Structuration du Pôle Compétitivité Fibres Grand'Est), local (Conseil Général des Vosges), regional (Région Lorraine), national (DRRT et FNADT) and European (FEDER) founds.

References

1. Ashby, R.: An introduction to cybernetics. Chapman and Hall, London (1957)
2. Beer, S.: The liberty machine. Futures 3, 338–348 (1971)
3. Beer, S.: Brain of the firm: the Managerial cybernetics of organization. Allen Lane the Penguin Press, London (1972)
4. Beer, S.: The viable system model: Its provenance, development, methodology and pathology. Journal of the Operational Research Society 35, 7–25 (1984)
5. Espejo, R.: The viable system model a briefing about organisational structure (2003), `http://maine.gov/dhhs/btc/PDF/ViableSytemModel.pdf`
6. Herrera, C., Thomas, A., Belmokhtar, S., Pannequin, R.: A viable system model for product-driven systems. In: Proceeding of the International Conference on Industrial Engineering and Systems Management, IESM 2011, Metz (to appear, 2011)
7. Herring, C., Kaplan, S.: Viable Systems: The Control Paradigm for Software Architecture Revisited. In: Australian Software Engineering Conference, pp. 97–105 (2000)
8. Herrmann, C., Bergmann, L., Halubek, P., Thiede, S.: Lean production system design from the perspective of the viable system model. In: Mitsuishi, M., Ueda, K., Kimura, F. (eds.) Manufacturing Systems and Technologies for the New Frontier, pp. 309–314 (2008)
9. Laws, A., Taleb-Bendiab, A., Wade, S.: Genetically modified software: Realizing viable autonomic agency. Innovative Concepts for Autonomic and Agent-Based Systems, 184–196 (2006)
10. Maturana, H., Varela, F.: Autopoiesis and cognition: The realization of the living. Reidel Publishing, The Netherlands (1980)
11. McFarlane, D.: Product identity and its impact on discrete event observability. In: Proceedings of the European Control Conference, ECC (2002)
12. McFarlane, D., Sarma, S., Chirn, J., Wong, C., Ashton, K.: The intelligent product in manufacturing control. Journal of EAIA 5464 (2002)
13. Mella, P.: The Holonic Revolution Holons, Holarchies and Holonic Networks. In: The Ghost in the Production Machine. Pavia University Press, Pavia (2009)

14. Morel, G., Valckenaers, P., Faure, J.M., Pereira, C.E., Diedrich, C.: Manufacturing plant control challenges and issues. Control Engineering Practice 15, 1321–1331 (2007)
15. Ribeiro-Justo, G., Saleh, A., Karran, T.: Intelligent reconfiguration of dynamic distributed components. Electron. Notes Theor. Comput. Sci. 180, 91–106 (2007)
16. Rosenkranz, C., Holten, R., Laumann, M.: Designing ic structures by variety engineering. In: SAC 2008: Proceedings of the 2008 ACM Symposium on Applied Computing, pp. 518–523 (2008)
17. Sarin, S., Jaiprakash, P.: Flow Shop Lot Streaming. Springer, New York (2007)
18. Thomas, A., Castagna, P., Blanc, P., Cardin, O., El Haouzi, H., Klein, T., Pannequin, R.: Simulation for Product Driven Systems, pp. 221–255. Wiley (2008)
19. Trentesaux, D.: Les systèmes de pilotage hétérarchiques: Innovations réelles ou modèles stériles? Journal Européen des Systèmes Automatisés 41(9-10), 1165–1202 (2007)
20. Truszkowski, W., Karlin, J.: A Cybernetic Approach to the Modeling of Agent Communities. In: Klusch, M., Kerschberg, L. (eds.) CIA 2000. LNCS (LNAI), vol. 1860, pp. 166–178. Springer, Heidelberg (2000)
21. Ueda, K., Vaario, J., Ohkura, K.: Modelling of biological manufacturing systems for dynamic reconfiguration. CIRP Annals-Manufacturing Technology 46(1), 343–346 (1997)
22. Valckenaers, P., Van Brussel, H.: Holonic manufacturing execution systems. CIRP Annals. General Assembly of CIRP 54(55), 427–432 (2005)
23. Warnecke, H.: The Fractal Company (1993)
24. Wong, C., McFarlane, D., Zaharudin, A., Agarwal, V.: The intelligent product driven supply chain. In: IEEE International Conference on Systems, Man and Cybernetics 2002, vol. 4, pp. 6–11. IEEE (2002)
25. Zins, C.: Conceptual approaches for defining data, information, and knowledge: Research articles. J. Am. Soc. Inf. Sci. Technol. 58, 479–493 (2007)

Speech to Head Gesture Mapping in Multimodal Human-Robot Interaction

Amir Aly and Adriana Tapus

Cognitive Robotics Lab, ENSTA-ParisTech, France
{amir.aly,adriana.tapus}@ensta-paristech.fr

Abstract. In human-human interaction, para-verbal and non-verbal communication are naturally aligned and synchronized. The difficulty encountered during the coordination between speech and head gestures concerns the conveyed meaning, the way of performing the gesture with respect to speech characteristics, their relative temporal arrangement, and their coordinated organization in a phrasal structure of utterance. In this research, we focus on the mechanism of mapping head gestures and speech prosodic characteristics in a natural human-robot interaction. Prosody patterns and head gestures are aligned separately as a parallel multi-stream HMM model. The mapping between speech and head gestures is based on Coupled Hidden Markov Models (CHMMs), which could be seen as a collection of HMMs, one for the video stream and one for the audio stream. Experimental results with Nao robots are reported.

Keywords: Coupled HMM, audio-video signal synchronization, human-robot interaction, signal mapping, robot services.

1 Introduction

Robots are more and more present in our daily lives and the new trend is to make them behave more natural so as to obtain an appropriate social behaviour and response. The work described in this chapter presents a new methodology that allows the robot to automatically adapt its head gestural behavior to the user's profile (e.g. the user prosodic patterns) and therefore to produce a personalizable interaction. This work is based on some findings in the linguistic literature that show that head movements (e.g., nodding, turn taking system) support the verbal stream. Moreover, in human-human communication, prosody expresses the rhythm and intonation of speech and reflect various features of the speakers. These two communication modalities are strongly linked together and synchronized. Humans use gestures and postures as a communicative act. McNeill in [1] defines a gesture as a movement of the body synchronized with the speech flow. The mechanism of the human natural alignment of verbal and non-verbal characteristic patterns based on the work in [2] shows a direct relationship between prosody features and gestures/postures, and constitutes an inspiration for our work.

T. Borangiu, A. Thomas, and D. Trentesaux (Eds.): SOHOMA 2011, SCI 402, pp. 183–196.
springerlink.com

Recently, there has been a growth of interest in socially intelligent robotic technologies featuring flexible and customizable behaviours. Based on the literature in linguistics and psychology that suggests that prosody and gestural kinematics are synchronous and therefore strongly linked together, we posit that it is important to have a robot behaviour that integrates this element. In this chapter, we describe a new methodology for speech prosody and head gesture mapping for human-robot social interaction. The gesture/prosody modelled patterns are aligned separately as a parallel multi-stream HMM model and the mapping between speech and head gestures is based on Coupled Hidden Markov Models (CHMMs). A specific gestural behaviour is estimated according to the incoming voice signal's prosody of the human interacting with the robot. This permits to the robot to adapt its behaviour to the user profile and to produce a personalizable interaction.

To the best of our knowledge, very little research has been dedicated to this investigation area. An attempt is described by the authors in [3] that present a robotic system using dance so as to explore the properties of rhythmic movement in general social interaction. Most of the existing works are related to computer graphics and interactive techniques. A general correlation between head gestures and voice prosody had been discussed in [4] and [5]. The emotional content of the speech can also be correlated to some bodily gestures. In [6], it is discussed the relation between voice prosody and hand gestures, while [7] discusses the relation between the verbal and semantic content and the gesture. In [8], which is somehow closed to the discussed topic on this research, the relation between prosody changes and the orientation of the head (Euler angles) is presented. Moreover, authors in [9], propose a mechanism for driving a head gesture from speech prosody.

Our work presents a framework for head gesture and prosody correlation for an automatic robot gesture production from interacting human user speech. The system is validated with the Nao robot in order to find out how naturalistic will be the driven head gestures from a voice test signal with respect to an interacting human speaker. The rest of the chapter is organized as follows: section 2 presents the applied algorithm extracting the pitch contour of a voice signal; section 3 illustrates the detection of head poses and Euler angles; section 4 describes speech and gesture temporal segmentation; section 5 presents the speech to head gesture coupling by using CHMMs; section 6 resumes the results; section 7 concludes the chapter.

2 Prosodic Features Extraction

In human-robot interaction applications, the human voice signal can convey many messages and meanings, which should be understood appropriately by the robot in order to interact properly.

Next, the methodology used for pitch extraction is described. Talkin [10] defined the pitch as the auditory percept of tone, which is not directly measurable from a signal. Moreover, it is a nonlinear function of the signal's spectral and temporal energy distribution. Another vocal characteristic, the fundamental frequency F0, is measured as it correlates well with the perceived pitch. Voice processing systems that estimate the fundamental frequency F0 often have 3 common processes: (1) Signal Conditioning; (2) Candidate Periods Estimation, and (3) Post

Processing. Signal preconditioning process is concerned by removing interfering signal components like noise and DC offset, while post processing process chooses the more likely candidate period in order to precisely estimate the fundamental frequency F0. Talkin in [10] developed the traditional (NCC) method in order to estimate reliably the voicing periods and the fundamental frequency F0 by considering all candidates simultaneously in a large temporal context. This methodology uses two pass normalized cross correlation (NCC) calculations for searching the fundamental frequency F0 which reduces the overall computation load with respect to the traditional (NCC) method. We express here the characterizing vector of the voice signal in terms of the pitch and the intensity of the signal.

3 Head Pose Estimation

During social human-robot interaction, robots should be able to estimate the human head pose. This can help the robot to understand the human focus of attention and/or the meaning of the spoken message. The authors in [11] present a survey on existing algorithms for head poses estimation and detection. We use for the detection of head poses is Viola and Jones algorithm [12]. After extracting the head region, the eyes are detected by the valleypoints' detection algorithm [13].

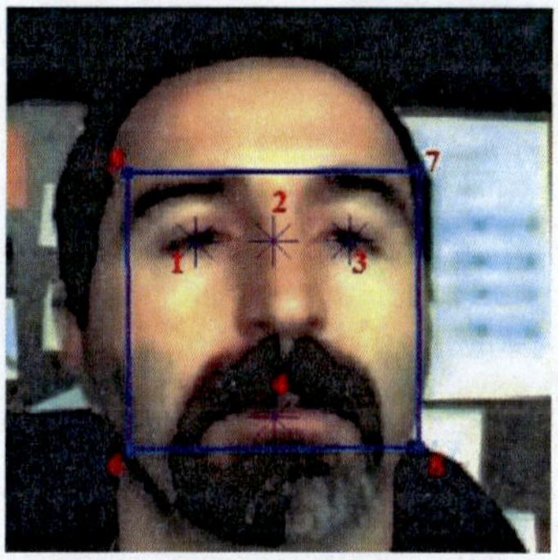

Fig. 1 Detecting the face rectangle that contains all salient points

After detecting the location of eyes, it is possible to detect the location of other salient points of the face using the face geometry [14]. For example, if the distance between the two eyes points (1&3) equals to D (see Fig.1), and point 2 is the midpoint between the eyes, then the mouth point 4 is located at a distance = 1:1D downwards from point 2.

The X -Y coordinates of the rectangle surrounding the salient points of the face (points 5, 6, 7, and 8) (see Fig.1) could be defined as following:

- The difference between the Y - coordinates of points (5&1 or 3&7) = 0:2 * 1:8D

- The difference between the X - coordinates of points (5&1 or 3&7) = 0 :225 * 1:8 D

After calculating the coordinates of points (5; 7), the coordinates of points (6; 8) are directly calculated based on the vertical distance between points (7&8 or 5&6), which is equal to 1:8D.

One of the problems that may appear when detecting the surrounding rectangle of the facial salient points is the rotation of the head clockwise and counter clockwise (see Fig.2).

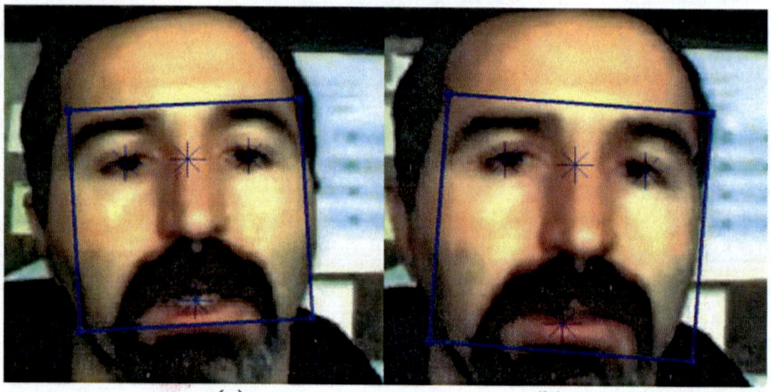

(a) (b)

Fig. 2 Tracking salient face details in different cases of rotation: (a) Rotation clockwise; (b) Rotation counter clockwise

Therefore, the (X, Y) coordinates of the eyes have to be rotated first to (X^{-}, Y^{-}) before following the previous steps in order to precise the points of the surrounding rectangle, because the above motioned relations are valid when the eyes coordinates are in the same plane of the face (i.e., if the face is rotated, the coordinate of the eyes have also to be located in the rotated plane). The direction of rotation will be detected by calculating the slope (i.e., rotation angle θ) of the line passing by the two eyes using their (X, Y) coordinates. The rotation of the axes is described by the following equations:

$$X^{-} = X \cos\theta - Y \sin\theta \quad (1)$$

$$Y^{-} = X \sin\theta + Y \cos\theta \quad (2)$$

After calculating the coordinates of a face salient point in the rotated (X ; Y) plane while the head is rotating clockwise or counter clockwise, it is important to make the inverse rotation each time to detect the location of this point in the (X ; Y) plane. The pose of the head is calculated in terms of Euler angles: Pitch, Yaw, and Roll. These angles are calculated using the previously detected 8 salient facial points and their relative positions based on the geometry and symmetry of faces (Fig.1) as following:

$$Yaw_i = b_1 [\frac{(P2_{x,i} - C1_{x,i}) + (P2_{x,i} - C2_{x,i})}{2D\_eyes_0)} - \frac{(P2_{x,0} - C1_{x,0}) + (P2_{x,0} - C2_{x,0})}{2D\_eyes_0)}] + \\ + b_2 [\frac{(P4_{x,i} - C1_{x,i}) + (P4_{x,i} - C2_{x,i})}{2D\_exys_0)} - \frac{(P4_{x,0} - C1_{x,0}) + (P4_{x,0} - C2_{x,0})}{2D\_exys_0)}] \tag{3}$$

$$Pitch_i = b_3 [\frac{(P2_{y,i} - C3_{y,i}) + (P2_{y,i} - C4_{y,i})}{2D\_eyes_0} - \frac{(P2_{y,0} - C3_{y,0}) + (P2_{y,0} - C4_{y,0})}{2D\_eyes_0}] + \\ + b_4 [\frac{(P4_{y,i} - C3_{y,i}) + (P4_{y,i} - C4_{y,i})}{2D\_eyes_0} - \frac{(P4_{y,0} - C3_{y,0}) + (P4_{y,0} - C4_{y,0})}{2D\_eyes_0}] \tag{4}$$

where:

- $P2_{x,i}$, $P4_{x,i}$: the x coordinates of the midpoint between eyes, mouth point, respectively (see Fig.1), in frame i of the video.
- $P2_{y,i}$, $P4_{y,i}$: the y coordinates of the midpoint between eyes, mouth point, respectively (see Fig.1), in frame i of the video.
- $P2_{x,0}$, $P4_{x,0}$: the x coordinates of the midpoint between eyes, mouth point, respectively (see Fig.1), in frame 0 which is the reference frame in the video (1st frame).
- $P2_{y0}$, $P4_{y0}$: the y coordinates of the midpoint between eyes, mouth point, respectively (see Fig.1), in frame 0 which is the reference frame in the video (1st frame).
- $C1_{x,i}$: the x coordinates of the centre point between point 5 and point 6 (Fig.1), in frame *i*.
- $C2_{x,i}$: the x coordinates of the centre point between point 7 and point 8 (Fig.1), in frame *i*.
- $C3_{y,i}$: the y coordinates of the centre point between point 5 and point 7 (Fig.1), in frame *i*.
- $C4_{y,i}$: the y coordinates of the centre point between point 6 and point 8 (Fig.1), in frame *i*.

The regression values b_1, b_2, b_3, and b_4 are constants throughout all the video frames. They are calculated by defining the absolute values of *Yaw* and *Pitch* angles in the 2nd and 3rd frames (as per the empirical test) as shown in Table 1. The substitution of the second and third values of *Pitch* and *Yaw* in the equations 3 and 4, leads directly to the computation of the values of the constants b_1, b_2, b_3, and b_4.

The calculation of *Roll* angle is straightforward, it depends on the coordinates of the midpoint between eyes (point 2) in frame *i* with respect to the reference frame [15], and it is clear that the value of *Roll* angle in the 1st reference frame equals to 0.

$$Roll_i = \tan^{-1}(\frac{P2_{y,i}}{-P2_{x,i}}) - \tan^{-1}(\frac{P2_{y,0}}{-P2_{x,0}}) \tag{5}$$

Table 1 Yaw and Pitch Initial Angles (Frames 1-3) Used for Calculation of Regression Values

	Frame1(reference)	Frame 2	Frame 3
Yaw	0	0.0016	0.0034
Pitch	0	0.00255	0.0075

4 Speech and Head Gesture Segmentation

The mapping between speech and head gestures is done by using the Coupled Hidden Markov Models (CHMMs), which could be seen as a collection of HMMs, one for the video stream and one for the audio stream. The advantage of this model over a lot of other topologies is its ability to capture the dual influences of each stream on the other one across time (see Fig.5).

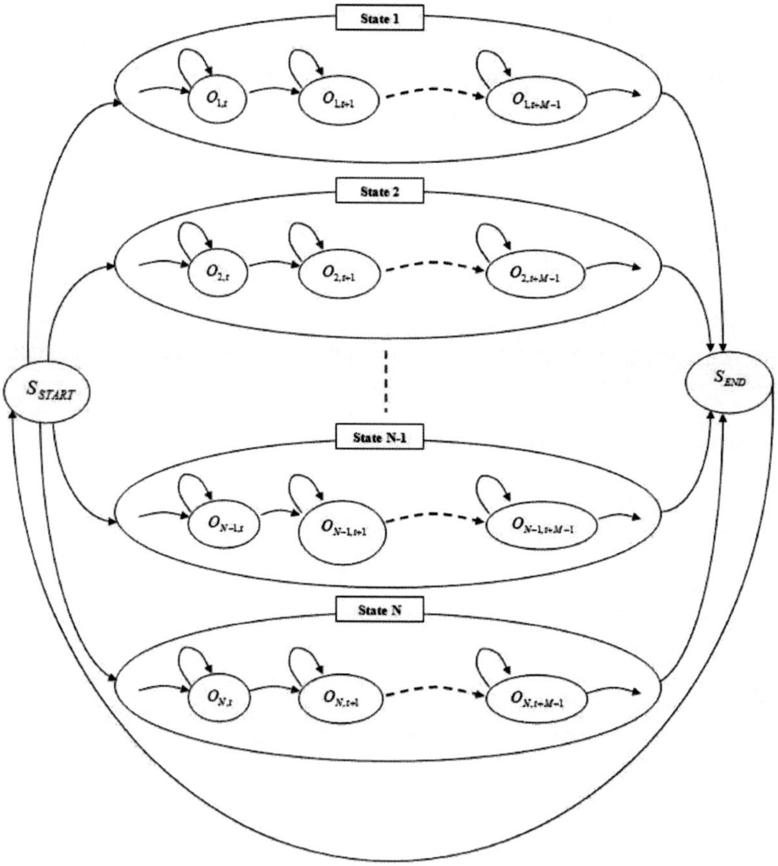

Fig. 3 HMM structure for gesture and prosody analysis

In the beginning, speech and head gestures streams are aligned separately as a parallel multi-stream HMM model. The mapping between speech and head gestures is performed in 2 main steps: (1) the first is modelling the gesture sequence and the associated voice prosody sequence (in terms of their characteristic vectors) into two separate HMMs; (2) then after training both models, a correlation between the two HMMs is necessary so as to estimate a final head gesture state sequence given a speech test signal.

The HMM structure used in analyzing gestures (and similarly voice prosody) is indicated in Fig.3. It is composed of N parallel states, where each one represents a gesture composed of M observations. The goal of the transition between states S_{END} to S_{START} is to continue the transition between states from 1 to N (e.g., after performing gesture state 1, the model transfers from the transient *end* state to the *start* state to perform any gesture state from 2 to N in a sequential way and so on). In order to be able to model gestures/prosody, it is necessary to make a temporal segmentation of the video content to detect the M number of observations in each state and the total number of states N.

4.1 Speech Temporal Segmentation

Speech is segmented as syllables presented by the states from 1 to N as indicated in Fig.3. The segmentation is performed by intersecting the inflection points (zeros crossing points of the rate of change of the curve) for both the pitch and intensity curves, beside the points that separate between the voiced and unvoiced segments of the signal (see Fig. 4 for an example of pitch and intensity curves). When comparing the two curves together, 5 different trajectory states can result [16] (see Table 2).

Table 2 Voice Signal Segmentation Labels

Trajectory Class	Trajectory State
1	pitch ↑ & intensity ↑
2	pitch ↑ & intensity ↓
3	pitch ↓ & intensity ↑
4	pitch ↓ & intensity ↓
5	Unvoiced segment

The goal is to code each segment of the signal with its corresponding pitch-intensity trajectory class (e.g., a voice signal segment coding could be: 5, 3, 4, 2, etc.). This segmental coding is used as label for CHMM training. The next step corresponds to segmenting the voice signal with its corresponding trajectory labelling into syllables. Arai and Greenberg in [17] defined the average duration of a syllable as 200 ms and this duration can increase or decrease according to the nature of the syllable as being short or long.

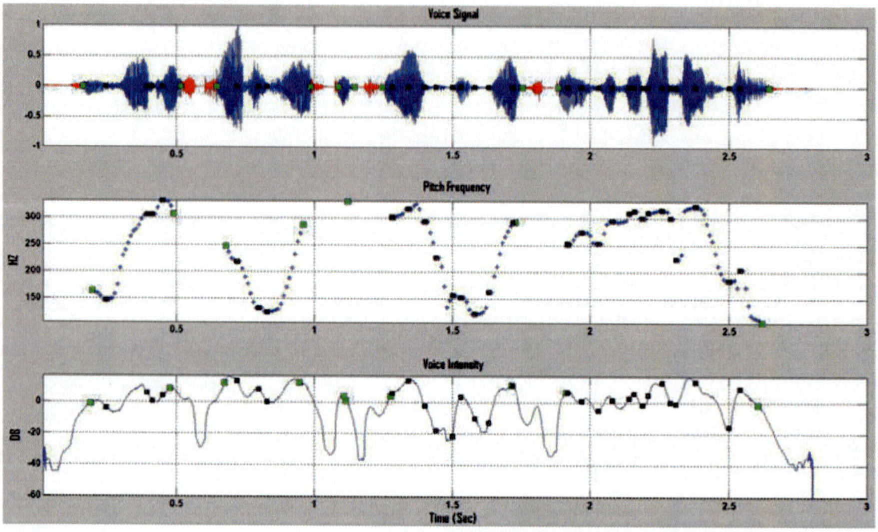

Fig. 4 Speech, Pitch and Intensity Curves (The red parts in the voice signal are the unvoiced parts, while blue parts are the voiced parts of the signal. The black points depict the inflection points of the signal, while green points represent the separating points between the unvoiced and the voiced segments.)

Practical tests proved that within a syllable of duration varying from 180 ms to 220 ms, the average number of trajectory classes in its corresponding pitch and intensity curves is around 5. Therefore, given the voice signal with its segments coded by the corresponding pitch-intensity trajectory labels, each 5 segments of the signal will create a syllable state (from 1 to N) and the corresponding 5 labels will be the observations M within the syllable state.

4.2 Gestures Temporal Segmentation

The average duration for making gestures, in general, varies between 0.1 to 2.5 seconds according to the speed and the performed gesture as being pointing or head gesture for example. In case of head gestures, the average duration of performing a gesture will be limited to 0.4 seconds [18, 19]. In our case, the camera used to capture the gestures had the ability of capturing 30 frames/second, and therefore we can estimate to 12 frames the average number of frames sufficient to characterize a gesture.

Similarly to the speech temporal segmentation (see Section 4.1), gesture temporal segmentation is performed by comparing the 9 trajectory classes according to the sinusoidal evolution of the extracted angles curves. However, the mechanical characteristics of our platform (NAO robot) are limited only to *pitch* and *yaw* movements, therefore introducing only 5 trajectory classes (see Table 3). In the context of the CHMM model each group of 12 frames will form a complete gesture state from 1 to N, and the corresponding coding labels will constitute the observations within the gesture state.

Table 3 Gesture Segmentation Labels

Trajectory Class	Trajectory State (Rate of Change)
1	Yaw ↑ & Pitch ↑
2	Yaw ↑ & Pitch ↓
3	Yaw ↓ & Pitch ↑
4	Yaw ↓ & Pitch ↓
5	No change

5 Speech to Head Gesture Coupling

A typical CHMM structure is shown in Fig.5, where circles present the discrete hidden nodes/states while rectangles present the observable continuous nodes / states, which contain the observation sequences of voice and gestures features.

According to the sequential nature of gestures and speech, the CHMM structure is of type lag-1 in which couple (backbone) nodes at time *t* are conditioned on those at time *t*-1 [20, 21, 22]. A CHMM model λ_C is defined by the following parameters:

$$\pi_0^C(i) = P(q_1^C = S_i) \tag{6}$$

$$a_{i|j,k}^C = P(q_t^C = S_i \mid q_{t-1}^{audio} = S_j, q_{t-1}^{video} = S_k) \tag{7}$$

$$b_t^C(i) = P(O_t^C \mid q_t^C = S_i) \tag{8}$$

where C $\epsilon\{audio, video\}$ denotes the audio and visual channels respectively, and q^C_t is the state of the coupling node in the c_{th} stream at time *t* [23, 24].

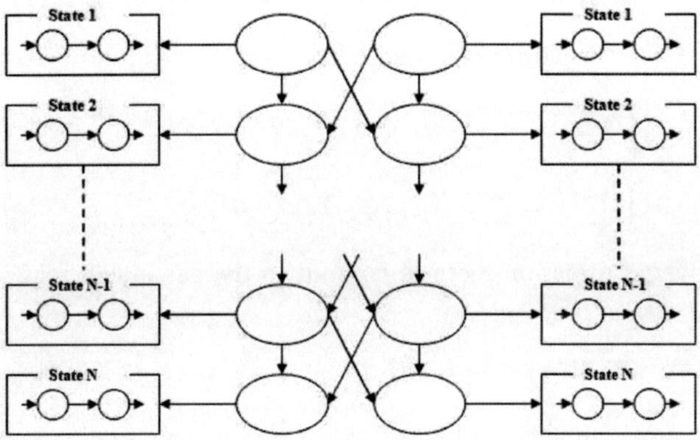

Fig. 5 Coupled Hidden Markov Model CHMM lag-1 Structure

The training of this model is based on the maximum likelihood form of the expectation maximization (EM) algorithm. Supposing there are 2 observable sequences of the audio and video states O = $\{A_{1..N}, B_{1..N}\}$ where $A_{1..N}=\{a_1,..,a_N\}$ is the set of observable states in the first audio sequence, and similarly $B_{1..N} = \{b_1, ..,b_N\}$ is the set of observable states in the second visual sequence, and S = $\{X_1..X_N\}$ is the set of states of the couple nodes at the first audio chain and the second visual chain respectively [21, 22]. The expectation maximization algorithm finds the maximum likelihood estimates of the model parameters by maximizing the following function [22]:

$$f(\lambda_C) = P(X_1)P(Y_1)\prod_{t=1}^{T} P(A_t \mid X_t)P(B_t \mid Y_t) \\ P(X_{t+1} \mid X_t, Y_t)P(Y_{t+1} \mid X_t, Y_t), 1 \leq T \leq N \tag{9}$$

where:

- $P(X_1)$ and $P(Y_1)$ are the prior probabilities of the audio and video chains, respectively,
- $P(A_t \mid X_t)$ and $P(B_t \mid Y_t)$ are the observation densities of the audio and video chains, respectively,
- $P(X_{t+1} \mid X_t , Y_t)$ and $P(Y_{t+1} \mid X_t , Y_t)$ are the couple nodes transition probabilities in the audio and video chains.

The training of the CHMM differs from the standard HMM in the expectation step (E) while they are both identical in the maximization step (M) which tries to maximize equation 9 in terms of the expected parameters [25]. The expectation step of the CHMM is defined in terms of the forward and backward recursion. For the forward recursion we define a variable for the audio and video chains at $t = 1$:

$$\alpha_{t=1}^{audio} = P(A_1 \mid X_1)P(X_1) \tag{10}$$

$$\alpha_{t=1}^{video} = P(B_1 \mid Y_1)P(Y_1) \tag{11}$$

The variable α is calculated incrementally at any arbitrary moment t as follows:

$$\alpha_{t+1}^{audio} = P(A_{t+1} \mid X_{t+1})\iint \alpha_t^{audio}\alpha_t^{video}P(X_{t+1} \mid X_t, Y_t)dX_t dY_t \tag{12}$$

$$\alpha_{t+1}^{video} = P(B_{t+1} \mid Y_{t+1})\iint \alpha_t^{audio}\alpha_t^{video}P(Y_{t+1} \mid X_t, Y_t)dX_t dY_t \tag{13}$$

Meanwhile, for the backwards direction there is no split in the calculated recursions which can be expressed as follows:

$$\beta_{t+1}^{audio,video} = P(O_{t+1}^{N} \mid S_t) = \iint P(A_{t+1}^{N}, B_{t+1}^{N} \mid X_{t+1}, Y_{t+1}) \\ P(X_{t+1}, Y_{t+1} \mid X_t, Y_t)dX_{t+1}dY_{t+1} \tag{14}$$

After combining both forward and backwards recursion parameters, an audio signal will be tested on the trained model, generating a synthesized equivalent gesture that most likely fits the model. The generated gesture sequence is determined when the change in the likelihood is below a threshold.

6 Experimental Results

The experimental test-bed used in this study is the humanoid robot Nao developed by Aldebaran Robotics. For the training and testing, we used the MVGL-MASAL gesture-speech Turkish database [9]. The database is composed of 4 videos of different durations that go from 6 to 8 minutes. It contains the audiovisual information of different subjects instructed to tell stories to children audience. We use one part of the database for the training of the models and the other part for the testing. The audio signals are extracted and then they are processed in order to extract the relevant prosodic characteristics. The proposed speech to gesture mapping methodology was tested on the database using cross validation algorithm. The system was trained on the audio/visual sequences of 3 videos from the database, and then tested on the audio sequence of the 4th video. The corresponding generated gestures are compared to the natural gesture sequence in the video of test and an average score of 62% was found in terms of the similarity of trajectory classes.

Table 4 depicts the confusion matrix between the original and synthesized gesture labels trajectories. The confusion matrix reveals that the trajectory state 5 in which there would be no change in the *Yaw* and *Pitch* angles is the dominant trajectory class. This can be a result of the smoothing processes and/or of the precision of Euler angles extracting algorithm; however this will not cause unnaturalness when the robot and the human are interacting in long conversations.

Table 4 Confusion matrix of the original and synthesized trajectories' classes

Synthesized/Real Gesture Classes	1	2	3	4	5
1	25	13	13	6	36
2	3	29	5	3	28
3	2	6	20	8	25
4	4	2	5	40	33
5	20	18	30	43	351

After calculating the score of similarity between the trajectory labels of the original and the synthesized signals, it is important to generate the corresponding *Yaw* and *Pitch* curves for the head motion and compare them to the original curves by calculating the total average root mean square (RMS) error between the corresponding curves points. The RMS errors found between the generated *Yaw* and *Pitch* curves with respect to the original curves are 10% and 12% respectively. In fact, the obtained score 62% and the RMS errors between the original and the synthesized curves can be considered a reasonable result, because the duration and the surrounding environment conditions of the test video and the training videos set were similar. Also, the speaker's tonality in all training and test videos were similar.

However, we don't know yet the score we will obtain in real applications where the robot will be tested under different conditions. The performed head gestures could differ in amplitude or direction from one person to another without hindering the transfer of the meaning of the gesture message between interacting humans and similarly, between the interacting robot and human.

Fig. 6 and 7 show a comparison between a part of the original and synthesized pitch and yaw curves of the test video from the database.

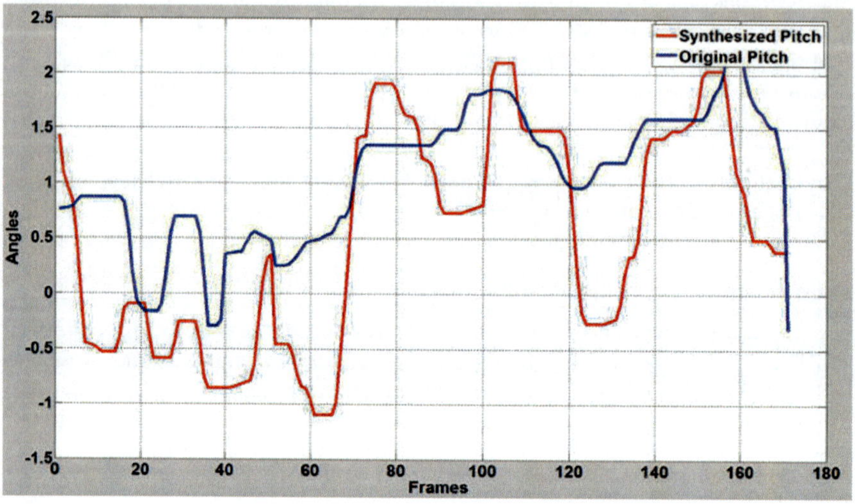

Fig. 6 View of the original (blue curve) and synthesized (red curve) Pitch angles of a part of the test video

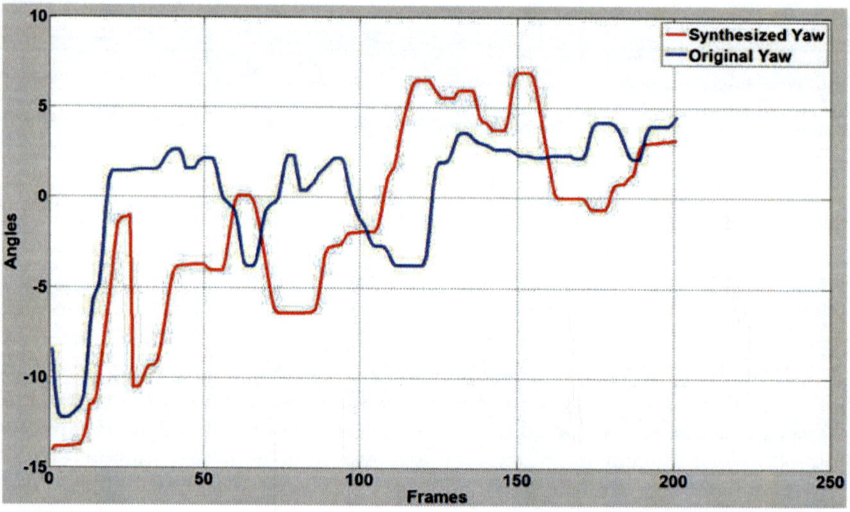

Fig. 7 View of the original (blue curve) and synthesized (red curve) Yaw angles of a part of the test video

7 Conclusions

This research focuses on synthesizing head gestures based on speech characteristics (e.g., pitch and intensity of the signal). Our mapping system is based on the Coupled Hidden Markov Model (CHMM) that tries to find a coupling joint between audio and visual sequences. The audio sequence is composed of parallel states presenting the syllables and each syllable is composed of a specific number of observations (M=5, in our case). Meanwhile, the video sequence has the same parallel construction where the states present the gestures and each state is composed of another specific number of observations determined experimentally (M=12, in our case). After training the CHMM on audio-visual sequences from a database, and when a test audio signal is generated, the system tries to find a corresponding sequence of gestures based on its own experience learnt during the training phase. The generated gesture sequence is the sequence that achieves the maximum likelihood estimation with the speech test signal. Our system shows a score of 62%, which measures the similarity between the original gesture sequence labels and the synthesized gesture sequence labels, over a test video of 8 minutes. This can be considered a good score. The proposed system is able to generate appropriate robot head gesture from speech input, which allows it to produce an automatic natural robot behavior that is almost completely absent from present-day human-robot interactions. Further work will focus on creating a triadic alignment between the speech, head gestures, and hand gestures in different human-robot interactional contexts that will allow the robot to interact naturally under different conditions.

Acknowledgment. This work is supported by the French National Research Agency (ANR) through Chaire d'Excellence program 2009 (HRIAA project).

References

1. McNeill, D.: Hand and mind: what gestures reveal about thought. University of Chicago Press, Chicago (1992)
2. Eyereisen, F.P., Lannoy, J.D.D.: Gestures and Speech: Psychological Investigations. Cambridge University Press (1991)
3. Michalowski, M.P., Sabanovic, S., Kozima, H.: Proceedings of the Human-Robot Interaction Conference, Arlington, USA, pp. 89–96 (2007)
4. Munhall, K., Jones, J.A., Callan, D.A., Kuratate, T., Vatikiotis-Bateson, E.: Psychological Science 15(2), 133–137 (2004)
5. Kuratate, T., Munhall, K.G., Rubin, P.E., Vatikiotis-Bateson, E., Yehia, H.: Proceedings of the 6th European Conference on Speech Communication and Technology (EUROSPEECH), pp. 1279–1282 (1999)
6. Valbonesi, L., Ansari, R., McNeill, D., Quek, F., Duncan, S., McCullough, K.E., Bryll, R.: Proceedings of the European Signal Processing Conference (EUSIPCO), vol. 1, pp. 75–78 (2005)
7. Quek, F., McNeill, D., Ansari, R., Ma, X., Bryll, R., Duncan, S., McCullough, K.: Proceedings of the ICCV, pp. 64–69 (1999)

8. Graf, H.P., Cosatto, E., Strom, V., Huang, F.J.: Proceedings of IEEE Int. Conf. Automatic Face and Gesture Recognition, pp. 381–386 (2002)
9. Sargn, M.E., Yemez, Y., Erzin, E., Tekalp, A.M.: IEEE Transactions on Pattern Analysis and Machine Intelligence 30(8), 1330–1345 (2008)
10. Talkin, D.: In: Kleijn, W.B., Paliwal, K. (eds.) Speech Coding and Synthesis, pp. 497–518. Elsevier (1995)
11. Chutorian, E.M., Trivedi, M.M.: IEEE Transactions on Pattern Analysis and Machine Intelligence 31(4), 607–626 (2009)
12. Viola, P., Jones, M.J.: International Journal of Computer Vision 57, 137–154 (2004)
13. Wong, K., Lam, K., Siu, W.: Signal Processing: Image Communication 18(2), 103–114 (2003)
14. Wong, K.W., Lam, K.I., Siu, W.: Pattern Recognition 34(10), 1993–2004 (2000)
15. Yip, B., Siu, W.Y., Jin, S.: Proceedings of IEEE Int. Conf. on Multimedia and Expo. (ICME), vol. 2, pp. 1183–1186 (2004)
16. Ringeval, F., Demouy, J., Chetouani, M., Robel, L., Xavier, J., Plaza, D.C.: IEEE Transactions on Audio, Speech and Language Processing 99, 1–15 (2010)
17. Arai, T., Greenberg, S.: Proceedings of Eurospeech, Rhodes, Greece, pp. 1011–1114 (1997)
18. Nickel, K., Stiefelhagen, R.: Proceedings of DAGM-Symposium, Magdeburg, Germany, pp. 557–565 (2003)
19. Moubayed, S.A., Beskow, J.: Proceedings of the International Conference on Auditory-Visual Speech Processing (AVSP), Norwich, UK (2009)
20. Rabiner, L.R.: Proceedings of the IEEE 77, 257–286 (1989)
21. Rezek, I., Sykacek, P., Roberts, S.J.: Proceedings of the International Conference on Advances in Medical Signal and Information Processing, MEDSIP (2000)
22. Rezek, I., Roberts, S.J.: Proceedings of the IEEE International Workshop on Neural Networks for Signal Processing (NNSP), Sydney, Australia (2000)
23. Nean, A.V., Liang, L., Pi, X., Liu, X., Mao, C.: Proceedings of the International Conference on Acoustics, Speech and Signal Processing, ICASSP, Orlando, USA, vol. 2, pp. 2013–2016 (2002)
24. Liang, L., Liu, X., Pi, X., Zhao, Y., Nean, A.V.: Proceedings of the International Conference on Multimedia and Expo. (ICME), Lausanne, Switzerland, vol. 2, pp. 25–28 (2002)
25. Penny, W., Roberts, S.: Technical Report TR-98-12, Imperial College London, UK (1998)

Myopia of Service Oriented Manufacturing Systems: Benefits of Data Centralization with a Discrete-Event Observer

Olivier Cardin[1] and Pierre Castagna[2]

[1] IRCCyN, 1 rue de la Noë, 44321 Nantes, France
[2] IUT de Nantes, 2 avenue du Pr J. Rouxel, 44475 Carquefou, France
olivier.cardin@irccyn.ec-nantes.fr,
pierre.castagna@univ-nantes.fr

Abstract. Service orientation paradigm is particularly well adapted to distributed manufacturing systems. The difficulty of such systems' production activity control deals with knowledge management. Indeed, the knowledge is distributed among each entity which is able to create, modify or communicate with other entities. An entity cannot have a full up-to-date access to all the system's data. At shop floor level, a convenient way to implement service-oriented manufacturing systems is to rely on the paradigm of Holonic Manufacturing Systems. The chapter introduces the possibility of specializing a resource holon in order to gather data from the whole holarchy and make these data available to any holon for a decision making. This holon is thus playing the role of a discrete-event observer. Upon positioning the service-oriented architectures, the HMS reference architecture PROSA is described, especially in terms of decision making. After decisions were defined, the problem of online decision making in a HMS is described, and an implementing solution for the observer and the forecasting tools in the architecture is exposed. Finally, two applications are presented, based on an industrial job-shop.

Keywords: SOA, Service Oriented Manufacturing, Online simulation, HMS, Production Activity Control, PROSA, Staff holon, Observer.

1 Introduction

Since the beginning of the 21st century, service oriented architectures developed, dedicated to interoperability of computer services in the companies. Along time, this architecture, mainly based on autonomy, negotiation and data distribution was transposed on the shop floor [10]. In the field of systems involving a high level of information and control distribution, Holonic Manufacturing Systems are more and more common in both academic and industrial worlds [2], and particularly well adapted to the implementation of SOA. Many decisions must be taken by individual holons during production. These decisions are based on the data that are retrieved on the HMS, but are generally insufficient for the operator to forecast the behaviour of the system. In addition, holons face the problem of myopia [1].

T. Borangiu, A. Thomas, and D. Trentesaux (Eds.): SOHOMA 2011, SCI 402, pp. 197–210.
springerlink.com

However, as shown in [13], efficiency of holonic architectures goes by the ability of holons to forecast the future behaviour of the underlying system. Thus, several authors suggest the use of ant colonies to predict the emergent behaviour of the system on short term [8, 12, 13].

The general framework of this work deals with the use of discrete-event simulation as online forecasting tool. This possibility was already exposed in [7] where it was stated that the use of tools supporting a discrete-event simulation of the production flows could ensure that no live-locks occur when various products are considered at the same time in a product-driven system. Regarding their behaviour, the production systems can mainly be seen as discrete-event systems. As a matter of fact, discrete-event simulation is a very powerful tool to model HMS.

This chapter introduces the possibility of specializing a resource holon with the objectives to gather the data from the whole holarchy and make these data available to any holon for decision making. The resource holon is thus playing the role of a *discrete-event observer*. This work is based on the PROSA holonic reference architecture, and describes the structure needed to integrate this observer, jointly with a forecasting solution in the holons called Staffs, based on online discrete-event simulation. After positioning service-oriented architectures in the first section, the HMS reference architecture PROSA is described, especially in terms of decision making. After defining decisions, the problem of online decision making in a HMS is described, and an implementing solution for the observer and the forecasting tools is exposed. Finally, two applications are presented, based on an industrial job-shop.

2 Service Oriented Manufacturing Systems

2.1 Service Orientation in the Context of Distributed Manufacturing Systems

Service orientation paradigm is particularly well adapted to distributed manufacturing systems. In this context, the system can be seen as a set of multiple autonomous entities, interacting and cooperating in a complex network to accomplish a number of tasks.

The difficulty of such systems production control resides in knowledge management. Indeed, the knowledge is distributed among entities, which are able to create, modify or communicate between them. As a matter of fact, any entity cannot have a full up-to-date access to all the data of the system.

This is why service oriented architectures (SOA) were suggested to handle the communication between the manufacturing entities. In [9] the basic conceptual model of the SOA architecture is defined; it consists of:

- *Service requesters*: typically product entities when they are realized as order entities. Order entities call on the services they require to be executed.
- *Service providers*: usually resource entities, which have the capabilities needed to provide the services requested.

- Service brokers: actors that contain the rules and logics of using the services; their function is to identify service providers for the requesters on the basis of criteria such as cost, quality, and time.

2.2 Service Oriented Manufacturing System Example

The studied HMS is a job-shop with automated transfers system, including six workstations, each of them being a transformation service provider with its own intelligence [3]. The items are carried out one-by-one by a set of 40 transporters equipped with smart tags, all considered as separate resources - thus transportation service providers. A transporter storehouse (storage service provider) is available to store unused transporters (Fig. 1).

When a new order is placed, a service requester is created, called here *order service requester*. A negotiation between this requester, the other main service requesters and the transportation service providers is established to determine how many transporters are dedicated to the order. When the decision is taken, the service requester is split into as many service requesters as items to be manufactured in the order, which are then called atomic service requesters. Then, service requesters act independently: they are able to negotiate with each transformation service provider in order to go on its recipe, until the production of the item they represent is finished, and the transportation service provider enters the storage.

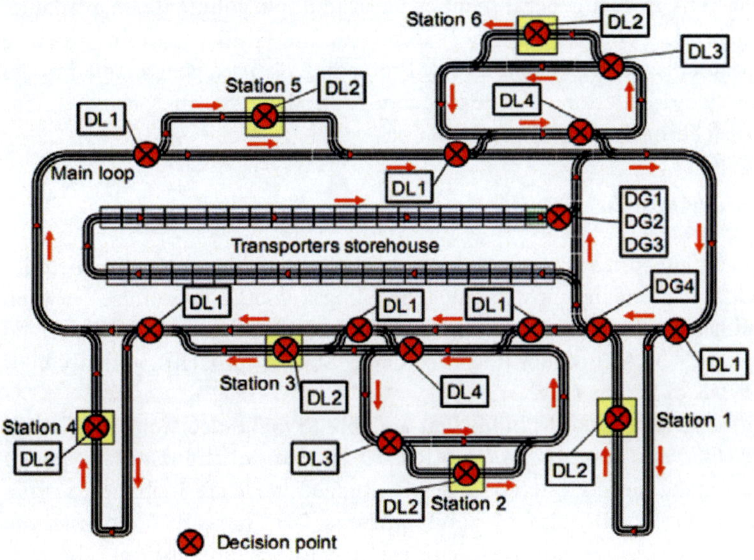

Fig. 1 The assembly line and the related decisions

As a matter of fact, to run an order, the main service requester needs to make three decisions by negotiating with the other main service requesters (Fig.1 shows the localization of all these decisions on an assembly line scheme):

- DG1: At which date will the production begin?
- DG2: How many transporters will be allocated to this order?
- DG3: Which priority is given to the order?

Atomic service requesters negotiate with the storage service provider to decide:

- DG4: Does the transporter located on the main loop at the entrance enter the storehouse or stay on the loop?

Along the production, atomic service requesters and transformation service providers negotiate to decide:

- DL1: Does the transporter located on the main loop at the entrance enter the station or stay on the loop? (particularly detailed in [4])
- DL2: At the end of an operation, is the atomic service requester authorized to step to the next operation of the recipe?

3 HMS and SOA

To implement SOA from a general point of view, multiple solutions are available. At shop floor level, a convenient way is to rely on the HMS paradigm. Existing negotiation protocols, such as Contract Net Protocol (CNP) are well adapted to the concept of service orientation. The next sections introduce the reference architecture which was chosen for this implementation.

3.1 PROSA Modelling

Many holonic architectures are exposed in the literature for the production activity control of HMS. One of the most famous, enabling a good compromise between hierarchy and heterarchy, is the reference architecture PROSA [14]. As expressed in the acronym, PROSA (Product Resource Order Staff Approach) is mainly built around four types of holons (Fig. 2).

There are only two types of holons that are able to make decisions in a HMS: *order holons* (OH) and *resource holons* (RH). It is to be noticed that the types of decisions they make are completely different. Indeed, RHs are inclined to make decisions on a very local point of view, whereas OHs have to cope with the global objectives of the system, and therefore make decisions not only on a local, but also on a global point of view. Furthermore, as each holon deals with fixed rules and local variants, the mechanisms of decision making are quite different.

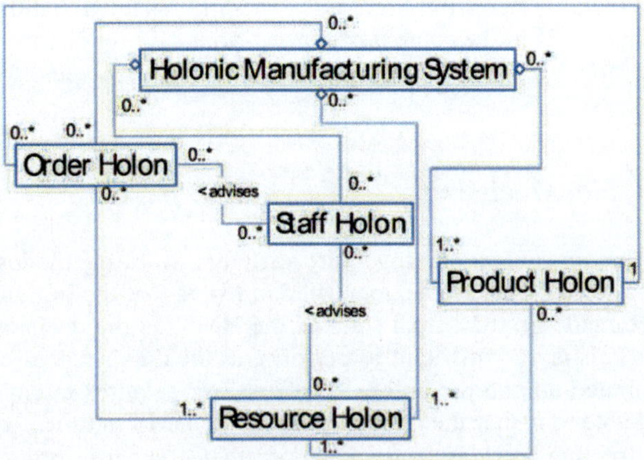

Fig. 2 Basic building blocks of a HMS and their relations, based on [11]

3.2 *PROSA in SOA*

In SOA, three main entities were identified: service providers, service requesters and service brokers. When looking carefully at PROSA, only Order and Resource Holons are making decisions. It seems also obvious to consider that Resource Holons might be identified to service providers, and Order Holons to service requesters. The service brokers are meant to support decisions. Their definition leads to an identification to Staff Holons (SH).

In order that staff holons are considered as services, they have to be accessible from different holons. Indeed, in the case of a staff holon only dedicated to the decision support of one only other holon, its definition as staff holon might be questionable. In this case, the problem is to know why would not this staff holon be integrated within the supported holon.

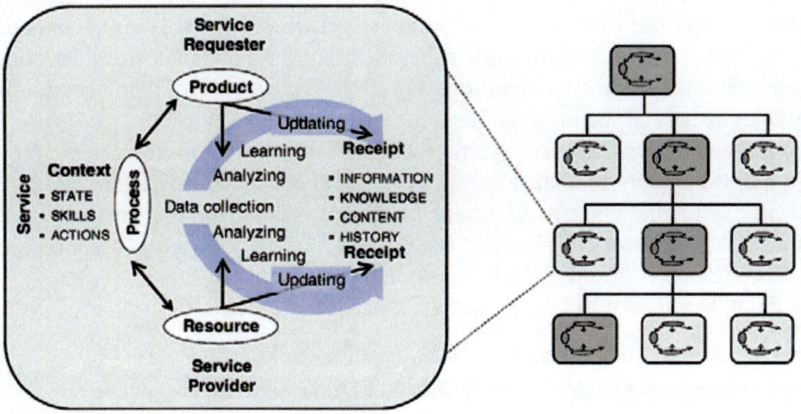

Fig. 3 An example of service in detail [9]

Concerning the Product Holons (PH) from Fig.3 it is generally assumed that the process represents the service itself in the context of service-oriented production.

In PROSA, the Product Holon representing the process data is well adapted for this analogy.

4 Centralizing Data, Not Decisions

There exist numerous forecasting tools having quite different working modes. However, the main problem for all these tools, identified in [5], is that the initialization of the tools must be made on the actual state of the HMS, as the decision horizon is generally short. This is very difficult to perform, as the data needed for this initialization are distributed among the holons. The first step requires to have at anyone's disposal a tool able to gather the updated state of the HMS in time.

The main contribution of this work is to demonstrate that a specific holon, centralizing a huge amount of data coming from all the other holons and reconstructing the missing ones, would be very helpful in the production control of HMS as support in the decision made by the other holons. This proposal is further detailed.

4.1 Gathering an Up-to-Date State of a HMS

The dynamic behaviour of autonomous decisional entities, such as those found in holonic and multi-agent systems, makes it hard to obtain performance guarantees. This difficulty is mainly due to the "myopic behavior" of distributed control systems [15]. In fact, this myopic behaviour is one of the major obstacles for using such systems. The analogy with myopia is justified since this condition causes a lack of visual acuity and can be extended to the lack of knowledge of a particular holon about the entire system.

This section only focuses on the state gathering of the HMS. The objective of this work is to derive a solution that could be applied to the widest class possible. As a matter of fact, the idea is to use as much as possible the real system's data to have as reliable as possible data. However, the set of data obtained from the control is generally not sufficient. Therefore, the idea, exposed in [6], is to use an observer to reconstruct all the missing data.

Fig. 4 presents the principles of such a solution. The observer is connected online with the entire holarchy so that it can retrieve all events occurring in the system. Then, its role is to continuously determine the missing data from the received ones. Using these data, it is also able to reset the deviation of its predictions. As a matter of fact, this observer is meant to run at the wall clock speed (real-time).

To sum up, the observer is meant to have three main specificities:

1. Gathering all the events happening on every other holons.
2. Reconstructing a probable behaviour of each holon between two occurrences of events.

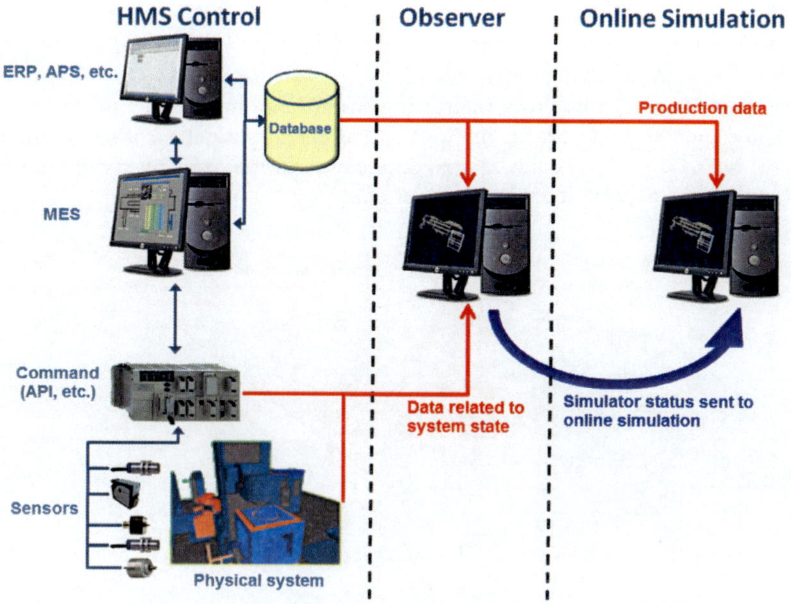

Fig. 4 Use of an observer for gathering the state of a HMS

3. Make available all or part of its state to any other holon which desires to gather data about the state of the holarchy.

A lot of technologies can be used to implement this observer, according to the objectives that were designed. To implement online simulation, the choice that should be made is to use discrete-event simulation. Indeed, a lot of simulation pieces of software meet the requirements of our study. First, it was widely used to model the behaviour of such production systems. As a matter of fact, the model that was eventually created for the design of the facility can be used again in the production phase. This implies a significant decrease of the investment time and costs. The available communication means are generally very well adapted to the communications inside such control architecture. Their graphical user interfaces enable a clear vision on the behaviour of the system for an operator. Finally, the state gathered by the observer is particularly well adapted to initialize online simulations in order to predict the future behaviour of the system.

4.2 *Integration in HMS*

As said before, online simulation tools are best situated in a staff holon. However, this staff holon may use several forecasting tools (ANN, Petri nets, etc.). These technologies might not be known by the other holons, which are only negotiating with the staff holon. It is thus necessary to split the staff holon into several resource holons, each of them representing a forecasting tool (Fig.5). In the most evolved decision making mechanisms, it is also possible to create order holons, representing the requests of the outer holons.

This organization is particularly interesting from the point of view of decision making delay. Indeed, each forecasting tool having its proper delays, a negotiation with a short lap of time before the desired application of the decision should use a different tool than a negotiation with a longer delay.

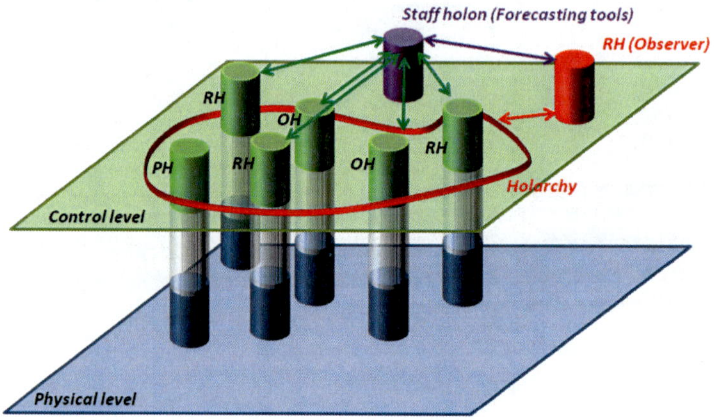

Fig. 5 Integration of online simulation in a HMS

The position of the observer is clearly that of a resource holon. Indeed, its role is to deliver the state of the system to any holon at any time. The next question is to know whether this resource holon should be included in the staff holon. Considering the necessary link between this resource and the whole holarchy for retrieving the control data, it does not seem optimal to include this into a staff holon.

Finally, with the observer considered as a resource outside the staff holon, placing an order inside the staff is modified. This means that an order's attribute is the considered state of the system, constant during the forecast inside the staff holon.

5 Applications

This section introduces two examples, showing some of the benefits for production control obtained from the use of an observer coupled with a HMS.

5.1 Application to Decision DG2

This section describes the use of an online simulation decision support tool on the HMS presented in section 2.2 to determine the best configuration of the orders placed on the system. This evaluation considers the total makespan criterion. As a matter of fact, the evaluation also gives the information of the estimated makespan of the orders, running or to be run. The singularity of this decision is that it is finally made by the operator in charge with production control of the system. The forecasting tools are only used to provide data to this operator, in order to help him make his decision.

The orders have several attributes, and among them:

- An ordered list of services needed for the completion of the item. This list is provided by the product holons.
- A number of items to treat, divided into :
  - A number of transporters resources ;
  - A number of items each transporters has to treat ;
- A running date.

The difficulty is to assign the parameters related to the number of items. Several combinations are possible, each one having positive and negative effects: for example, a low number of transporters increases the makespan avoiding the operations to happen simultaneously, whereas a high number of transporters decreases the makespan of this order, but only a few are left for the others (transporters are in finite capacity).

Fig. 6 describes the communications between holons during a simulation-based evaluation of the total makespan on the occurrence of a new order with a given parameter in a sequence diagram. This sequence is repeated as many times as necessary to evaluate every desired parameter sets.

After it was placed, the order triggers the calculation of the makespan according to the different scenarios. This calculation is limited in time by a due date parameter, which prevents the staff holon to spend too much time. Then, the staff holon gathers the state of the system and creates an atomic order holon, corresponding to the request of the original holon. This order holon negotiates within the staff holon to obtain the data before due date. When decided (in this example in favour of online simulation), the replications are run. Finally, the results are propagated to the order holon.

5.2 *Application to Decision DL1*

This section describes the use of the observer coupled with the HMS presented in section 2.2. This application deals with the use of the data available in the observer during the negotiations between holons.

The observer deployed has historical functionalities. It is able to memorize specific events, such as items treated on each station. Then, it is possible to measure the load of each station, with the objective of balancing this load. This balance is made for maintenance purpose: if two stations are able to provide the same services, the maintenance costs tend to decrease when reaching a relative balance between the stations. The section presents an academic study about the pertinence of using these data to balance the loads between stations performing similar services.

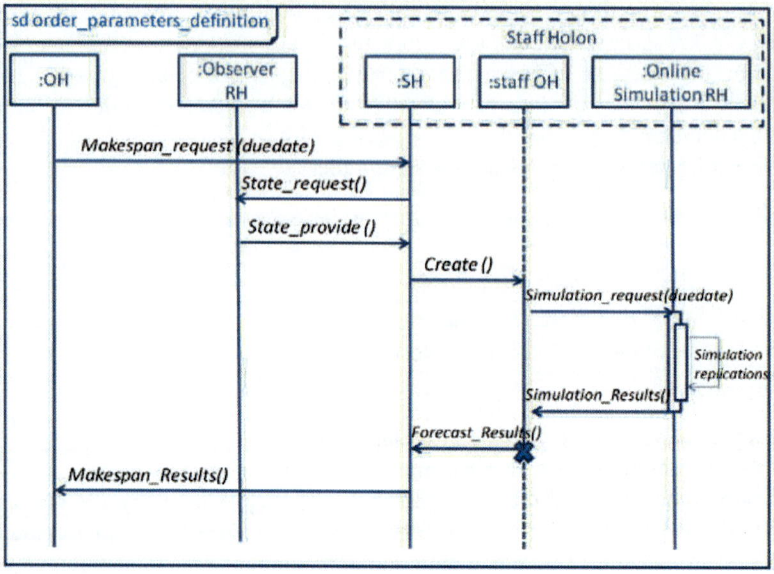

Fig. 6 Sequence diagram of the makespan calculation

Problem Definition

Table 1 presents the recipes of each item to handle.

Table 1 Recipes used for DL1 scenarios

Recipe number	Service n°1	Service n°2	Service n°3	Service n°4	Service n°5	Service n°6
1	10	20	30	40	50	60
2	50	60	60	20	30	60

Table 2 presents the data relative to the services needed by the recipes.

Table 2. Services performed for DL1 scenarios

Service number	Setup Time	Manufacturing Time
10	30	5
20	60	20
30	10	4
40	160	10
50	10	30
60	**80**	**10**

Finally, Table 3 presents the placed orders. By convention, the origin of the timeline (i.e. t=0) corresponds to the beginning of the manufacturing of these orders.

Table 3 Orders placed for DL1 scenarios

Order Number	Number of items	Number of items by transporter	Recipe
1	100	5	1
2	100	5	2

Scenarios

Three different scenarios were tested and are described in Table 4. The first scenario is a reference scenario, where all the services are exclusively provided by one station. The difference between scenarios 2 and 3 comes from the presence of the load balance objective in the negotiation between requesters and providers.

Table 4 Scenarios of DL1

Scenario Number	Load balance	Services performed by station					
		1	2	3	4	5	6
1	No	10	20	30	40	50	60
2	No	10,60	20	30,50	40	50	60
3	Yes	10,60	20	30,50	40	50	60

Results and Discussion

The results presented in Table 5 are expressed in seconds. The percentages relative to the total makespan are not expressed to ease the reading of the data.

Table 5 Results of the simulation

Performance indicator	Scenario Number		
	1	2	3
Load of station 1	530	1020	3460
Load of station 2	4060	4060	4060
Load of station 3	810	3970	3970
Load of station 4	1160	1160	1160
Load of station 5	6010	3820	3820
Load of station 6	4080	4030	3460
Cmax	6710	4784	4908

The results of scenario number 1 indicate that stations 1, 3 and 4 have a load strongly inferior to stations 2, 5 and 6. For the purpose of this study, a choice was made to try and balance the load of stations 1 and 6 on one side, and stations 3 and 5 on the other side (Table 4).

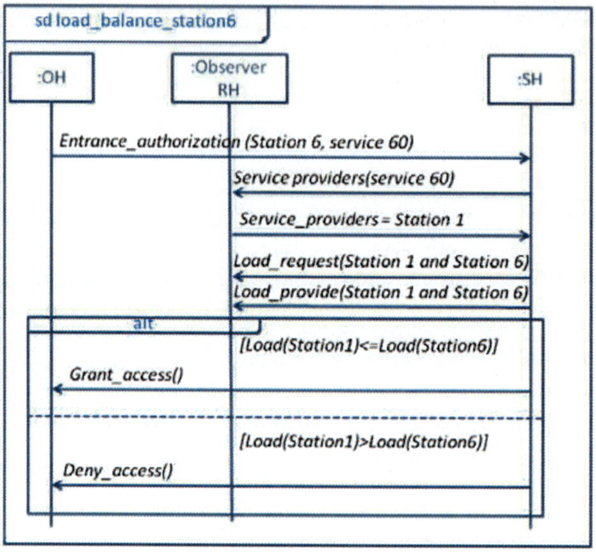

Fig. 7 Sequence diagram of the load balance negotiation mechanism

This balance is first performed by simply declaring that stations 1 and 3 are respectively able to perform services 60 and 50 (scenario number 2). Looking at the results of this scenario, the balance between stations 3 and 5 is globally acceptable, whereas the results of station 1 are not sufficient (still 4 times less load). Moreover, looking carefully at the results, it is possible to determine that the time saved on station 6 was almost integrally spent on station 1 for setups.

The difference between these balances can be explained by the topography of the HMS. Indeed, station 6 has a bigger buffer than station 5. Thus, very few items are not accepted by station 6 in the usual control, when a lot of items have to make supplementary loops on the central loop because of the lack of space in the buffer of station 5.

Thus, the decision was taken to implement a load balance decision rule in the negotiation between the service requesters and providers. This decision (Fig.7) is made after the provider (materialized by a resource holon) granted its access to the requester (materialized by an order holon).

The chosen rule was the most simple: when the load of station 6 is greater than station 1, the items are not accepted in station 6. The results show that this rule is very efficient for the load balance (scenario number 3). However, the Cmax indicator shows that some performance indicators are obviously degraded. This can be explained by the simplicity of the rule, which does not take into account anything else than the load balance.

6 Conclusion and Future Works

This chapter highlights the benefits of data centralization instead of decision centralization for the production activity control of service oriented manufacturing systems, more precisely their implementation as Holonic Manufacturing Systems. This data centralization is suggested to be performed using a discrete-event observer.

The concept of observer, enabling for example the forecasting tools to initialize on the actual state of the system was presented, and its integration in the HMS was explained. This observer can be seen as a specific resource of the holarchy, communicating with all the other holons in order to retrieve all the data necessary to reconstruct the complete actual state of the HMS.

Furthermore, an example of the benefits that can be provided in terms of production activity control from the use of an observer coupled with a HMS on an industrial FMS was developed, in order to exhibit the timed communications between all the holons involved in the production activity control (decision makers and decision supports). With very simple rules, a second example also shows how this data centralization can make the load balancing between several machines of a job shop easy and efficient.

The future developments of this work will deal with the generalization of the approach to other control architectures, such as hybrid energy production systems. Indeed, the reconstruction of the state between two consecutive events is quite different, and makes the behaviour of the observer more difficult to model.

References

1. Adam, E., Zambrano, G., Pach, C., Berger, T., Trentesaux, D.: Myopic Behaviour in Holonic Multiagent Systems for Distributed Control of FMS. Adv. Soft. Comp. 90, 91–98 (2011)
2. Blanc, P.: Holonic control of a laminated security glass production system, PhD Thesis, Université de Nantes (2006)
3. Cardin, O.: Contribution of online simulation to production activity control decision support – Application to a flexible manufacturing system, PhD Thesis, Université de Nantes (2007)
4. Cardin, O., Castagna, P.: Handling uncertainty in production activity control. In: Proceedings of 12th IFAC Symposium on Information Control Problems in Manufacturing, INCOM 2006, Saint-Etienne, pp. 579–584 (2006)
5. Cardin, O., Castagna, P.: Using online simulation in holonic manufacturing systems. Engineering Applications of Artificial Intelligence 22, 1025–1033 (2009)
6. Cardin, O., Castagna, P.: Proactive production activity control by online simulation. Int. J. Sim. Process. Model. (2011)

7. Gouyon, D., Pétin, J.-F., Morel, G.: A product driven reconfigurable control for shop floor systems. Stud. Inform. Control 16 (2007)
8. Karuna, H., Valckenaers, P., Saint-Germain, B., Verstraete, P., Zamfirescu, C.B., Van Brussel, H.H.: Emergent Forecasting Using a Stigmergy Approach in Manufacturing Coordination and Control. In: Brueckner, S.A., Di Marzo Serugendo, G., Karageorgos, A., Nagpal, R. (eds.) ESOA 2005. LNCS (LNAI), vol. 3464, pp. 210–226. Springer, Heidelberg (2005)
9. Nylund, U., Andersson, P.H.: Simulation of service-oriented and distributed manufacturing systems. Robot Cim.-Int. Manuf. 26, 622–628 (2010)
10. Trentesaux, D.: Distributed control of production systems. Engineering Applications of Artificial Intelligence 22, 971–978 (2009)
11. Valckenaers, P., Van Brussel, H., Wyns, J., Bongaerts, L., Peeters, P.: Designing Holonic manufacturing systems. Robot Cim.-Int. Manuf. 14, 455–464 (1998)
12. Valckenaers, P., Van Brussel, H.: Holonic Manufacturing Execution Systems. CIRP Ann.-Manuf. Techn. 54, 427–432 (2003)
13. Valckenaers, P., Karuna, H., Saint Germain, B., Verstraete, P., Van Brussel, H.: Emergent short-term forecasting through ant colony engineering in coordination and control systems. Adv. Eng. Inform. 20, 261–278 (2006)
14. Van Brussel, H., Wyns, J., Valckenaers, P., Bongaerts, L., Peeters, P.: Reference architecture for holonic manufacturing systems: PROSA. Computers in Industry 37, 255–274 (1998)
15. Zambrano, G., Pach, C., Adam, E., Berger, T., Trentesaux, D.: Myopic Behaviour in Heterarchical Control of FMS. In: Proceedings of Int. Conf. on Ind. Eng. and Syst. Man, IESM 2011, Metz (2011)

A Multi-agent Model for Job-Shop Scheduling

Gabriel Neagu

National Institute for Research and Development in Informatics - ICI, Department of Advanced Research and Modelling, B-dul Averescu 8-10, 011455 Bucharest, Romania,
e-mail: gneagu@ici.ro

Abstract. The chapter is devoted to a multi-agent solution for the job-shop scheduling problem and its current migration to a service oriented platform. The co-ordination mechanism for the agent population and the Decision Control Petri Nets formalism for the behaviour modelling at agent level are introduced. In the framework provided by the Generic Prototyping Approach, the chapter presents the overall orientation of the manufacturing shop control system and identifies the generic prototypes devoted to decision support for resource allocation and activity monitoring at the shop floor level. The design specifications for the job-shop scheduling particular prototype developed according to the multi-agent paradigm are further detailed. Finally, in the context of building up a Grid computing platform for workflow oriented applications, the feasibility of the service oriented implementation of this particular prototype is evaluated.

Keywords: Multi-agent system, job-shop scheduling, DCPN, Service Orientation.

1 Introduction

The multi-agent paradigm has been widely used in the analysis, modelling and design of complex systems with intensely interacting components. A multi-agent system is composed of a set of agents, evolving in a dynamic environment that allows the coordination of their activities. As well as the object-oriented paradigm, the multi-agent paradigm has emerged from the software engineering field, where the agent concept is considered synonymous with the actor model in the concurrent object-oriented programming [1]. The agent was defined as an entity provided with the ability to perceive and represent the environment in which it is placed. It interacts with other agents and has an autonomous behaviour, depending on its observations, resources and interactions with others [3].

In manufacturing, the utilization of the multi-agent approach has been frequently reported since the early '90s. A survey of multi-agent systems applications in production design and development, production planning and control and supply chain management is provided in [8]. More specifically, the advantages offered by agent-based approaches for distributed manufacturing process-planning and scheduling systems, like modularity, reconfiguration, scalability, upgradeability, and robustness are emphasized in [17]. For dynamic scheduling in

T. Borangiu, A. Thomas, and D. Trentesaux (Eds.): SOHOMA 2011, SCI 402, pp. 211–225.
springerlink.com

manufacturing, a comparative study of several dynamic scheduling methods lists the advantages of multi-agent systems [16]. More recently, the advantages of the service orientation in implementing the agent-oriented systems have been reported [18].

The purpose of this chapter is two-fold: to present a job-shop scheduling solution based on the multi-agent approach, and to propose a solution for implementing this approach on a service oriented platform.

Section 2 presents two formerly proposed building blocks of the multi-agent approach: the multi-agent coordination mechanism MACOR [12] and the Decision Control Petri Nets (DCPN) formalism [13]. MACOR is based on the blackboard paradigm and plays the role of simulation engine for MASIM (Multi-Agent SImulator in Manufacturing), devoted to generating scheduling solutions adapted to the job-shop functional specificity. The DCPN formalism is used for the event oriented modelling of agent behaviour. Section 3 introduces the methodological framework provided by the Generic Prototyping Approach (GPA) [10, 11]. An overview of the Decision Support for Manufacturing Control (DSMC) generic system is then provided. Two decision support oriented generic prototypes are identified, dedicated to resource allocation problems and to monitoring. For the first generic prototype, the design specifications of the object model and the functional assumptions for the job shop (including resource allocation) are outlined, and the criteria for identifying particular prototypes are introduced. Section 4 presents the DCPN-based design specifications of the control model for a particular prototype for job-shop scheduling. To illustrate the solution a detailed case study for job object class is provided. Section 5 is dedicated to service oriented implementation prospects of this multi-agent model. The chapter ends with some concluding remarks and orientations for further work in the context provided by the recent installation of a powerful cluster for Grid and cloud computing [4].

2 Building Blocks of the Multi-agent Solution

2.1 MACOR - a Multi-agent Co-ordination Mechanism

The MACOR solution is based on the blackboard paradigm. According to this paradigm, a centralized data structure serves as communication support among modules (knowledge sources) for the cooperative problem solving. The main components of a system build up according to this paradigm are the following: (1) the blackboard playing the role of both the problem solving model and communication medium; (2) knowledge sources that incorporate the problem solving expertise; and (3) control strategy for the knowledge sources selection and activation according the current state of the blackboard. In a multi-agent environment, agents act as knowledge sources, as soon as their behaviour and related decision resources encapsulate the domain-specific problem solving knowledge. The control strategy defined for the agents community plays the role of the simulation controller working on a shared data structure and interacting with the agents.

Considering the specificity of discrete event processes behaviour, the event oriented schema was selected as the control strategy. According to this schema, one or more knowledge sources (agents) are activated depending on the current state of the blackboard. As a result, one or more state transition events are generated. A new event is selected in the current context and knowledge sources matching its significance are activated. Knowledge sources are considered schedulable entities.

In the adaptation of this control mechanism the focus was put on the compatibility with the DCPN formalism describing the agent behaviour.

To explain the proposed coordination mechanism it should be mentioned first that depending on the functional complexity of an object class identified in the object oriented (structural) model of the system, there may be more than one agent corresponding to it in the control model. The extension of the multi-agent paradigm facilitates the coordination of parallel processes identified in the behaviour of the respective object class, which are synchronized by specific events.

The coordination mechanism is based on the *stimulus* concept – a set of events linked by logical operators. The following criteria have been defined to structure the set of the relevant events:

- *event visibility*, to characterize its influence on the agent population: agent (local influence), object (set of agents modelling the same object), and system (the whole agent population);
- *event role*: active or passive role depending on its contribution to the system evolution;
- *event time*: past, current or future events.

Object and system events are active (stimulus type) events. Active events, irrespective of their visibility level, are included in the stimuli structure and stored at the coordination (blackboard) level. The passive events have limited visibility.

Each stimulus includes one or more condition events and one control event. At the moment of next stimulus identification the condition events are past events included in the blackboard sequence of events, which describes the blackboard current status as a stimuli activation context. The control events are included in the future events list.

The procedure of the stimulus activation starts with the identification of the proximate control event in the list of future events. Then, for all stimuli configurations built around this type of control event the availability of condition events in the blackboard current status sequence is checked. If more than one candidate stimulus is identified, the control strategy of the blackboard has to implement specific selection rules. For example, the stimulus with the largest number of condition events is selected, as this is the most relevant for the current blackboard state.

All events of the activated stimulus are switched in the status of current event. Under the action of the current control event, one or more agents are changing their status. New condition events are generated accordingly, which are included in the blackboard current status sequence. Then, all current condition events (belonging to the active stimulus) are removed from the blackboard current state sequence and are stored in the sequence of past condition events describing the system evolution. Finally, a new control event is generated and included in the future event list, while the current control event is transferred to the sequence of past control events representing the aggregated simulation solution at the system level.

In case of more candidate stimuli, the next one is selected, activated and fired according to the same procedure. Then, the next control event in the future event list is selected. If no candidate stimulus is available for activation the control event is removed form the future event list and its control capability is reconsidered.

Based on the sequence of past condition events various views on the system evolution may be generated depending on the intended level of description granularity or the target agent population. To detail this evolution at the agent level, additional passive events may be generated based on the existing condition events.

2.2 *The DCPN Formalism*

The *Decision Control Petri Nets* formalism has been defined as an extension of the Control PN model [9] to meet specific requirements for the DSS development:

$$DCPN = (CdPN, E, OP, f_K, f_T, R_D, T_D, f_D) \quad (1)$$

where: *CdPN* is a coloured PN; *E* is a finite set of events; *OP* is a finite set of process execution and control operators; f_K: $K \rightarrow E$ is a token function (with *K* the finite set of tokens moving in the net); and f_T: $T \rightarrow OP$ is a transition function (with *T* the finite set of transitions). These elements define a Control PN. R_D is a finite set of decision resources (belonging to problem solving knowledge), for the identified token, transition or place conflicting situations; T_D is a finite subset of decision transitions ($T_D \subset T$) related with these conflicting situations; f_D: $T_D \rightarrow R_D$ is a function associating to each decision transition the decision resources required to solve corresponding conflicting situations.

In the DCPN model, the transitions are fired by stimuli activated at the coordination level. When a transition is fired, all associated operators (including decision resources) are executed and new events are generated as candidates for future stimuli.

In the framework of the PN formalism the conflicting situations that could arise during a transition firing are characterized as token, transition or place conflicts.

To cope with these conflicting situations the DCPN model proposes a class of *decision transitions*. For these transitions the set of associated operators includes specific decision resources (algorithms, rules) for solving related conflicts. A specific graphical representation is used for each type of decision transition.

3 Generic Prototyping in Manufacturing Control

3.1 *Methodological Framework*

The *Generic Prototyping Approach (GPA)* belongs to the class of modelling approaches having in common the stress put on the abstract description of the future system as a major support of the development process. The approach complies with major orientations in the complex system engineering regarding: the end-user involvement into the system development process starting with the early requirement definition phase; the open system architecture as a prerequisite for gradual improvement of system functionality and performances; the object-oriented analysis and design with major benefits regarding the natural matching between the reality and the model, the abstraction power, the structuring flexibility and interaction flexibility, the robustness related to later changes in system requirements.

The *GPA life-cycle* includes the prototyping, integration and instantiation processes. The GPA prototyping and integration processes deal with the generic prototype (GP) development, while the GPA instantiation process refers to GP tuning, deployment and maintenance. The first two processes, which are of main interest for this work, are represented in Fig. 1.

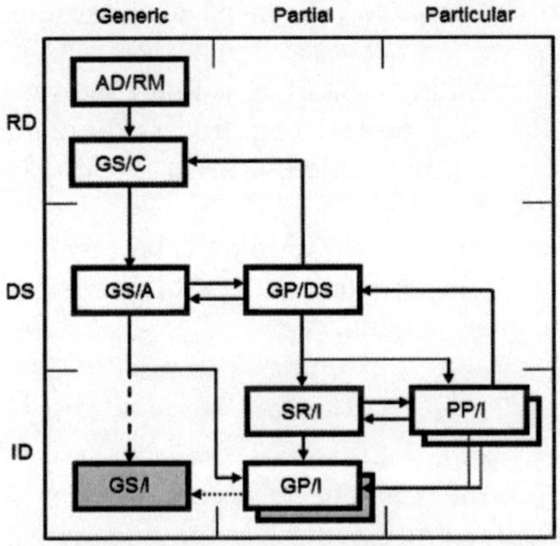

Fig. 1 GPA prototyping and integration processes

The GPA processes are defined in the framework provided by life cycle dimension and generic dimension of the of the CIMOSA methodology [7], where the first one is structured on requirement definition (RD), design specification (DS) and implementation description (ID) phases, while for the second one the generic, partial and particular stages are defined.

The *GPA prototyping process* has diagonal orientation as reported to CIMOSA framework, with emphasis put on the object oriented approach for the RD and DS phases. The process starts with the RD phase, which is placed at the generic stage and is structured on three main activities: (a) functional analysis of the application domain; (b) conceptual modelling of the generic system (GS/CM); and (c) identification of generic prototypes. The main results of the first activity are the reference model of the application domain (AD/RM), including the functional decomposition diagrams, and list of entities of the future GS. The second activity provides the conceptual model of the generic system (GS/CM), including the structural and behavioural sub-models according to the object oriented approach. The third activity aims at defining the development strategy of future GS through its structuring in generic prototypes and defining their implementation priority.

The identification of GPs is based on their functional relevance and consistency as regard the GS functionality. The implementation priority is defined according to the mutual conditionality between functional profiles of various GPs.

The aim of the second (DS) phase is two-fold: provide the overall architecture of the generic system (GS/A) as a prerequisite for achieving integrating capabilities of GPs, and to elaborate the detailed design specifications of the selected generic prototype (GP/DS).

During the third (ID) phase, the structural (data) resources are implemented at partial stage, considering their integration capabilities for the GP functional components. Then, these components are developed at the particular stage. They are called particular prototypes (PPs) and play the role of the GP building blocks that validate the implementation solutions for the high priority GP modules. The back arrows in Fig. 1 illustrate the opportunity of an iterative refinement of the results generated by the prototyping process.

The *GPA integration process* aims at implementing the GP, based on a relevant set of the available PPs developed during the GPA prototyping process. Further, the GS implementation phase of this process may be tackled, using a set of GPs compliant with the existing GS/CM and GS/A.

The decision of developing and integration of a new GP should be justified based on the analysis of the cost/performance gain reported to the user's satisfaction and requirements generated by the current GS version.

3.2 DSMC_A Generic Prototype

The *Decision Support for Manufacturing Control* is a generic system devoted to providing specific decision support for manufacturing control at the operational level. According to the GPA framework, the conceptual model of the generic system is the result of the requirement definition phase of the GPA prototyping process. In case of the DSMC GS, the conceptual model is structured on Job execution planning, Job execution dispatching and Manufacturing shop supervision components. For each component, the specific decision problems are investigated, i.e. *job execution scheduling*, *solving unpredictable events* and *activity monitoring*, respectively.

The aim is to identify specific decision support resources and, on this basis, to delimit generic prototypes which are subject of design specification at the second stage of the GPA prototyping process.

According to this taxonomy of decisional problems, two generic prototypes were identified: *DSMC_A for resource allocation problems* (scheduling, dispatching rules) and *DSMC_M monitoring problems* (identification and diagnosis of unpredictable situations, proposal of corrective actions).

During the GPA design specification phase, two main objectives have to be achieved: the object (structural) model for the generic prototype and the control models for particular prototypes.

For the DSMC_A GP the design specifications for the object model are structured on four views regarding the manufacturing shop: configuration, activity profile, functioning scenarios, and performance evaluation.

The adopted functional assumptions for the manufacturing shop aim to cover the entire *diversity of resource allocation decisions*: an idle machine at the group level allocated to the arrived job, a waiting job in the input queue allocated to the released machine, an operation waiting for execution allocated to the released robot working for the given machine, a machine group (including the current one among the alternatives for the next operation) allocated to a job with accomplished current operation, an idle transport device allocated to the job requesting transportation, a job waiting in the output queue allocated to the released transport device.

Also, it is assumed that the operation time is divided in loading (including set-up), execution and unloading phases. In case of a robotized operation, the robot allocated to the given machine has to be used for loading and unloading phases (or for entire operation as a particular case).

To expedite its transportation for the next operation, the job formulates the request for transport device at the beginning of the unloading phase for the current operation. Transport devices are administrated in a centralized manner. A released transport device goes to the parking place if there are no more requests to be processed.

In case of the scheduling problem, the manufacturing shop type (e.g. job-shop, flow-shop, open-shop) is relevant for PPs identification. Each PP is dedicated to a specific manufacturing shop.

Their integration in the DSMC_A GP aims at covering the full functional diversity of the manufacturing shop and is supported by the common object (structural) model at generic prototype level.

For the job-shop scheduling PP the design specifications of the control model are presented in the next subsection.

4 Job-Shop Scheduling Particular Prototype

4.1 Design Specification of the Control Model

For the GPA particular prototypes, the relevance of the control model is generated by its capability to reflect the PP specificity at the generic prototype level. The control component specifications include the DCPN diagrams for all relevant object classes identified in the GP object model: machine, machine group, transport device (e.g. AGV), robot (manipulator), input queue, output queue, storage cell, job, technological operation. According to the implementation specificity mentioned in Sect. 2.1, six additional agents were identified:

- *m2* - the allocated machine setting to waiting status the start of a planned maintenance period;
- *td2* - the allocated transport device setting to waiting status all coming transport requests;
- *td3* - the allocated transport device setting to waiting status the start of a planned maintenance period;
- *r2* - the allocated robot (manipulator) setting to waiting status all coming manipulation requests;

- *r3* - the allocated robot setting to waiting status the start of a planned maintenance period;
- *j2* - the job in the unloading status after the execution of the current technological operation waiting for the allocation decision regarding the next operation.

The behaviour synchronization of different agents belonging to the same object class is implemented at the transition level through common (convergence) output places.

A synthesis of the job-shop control design specifications is given in Table 1.

From the total of 94 transitions there are 8 decision transitions at the overall model level: 2 (token type conflict) for the first machine agent, 2 (token type) for the first transport device agent, one (token type) for the first robot agent, one (token type) and one (transition type) for the job agent 1, one (place type) for the job agent 2.

The overall number of places at the job-shop level is less than the sum of places for each agent because of the convergence places mentioned above.

The amount of the 98 events generated by transition firing is structured as following: 35 passive events and 63 active events (16 control events, 37 condition events and 10 control/condition events depending on the stimulus configuration). To the number of condition events should be added 4 additional external events, generated at the upper level of production control: job processing in the job-shop started, planned machine maintenance period stared, planned robot maintenance period stared, planned transport device maintenance period stared. From the visibility point of view there are 16 object level events and 47 system level events.

Table 1 Major parameters of the job-shop scheduling design specifications

Object	Agents	Places	Transitions (decision transitions)	Accepted stimuli	Generated events
Machine	*m1*	11	18(2)	30	28
	m2	2	3	4	2
Machine group	*mg*	2	2	2	2
Transport device	*td1*	8	16(2)	21	19
	td2	2	5	3	2
	td3	1	1	1	1
Robot	*r1*	3	4(1)	5	5
	r2	3	5	4	2
	r3	1	1	1	1
Input queue	*iq*	4	4	3	4
Output queue	*oq*	4	4	2	4
Storage cell	*sc*	2	2	4	2
Job	*j1*	14	12(2)	15	13
	j2	15	12(1)	8	13
Technological operation	*to*	6	5	10	5
TOTAL	15	72	94	86	98

86 stimuli have been configured that emphasize the control model complexity. Each stimulus includes up to 6 events. The level of interdependencies between PP components is illustrated by the number of stimuli acting on more than one agent: there are 4 stimuli acting on 2 agents belonging to the same object class, 12 stimuli acting on 2 agents belonging to 2 different object classes, 2 stimuli acting on 3 agents from different object classes, and 2 stimuli acting on 3 agents from different object classes.

4.2 A Case Study: The Job Object Class

The DCPN diagrams for the agents *j1* and *j2* are presented in Fig. 2 and Fig. 3, respectively. The decision transitions (see Sect. 2.2) have specific graphical representations. In Fig. 2, the output places for the transition t^{j1}_8 are alternative ones: the right one is selected according to the diagram for the *j2* agent (see Fig. 3).

The description of places, transitions and events for the *job* object class is respectively provided in the Table 2, Table 3 and Table 4. In these tables the term *preliminary* refers to the period of unloading the job from the current machine after the execution of the current technological operation.

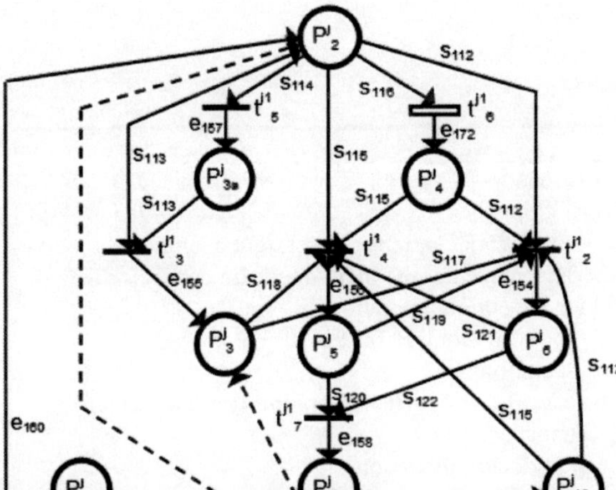

Fig. 2 The DCPN diagram for the *j1* agent

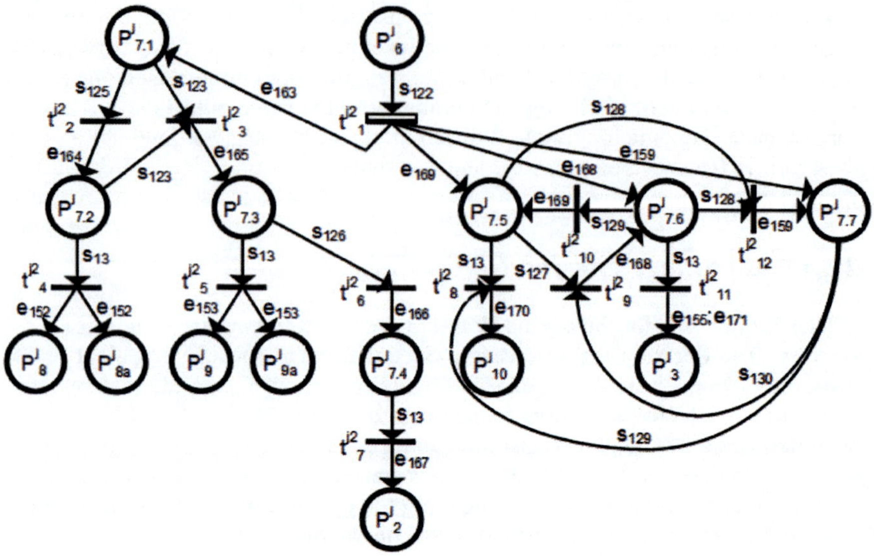

Fig. 3 The DCPN diagram for the *j2* agent

Table 2 List of places for the *job* object class

Place no	Place description
P^j_1	waiting for the first operation
P^j_2	in transit for the next operation
P^j_3	waiting in the input queue the work centre for the current operation
P^j_{3a}	waiting on the transport device to enter the input queue (in case it is full)
P^j_4	with machine allocated for the current operation
P^j_5	waiting for the available robot (if required for the current operation)
P^j_6	execution of the current operation
P^j_7	being unloaded after the execution of current operation (generic status)
$P^j_{7.1}$	and after requesting a transport device
$P^j_{7.2}$	and waiting for transport device allocation
$P^j_{7.3}$	and waiting for the allocated transport device
$P^j_{7.4}$	with available allocated transport device
$P^j_{7.5}$	and preliminarily reallocated on the current machine with free input queue
$P^j_{7.6}$	and preliminarily waiting for the reallocation on the current machine
$P^j_{7.7}$	and preliminarily reallocated on the current machine with full input queue
P^j_8	waiting for transport device allocation
P^j_{8a}	same, in case of the last operation
P^j_9	waiting for the allocated transport device
P^j_{9a}	same, in case of the last operation
P^j_{10}	reallocated on the same machine for the next operation
P^j_{11}	finished

Table 3 List of transitions for the *job* object class

Transition no	Transition description
$^{*}t^{j1}_{1}$	job enters the shop
t^{j1}_{2}	start job execution of the current operation
t^{j1}_{3}	start job waiting for the current operation
t^{j1}_{4}	start job waiting for the robot allocated to the current operation
t^{j1}_{5}	start job waiting on the transport device in front of the busy input queue
$^{*}t^{j1}_{6}$	select available machine from the machine group
t^{j1}_{7}	end of the execution of the current operation and start unloading the job
t^{j1}_{8}	end of the current operation
t^{j1}_{9}	start job waiting for the transport device to the next operation
t^{j1}_{9a}	start job waiting for the transport device to the shop exit
t^{j1}_{10}	start job transit to the next operation
t^{j1}_{10a}	start job transit after the last operation
$^{*}t^{j2}_{1}$	select machine / group for the next operation – simultaneous with t^{j1}_{7}
t^{j2}_{2}	start job preliminary waiting for transport device allocation
t^{j2}_{3}	start job preliminary waiting for the allocated transport device
t^{j2}_{4}	start job waiting for transport device allocation - simultaneous with t^{j1}_{8}
t^{j2}_{5}	start job waiting for the allocated transport device - simultaneous with t^{j1}_{8}
t^{j2}_{6}	start availability of the allocated for the next operation transport device
t^{j2}_{7}	start job transit on waiting available transport device - simultaneous with t^{j1}_{8}
t^{j2}_{8}	job reallocation on the current machine - simultaneous with t^{j1}_{8}
t^{j2}_{9}	start job preliminary waiting for reallocation on the current machine
t^{j2}_{10}	job preliminary reallocation on the current machine with free input queue
t^{j2}_{11}	start job waiting reallocation on the current machine - simultaneous with t^{j1}_{8}
t^{j2}_{12}	job preliminary reallocation on the current machine with full input queue

$^{*}t^{i}_{j}$ – decision transition

Events and stimuli are not numbered sequentially at the system level: to improve flexibility and clarity in design specification description, larger index sequences have been pre-allocated for each object class.

Table 4 List of events generated for the *job* object class

Event no	Event description
$\underline{e}_{151}$	job processing in the shop started (external event)
$\underline{e}_{152}$	job waiting for the transport device allocation started
$\underline{e}_{153}$	job waiting for the transport device started
e_{154}	job execution of the current technological operation started
e_{155}	job waiting for the execution of the current operation started
e_{156}	job waiting for the robot allocated to the current operation started
e_{157}	job waiting for the execution of the current operation being located on the transport device started
e_{158}	job unloading for the current operation started
e_{159}	job preliminary reallocated on the current machine with full input queue
e_{160}	job transit to the next operation after waiting for the transport device started

Table 4 (*continued*)

e_{161}	job waiting for the transport device in case of transit to the shop exit started
e_{162}	job transit to the shop exit started
$\underline{e}_{163}$	job request for transport device sent
e_{164}	job preliminary waiting for the allocation of the transport device started
e_{165}	job waiting for the allocated transport device started
e_{166}	allocated transport device available
$\underline{e}_{167}$	job transit to the next operation started
e_{168}	job preliminary waiting for reallocation on the current machine started
e_{169}	job reallocated on the current machine with free input queue
$\underline{e}_{170}$	job reallocated on the current machine
$\underline{e}_{171}$	unloaded job entered in the input queue
$\underline{e}_{172}$	job selected an available machine from the current group

$\underline{e}_i$ – control events

The diagrams in Fig. 2 and Fig. 3 are illustrative for all decision transition types defined in Sect. 2.2 (see Table 3). The transition $t^{j1}{}_{1}$ is a token type decision transition, as the priority job is selected for transportation to the machine group (if the transport device is available) or for allocation of the busy transport device. The transition $t^{j1}{}_{6}$ is a decision transition for the transition type conflict, as the job has to select the machine for the current operation among several available ones. The transition $t^{j2}{}_{1}$ is a place type decision transition in case the list of machine groups for the next operation includes the group for the current operation.

According to Table 4, 36.3% of the generated events for the *job* object class are control events. This proportion confirms the relatively passive role of this object class in the job-shop scheduling execution as compared with classes representing the job-shop configuration: *input queue* for machine group (100% control events), *machine* (64.5%), and *transport device* (60.8%).

5 Towards a Service Oriented Implementation

5.1 PEGAF Platform

The PEGAF platform provides the support for the implementation of DSMC_A GP in a distributed environment according to the service oriented paradigm. The platform is the result of the research project "*Experimental Grid platform for development of workflow oriented applications with dynamic allocation of resources*", aiming at designing and implementing a service based environment, on the Grid infrastructure. The main objective of the project is to support e-science workflow management and Grid service orchestration and scheduling.

The Grid service concept extends the Web service technologies by introducing life cycle management capabilities, statefulness (maintain state between invocations), a naming based identification and dynamic instantiation. A Grid service is basically a Web service able to store state information persistently rather than transiently beyond the lifetime of a single request. The Open Grid Services

Architecture (OGSA) is a set of standards defining the architecture for service-oriented Grid computing [5].

The software architecture implemented in the PEGAF project consists in a platform fulfilling the following functional requirements: service orientation, coverage of the entire workflow life cycle, usage of open source solutions, compliance with gLite middleware.

The main components of the platform are: the workflow specification component, the workflow execution engine component, and an OGSA compliant service environment dedicated for Grid service discovery, monitoring, execution and life cycle management.

The last component is of main interest for the DSMC_A GP implementation. Its architecture includes a Service Orchestration module, a Binding module for the integration with the Grid middleware, and an Information Model Persistence module. The Service Orchestration module of the PEGAF platform was implemented as a service container conforming to the OGSA basic principles. The proposed architecture integrates an OGSA information model manager and a stateful WSRF compatible service container. The stateful service container is built using the Apache Tomcat servlet server, Axis2 SOAP server and the WSRF and WS-* stack implementation provided by the Apache Muse framework. Apache Orchestration Director Engine (ODE) [http://ode.apache.org/] is an open-source workflow engine under the patronage of the Apache Software Foundation that executes business processes written following the WS-BPEL standard.

The Binding module is based on the Basic Execution Service (BES) [6]. There are already some solid implementations of this model, like Globus BES, UNICORE BES and the gLite compatible CREAM-BES.

The OGSA information model for Grid resources and services is provided by the GLUE 2.0 Specification [2]. The Information Provider Service is aware of the state of the manageable resources (CREAM-BES), is able to publish advertisements regarding this state and to provide the information in the GLUE Schema database to the other components of the platform.

5.2 Implementation Feasibility

Due to the direct compliance between the agent and the software service concepts, the service oriented implementation of the design specifications for the DSMC_A GP is straightforward. The only major impact of this implementation paradigm shift regards the co-ordination mechanism. The centralized MACOR solution should be reconsidered in the sense that the co-ordination responsibilities should be distributed at the service level, on the support of message based interaction.

To solve a concrete job-shop scheduling problem, all software services implementing the relevant agents for this problem are discovered and orchestrated by the workflow engine. Then, the workflow execution is driven by interaction between agents. There are two types of messages supporting this interaction: information messages and control messages. To identify them the list of active events provided by the PP design specifications is used.

An information message corresponds to a condition type event. It is generated by an agent at the moment its state changes and is sent to all other agents that are sensitive to this change. A control message corresponds to a control type event, while the role of condition events that enable its activation is played by the formerly received information messages. The following coordination rules have been defined:

- a control event generates a control message which is sent to all target agents;
- each recipient agent changes its current status according to the control message;
- at this moment, an information message is generated and is sent to all other agents which have to be aware about this status change;
- each control message is sent also to an archiving service which provides at the end the scheduling solution.

6 Conclusions

The proposed GPA methodological approach facilitates the gradual development of a generic system for manufacturing shop control aiming to cover the decision support requirements at this level. This gradual implementation is based on generic prototypes identified at the conceptual model phase, and their functional modules implemented as particular prototypes. To illustrate the approach the generic prototype devoted to resource allocation problems at the shop-floor level is used with the main focus on the particular prototype for the job-shop scheduling problem, implemented according to the multi-agent paradigm.

The decision to implement this generic prototype in a service oriented environment is compliant with the current research objective of the Distributed Computing Research Team in ICI Bucharest for implementing and using the Grid oriented infrastructure as support for e-science services. This activity is based on the National Grid Infrastructure supporting the first national virtual organization on modelling, simulation and optimization, which was set-up within the national research project GridMOSI [15]. Also, a structural fund project devoted to a significant upgrade of our Grid cluster was recently finalized [14]. The cluster currently integrates a powerful computing configuration (264 cores computing capacity, 20 TB storage) and an extended mathematical modelling platform for distributed computing based on MATLAB® and Simulink® product families, which are opening consistent opportunities for providing virtualized services in the benefit of science and industry.

References

1. Agha, G.A.: ACTORS: A model of concurrent computation in distributed systems. The MIT Press, Cambridge (1986)
2. Andreozzi, S., Burke, S., Ehm, F., Field, L., Galang, G., Konya, B., Litmaath, M., Millar, P., Navarro, J.P.: Glue specification v. 2.0. Open Grid Forum, Tech. Rep., March, http://www.ogf.org/documents/GFD.147.pdf (accessed August 12, 2009)
3. Bond, A.H., Gasser, L. (eds.): Readings in Distributed Artificial Intelligence. Morgan Kaufman Publishers, San Mateo (1988)

4. Florian, V., Neagu, G., Preda, S.: An OGSA Compliant Environment for eScience Service Management. In: Xhafa, F., Barolli, L., Nishino, H., Aleksy, M. (eds.) Proceedings, International Conference on P2P, Parallel, Grid, Cloud and Internet Computing (3PGCIC), pp. 4–6, 381–386. Fukuoka Institute of Technology, Fukuoka (2010), doi:10.1109/3PGCIC.2010.64
5. Foster, I., Kesselman, C., Nick, J.M., Tuecke, S.: The physiology of the Grid: An open grid services architecture for distributed systems integration. Open Grid Service Infrastructure WG, Global grid forum (2002), http://www.globus.org/alliance/publications/papers/ogsa.pdf (accessed January 25, 2004)
6. Foster, I., Grimshaw, A., Lane, P., Lee, W., Morgan, M., Newhouse, S., Pickles, S., Pulsipher, D., Smith, C., Theimer, M.: OGSA Basic Execution Service v.1.0. Open Grid Forum, OGSA-BES WG, GFD-R.108, November 13, http://www.ogf.org/documents/GFD.108.pdf (accessed February 12, 2009)
7. Kosanke, K., Vernadat, F., Zelm, M.: CIMOSA: Enterprise Engineering and Integration. Computers in Industry 40(2,3), 83–97 (1999), doi:10.1016/S0166-3615(99)00016-0
8. Lee, J.H., Kim, C.O.: Multi-agent systems applications in manufacturing systems and supply chain management: a review paper. International Journal of Production Research 46(1), 233–265 (2007), doi:10.1080/00207540701441921
9. Mercier des Rochettes, R.: Sur l'utilisation des réseaux de Petri pour la commande des systèmes de production. Ph.D. Thesis, INPG, Grenoble (1988)
10. Neagu, G.: Generic modelling vs. prototyping: an object-oriented approach to the decision support at the shop floor level. In: Sodhi, R.S. (ed.) Advances in Manufacturing System: Design, Modelling and Analysis. Manufacturing Research and Technology, vol. 22, pp. 19–26. Elsevier (1994)
11. Neagu, G.: Conceptual modelling in Generic prototyping approach. In: Kopacek, P. (ed.) Preprints of IFAC Workshop on Manufacturing Systems: Modelling, Management and Control, Vienna, pp. 373–377 (February 1997)
12. Neagu, G.: A multi-agent simulation model in manufacturing. In: Florea, A.M. (ed.) Proceedings, Summer School on Agent-based Approaches in Intelligent Systems, University "Politehnica" of Bucharest, June 13-16, pp. 155–164 (1998)
13. Neagu, G.: Petri net based modeling of job-shop resource allocation in Generic Prototyping Approach. In: Proceedings of IEEE Int. Conference on Systems, Man and Cybernetics - SMC 1998, San Diego, October 11-14, vol. 4, pp. 96–100 (1998)
14. Neagu, G., Stanciu, A.: Grid Infrastructure Development as Support for e-Science Services. WSEAS Transactions on Computers 9(10), 1181–1190 (2010)
15. Neagu, G., Andrei, N., Sima, V., Cristea, V., Nae, C., Potolea, R., Petcu, D., Stanciu, A.: Grid Enabled Applications for Modelling, Simulation and Optimization. In: Vasiliu, N., Lanyi, S. (eds.) Proceedings of Conference CEEX 2007: Excellence Research as a way to E.R.A. Brasov, October 25-26, pp. 1–6. Technical Publish. House, Bucharest (2007)
16. Ouelhadj, D., Petrovic, S.: A survey of dynamic scheduling in manufacturing systems. J. Scheduling 12, 417–431 (2009)
17. Shen, W., Wang, L., Hao, Q.: Agent-Based Distributed Manufacturing Process Planning and Scheduling: A State-of-the-Art Survey. IEEE Transactions on Systems, Man, and Cybernetics - Part C: Applications and Reviews 36(4), 563–577 (2006)
18. Zhang, Y., Huang, G., Qu, T., Ho, O.K.: Agent-based Smart Gateway for RFID-enabled real-time wireless manufacturing. International Journal of Production Research 49(5), 1337–1352 (2011)

Services for Competitive and Sustainable Manufacturing in the Smart Grid

Vittaldas V. Prabhu

Marcus Department of Industrial and Manufacturing Engineering, Penn State University,
310 Leonhard Building, University Park, PA 16802 USA
prabhu@engr.psu.edu

Abstract. With the growing emphasis on the triple bottom line of people, planet and profits, innovative services will be enablers and drivers of growth of next generation of manufacturing enterprises that are competitive and sustainable. This chapter presents a desiderata for such services for architecting highly distributed intelligent operation of manufacturing enterprises consisting of production processes, lighting and HVAC systems, renewable energy systems, and energy storage systems to cooperatively achieve load control, implicitly thereby emission management, in real-time. Such distributed architectures will leverage emerging developments in service oriented architectures and in interoperability standards for the smart grid. Research challenges for engineering such systems include understanding their dynamics, ensuring responsiveness, and ensuring humanistic process automation. The work extends recent developments in intelligent product-driven production for combining distributed production scheduling and power consumption scheduling. Preliminary results of an analysis and simulation of the distributed system dynamics are also presented.

Keywords: innovative services, manufacturing enterprise, Smart Grid, HVAC.

1 Introduction

With the growing emphasis on the triple bottom line of people, planet and profits, innovative services will be enablers and drivers of growth of the next generation of manufacturing enterprises that are competitive and sustainable. This in turn could stimulate innovative application of advances in information technologies (IT) such as cloud computing, service oriented architecture (SOA), and mobile computing as depicted in Fig.1. For example, web services can be used to support new servicisation business models for cutting tool supply chain [4]. Here "people" is used in a broad sense to include customers, consumers, suppliers, employees, shareholders and the society at large. From a corporate social responsibility (CSR) perspective, this would include fair trade practices for procuring agricultural products from small and marginal farmers. Similarly "planet" is used in a broad sense to include use of natural resources, energy, and emissions. "Profit" is used to indicate short-term and long-term accounting profits, and the risk to these arising from

T. Borangiu, A. Thomas, and D. Trentesaux (Eds.): SOHOMA 2011, SCI 402, pp. 227–240.
springerlink.com

regulatory compliance. With increasing regulations for limiting carbon and green house gas emissions, managing energy consumption is imperative for manufacturing enterprise management, and can provide a substantial competitive advantage by reducing waste [6].

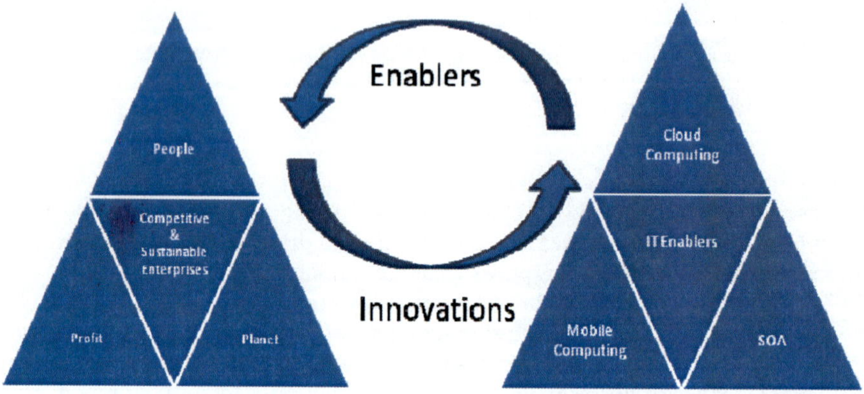

Fig. 1 Competitive and sustainable manufacturing enabled by information technologies

Cloud computing provides economies of scale to rapidly deploy and scale-up compute intense applications without major capital outlays, which makes it especially attractive for small and medium manufacturing enterprises [1]. An ERP solution for small and medium enterprises (SME) is among the early applications launched on the Amazon Cloud. Complimenting this, SOA provides loosely coupled, fine-grained services, that can be "mashed" together to build new applications independent of platform. In this chapter the term "services" is used to denote web services, which is the prevailing technology for implementing SOA at present. Cloud computing can be viewed as the outsourced part of a SOA, making them complimentary to each other. Mobile computing can play an important role in coordinating movement of people and goods thereby lowering cost and enhancing competitiveness. Mobile computing can also play a crucial role in sustainability by proving a means to delivering services to suppliers and customers in underdeveloped regions [17].

Manufacturing enterprises offer significant opportunities for combining innovations in information technologies with innovations in energy technologies to become more competitive and sustainable. The manufacturing sector is one of the largest consumers of energy, particularly electrical power. For example, U.S. manufacturing consumed over 800 billion kWh of electricity in 2006. This not only represents a significant cost but also impacts the planet's environment because every kWh of electricity generated results in an estimated emission of 1.3 pounds of CO2. This is exacerbated during peak load which is usually met using much more expensive and potentially less cleaner sources such as gas turbines. The cost of expensive peak power is reflected on the demand side through dynamic pricing of electricity. Unfortunately only about 20% of manufacturing companies in the U.S. use load control. Many heavy users of electricity opt for variable priced interruptible power in which the utility can shut down power with a certain minimum notice time, and in exchange for this the power rate is significantly reduced. Penalties for using electricity during a shutdown are usually significant. This will be most pronounced in SMEs where the operations managers may not have the expertise or resources to cope with these challenges.

Remarkably, even though the U.S. National Academy of Engineering considers electrification as the foremost among the greatest engineering achievements of the 20th Century, the electric grid has essentially remained an electromechanical system with little intelligence, especially to integrate new and renewable energy sources. In the U.S., one of the key R&D emphases in smart grid is advanced metering infrastructure (AMI) for effective demand response (DR) to reduce prices through competition and increasing awareness of energy usage thereby aligning prices with the value customers place on electric power [8].

The anticipation is that electricity will be priced dynamically therefore future manufacturing enterprise operation will require controlling power consumption through good management of local renewable energy and storage resources based on visibility into future grid conditions and prices. Such a load shaping strategy will be a major departure from the past, and the resulting consumption behaviour may be difficult to predict [7, 8]. Neural networks [10, 19] and multi-agent systems [12] have been suggested as a mechanism to adapt to changing conditions. But such techniques may not be rigorously predictable for critical infrastructure such as the electric grid. There are several technological developments underway such as distribution in the smart grid [3], home power consumption management using cell phones [9], wireless protocols to support Home Area Network (HAN) and Neighbourhood Area Network (NAN) for AMI [2], and Roadmap for Smart Grid Interoperability Standards [11]. At present much of the efforts in smart grid have focused on distribution, and the focus on the demand side seems to be limited to smart appliances in the residential segment. For the industrial segment, lighting and HVAC have been the main foci of smart grid efforts. There have been several concerted efforts to improve energy efficiency in a range of industrial processing technologies, which are also a way to boost productivity and competitiveness [18]. The research opportunity for manufacturing enterprises presented by developments in the smart grid is to optimize their operations by controlling their demand side load dynamics, which could be achieved by smartly controlling loads, generation, and storage [15].

This chapter builds on [15] to present a service oriented perspective of competitive and sustainable manufacturing in the smart grid. Section 2 presents a desiderata for such services for architecting highly distributed intelligent operation of manufacturing enterprises consisting of production processes, lighting and HVAC systems, renewable energy systems, and energy storage systems to cooperatively achieve load control in real-time. Research challenges for engineering such systems include understanding their dynamics, ensuring responsiveness, and ensuring humanistic process automation is also discussed in Section 2. Section 3 extends recent developments in intelligent product-driven manufacturing for combining distributed production scheduling and power consumption scheduling using a feedback control approach. Section 4 presents preliminary results of an analysis and simulation of the dynamics of the proposed distributed control system. Section 5 concludes the chapter with possible directions of future research.

2 Desiderata of Services

Fig. 2 illustrates the current opportunities and challenges in services for competitive and sustainable manufacturing. Every major function in manufacturing enterprises presents new opportunities for services: product design, manufacturing operations, and supply. These can be summarized as follows:

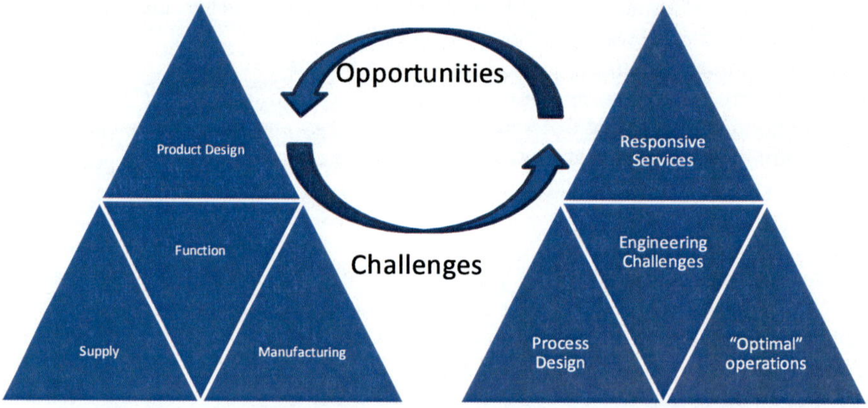

Fig. 2 Opportunities and challenges in services for competitive and sustainable manufacturing

- ***Design:*** CAE services to estimate power and energy profiles for components for specific combination of materials and manufacturing processes selected during "green design". Recently OMG PLM Services 2.0 has been developed for collaborative design, which could be enriched to include energy and emission data.
- ***Design/Manufacturing Interface:*** Digital provenance of products, e-Provenance, can be developed by keeping track of "actual" emissions and energy consumed during manufacturing operations. This service could be used for product differentiation and differential pricing, opening prospects of new business models.
- ***Manufacturing/Supplier Interface:*** Suppliers track the "actual" emissions and energy consumed during processing and transport of components/subassembly, and such accumulation continues all the way up to point of sale and through to end of life. With RFID technologies gaining wider use at item level tracking, e-Provenance are technologically viable.

- ***Supply Logistics:*** Traffic forecast services can be used to identify congestions and delays in real-time to better synchronize deliveries by choosing routes and modes that have desired energy and emission footprint. This could also be used to reduce "empty miles" in logistics, and improve delivery performance towards just-in-time "green logistics".
- ***Financial Supply Chain:*** Flow of finance is critical for the operation of material supply chains. One of the main "frictions" in global trade is paper-driven process and the resulting errors. Services for smooth and efficient flow of finance can clearly improve the competitiveness of the manufacturing enterprises involved.
- ***Micro-finance Services:*** When suppliers/consumers are in underdeveloped regions of the world, micro-finance services such as loans and insurance can increase the sustainability of the manufacturing supply chains. These could be delivered over mobile phones which are ubiquitous and affordable even in rural underdeveloped regions.
- ***Weather Forecast:*** Local weather forecast services such as sunshine and wind can be used to predict power generated using renewable sources such as solar PV and windmills; likewise temperature and sunshine can be used to predict load for HVAC and lighting.
- ***Price Forecast for Energy and Emissions:*** In situations where manufacturing campuses may have significant renewable energy assets installed sometimes it may be more profitable to sell energy and emission permits than to actually manufacture. Conversely, when manufacturing demand has to be satisfied in a timely manner the cost of energy and emissions may be significant.

In high energy manufacturing operations such as foundries, laser processing, and electron beam processing, managing grid demand and peak demand could easily become a dominant issue. Detailed process models coupled with production schedules could be useful to estimate power profile and energy consumption schedule. The envisioned system would be composed using fine grained services, in which detailed models will be hosted on a cloud platform. Besides the core manufacturing processes, HVAC and lighting are other important factors that determine electrical energy consumption, which would be largely determined by local weather, safety and comfort of the work environment. It should be noted that HVAC and lighting could present relatively easier opportunities to reduce energy consumption. The key point is that there are numerous opportunities for services in manufacturing enterprises.

The major engineering challenges in fully realizing the potential of services can be summarized as follows:

- ***Responsiveness:*** When services are scaled up through increased usage within enterprises and across enterprises, the infrastructure should smoothly scale up to provide adequate responsiveness. The fundamental issue is to be able to systematically characterize the parameters for meaningful service level agreement (SLA) without resorting to over provisioning of computing and networking resources.

- ***Optimizing Operations:*** The value of services can be realized when they lead to good, optimizing, decisions in the enterprise. The main issue here is to ensure that the services provide accurate information for analytics which support optimizing of operations in very dynamic environments, and lower the burden on the decision-makers.
- ***Humanistic Automation:*** Automated business processes and decision-making processes driven by services should be designed with due consideration to human computer interaction (HCI) issues such as responsiveness and cognitive complexity of the tasks. Many large software efforts in the past have failed in fully realizing their benefits because of shortcomings in this regard.

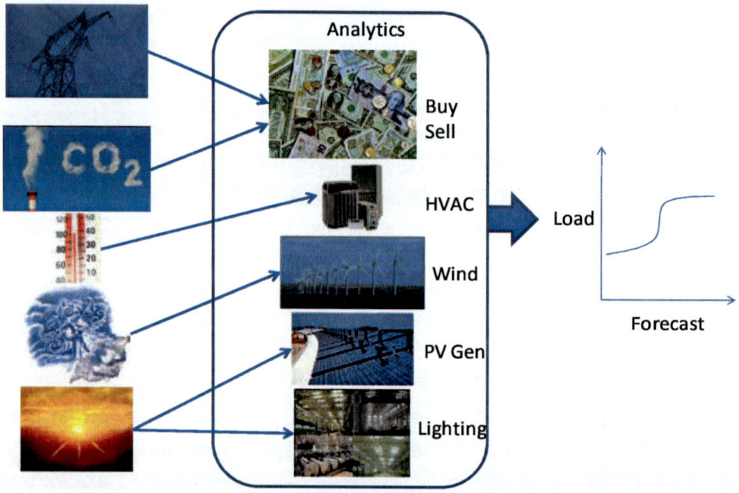

Fig. 3 Services for load forecasting analytics

There could be a variety of services that will be consumed for load forecasting analytics, as illustrated in Fig.3. Natural conditions forecast services will drive the analytics for predicting power generated using renewable sources and power consumed for HVAC and lighting.

Market conditions forecast of energy and emission prices will drive the analytics for buying/selling energy and emission permits. Similarly for logistics of supply and delivery, cost of fuel and emissions coupled with traffic congestion will drive the decisions regarding vehicle routing and delivery performance. Power load and emissions generated by a manufacturing enterprise will be largely determined by its production requirements such as those in the master production schedule generated by its ERP system. Therefore there is a *need for effective scheduling* of production, energy consumption, energy storage, energy sale into the grid, and setting HVAC and lighting levels under dynamically varying natural and market conditions.

3 Simulation-Based Distributed Feedback Control

Spatially distributed decision-making and autonomous control approaches developed for manufacturing can be extended to include additional functions of energy generation and storage [5, 16].

Fig.4 illustrates the overall architecture for simulation-based distributed feedback control in which each entity is capable of autonomous decision-making and control. This expands the concept of *part driven manufacturing* production to include logistics and power resources in which control is distributed among entities in manufacturing systems, generation systems, and storage systems. For example, in such a system each part seeks manufacturing and power resources required to complete its production while considering manufacturing performance and dynamically varying power cost in making decisions.

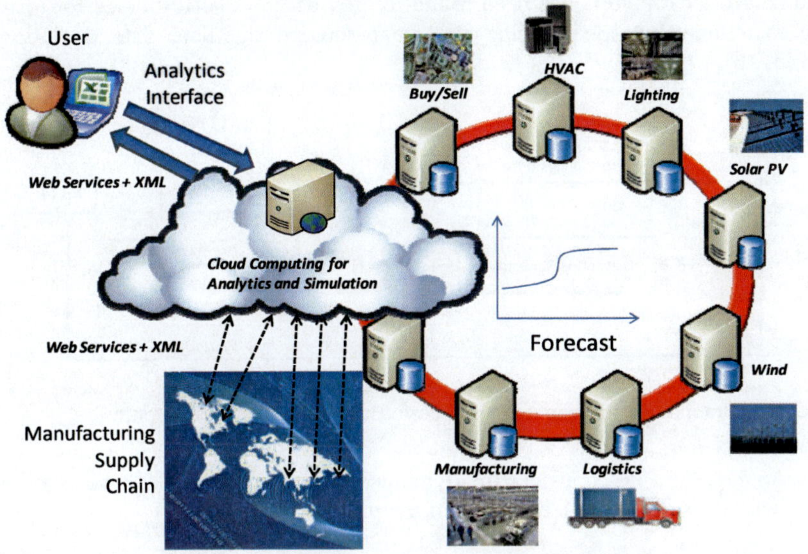

Fig. 4 Architecture for simulation-based distributed feedback control

Parts use simulation to assess the impact of the time at which they seek resources and iteratively try to improve their individual performance. In such simulations manufacturing and power entities make simultaneous decisions for scheduling manufacturing production, scheduling power consumption, and scheduling power storage from renewable energy resources [15].

In the envisioned manufacturing enterprise, production demand is taken as input along with time varying price based on prevailing grid conditions, time varying power generated from local renewable energy sources such as wind and solar PEV, and storage capacity, HVAC and lighting loads. Under dynamically varying conditions, the control system will adjust decisions in real-time to maximize competitiveness and sustainability. The challenge of engineering such systems is to ensure good "behaviour" and predictability [13, 14, 16], which will be discussed next.

Overall manufacturing performance will depend on manufacturing operations and the cost of power consumed in the resulting operations.

It is important to understand the combined dynamical interaction between manufacturing performance and power price. Fig.5 shows the controller topology for combined manufacturing performance and power cost controller suitable for part-driven manufacturing. For the ith part the notation used is $d_i(t)$, $z_i(t) = (d_i(t)-c_i(t))$, $a_i(t)$, $q_i(t)$, $p_i(t)$, and $c_i(t)$, are the due-date, due-date deviation, arrival time, queuing time, processing, and completion times, respectively. The arrival time for each part is manipulated by its manufacturing performance controller (MPC) and power price controller (PPC). The overall dynamics of the system could potentially consist of 1000s of such feedback control loops interacting with each other. Integral controller for MPC has been found to provide good performance for just-in-time manufacturing by reducing the mean-squared due-date date deviation (MSD) [5, 16].

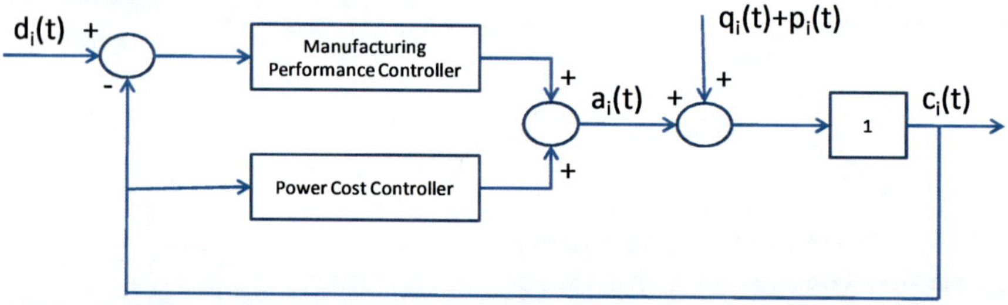

Fig. 5 Controller topology for manufacturing performance and power cost controller

To formulate the PPC, consider the dynamically varying price of power shown in Fig.6, where c is the completion time of the part [15].

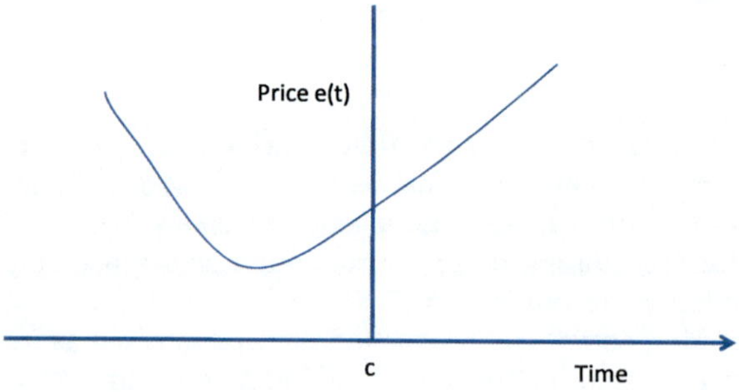

Fig. 6 Dynamically varying price of power

A possible PPC approach would be to "do a gradient search" by adjusting the arrival time of the part based on the direction and magnitude the rate of change of price of power at c. One such PPC, combined with MPC, can be expressed as follows [15]:

$$a_i(t) = \left[k_i \int_0^t z_i(\tau)\, d\tau \right] - \left[k_i^{'} \frac{de}{dt} \right] + a_1(0) \tag{1}$$

Intuitively, MPC and PPC would adjust arrival time such that there is a "balance" between manufacturing performance and power price depending on parameters k_i and $k_i^{'}$. Furthermore, prior work has established some of the nonlinear behaviour in MPC, which will be modified by the addition of PPC dynamics.

Specifically, MPC has discontinuous dynamics because infinitesimal changes in arrival times that change the part processing sequence cause large changes in completion times. The solution of such a system of differential equation is found using Fillipov's technique by taking the intersection of the discontinuity surface with the closure of the convex hull of all possible derivative vectors in the neighbourhood of the discontinuity. The velocity of the trajectory may not be uniquely determined by the state of the system on the discontinuity surface but belongs to a set of feasible values that depends "loosely" on the state. Fillipov's solution at a discontinuity point x(t) is prescribed as the set F(x(t)) where

$$F(\mathbf{x}(t)) = \bigcap_{\delta>0} \bigcap_{M} \overline{Conv}\, f(\beta(\mathbf{x}(t), \delta) - M) \tag{2}$$

M is the discontinuity region, β is a ball centred at x(t) with a radius δ. Details of applying this technique for analyzing distributed control system can be found in earlier works [13, 14, 16].

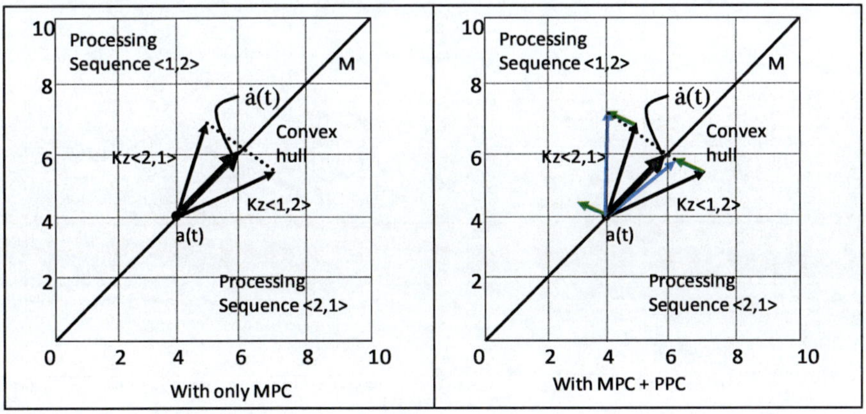

Fig. 7 Discontinuous dynamics with MPC only and with PPC

Fig.7 illustrates application of Equation 2 to a 2-part case with MPC only, and with MPC and PPC. Consider the system with only MPC. The z vectors are due-date deviation errors from Fig. 5. In this case there are 2! sequences possible in the vicinity of the discontinuity point, and the convex hull of the kz vectors intersects the discontinuity line, M, at a unique point, therefore the velocity, $\dot{a}(t)$ is unique. When MPC and PPC are combined, $-k'\dot{e}(t)$ is added to the kz vectors, and the convex hull is formed. In this particular illustration the resulting convex hull also intersects the discontinuity line at a unique point therefore the velocity, $\dot{a}(t)$ remains unique for combined MPC and PPC.

4 Simulation Results

To explore the dynamics of the interplay between the MPC and PPC, a simulation model was built in `Matlab`. The main objectives were to get some insights into the impact of gain parameters of the controllers in Equation 1, and to explore the dynamics as the system size is scaled up. Also, the time varying price of power was modelled as a sinusoidal function, -sin(wt) +1.5 with frequency w is set to 0.0145 rads/sec; the offset of 1.5 is used to ensure that the price is never negative. This power price function is chosen because of its simplicity to implement and because it is a non-convex function with multiple minima.

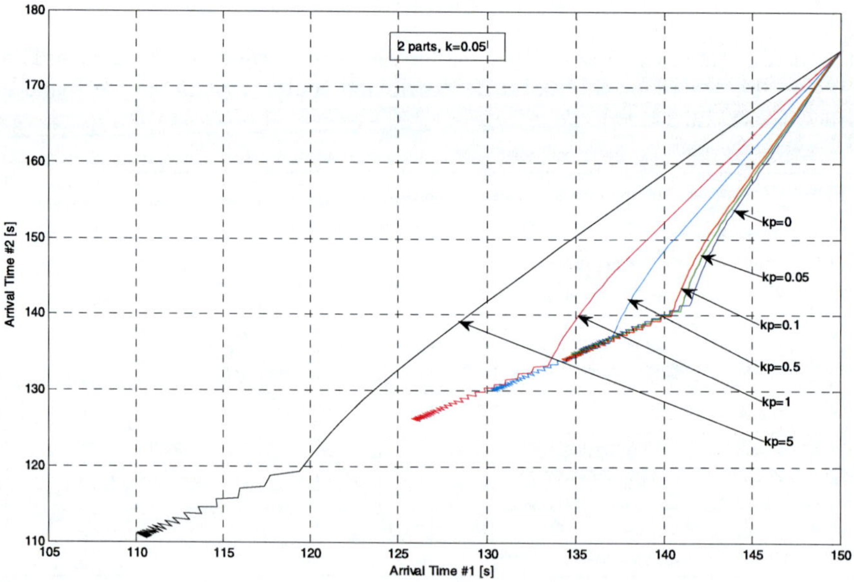

Fig. 8 Effect of changing controller gain on steady state arrival times

Fig.8 illustrates the dynamic response of a simple system with two parts with infeasible due-dates and k = 0.05. As the PPC parameter k'= kp is increased (0, 0.05, 0.1, 0.5, 1, and 5), the steady state arrival times decrease and the Mean Power Cost (MPC) decreases, thus illustrating the effect of weighting k and k' in Equation 1.

This can be interpreted in the context of Fig. 7 by noting that as k' increases, $-k'\dot{e}$ vector increases, causing the convex hull geometry to skew more towards decreasing the cost of power.

Fig. 9 illustrates the impact of controller parameters on manufacturing performance cost measured by MSD, and power cost measured by mean power cost for a system with 50 parts, with due dates and processing times randomly selected from [0, 100] uniformly. In this case k was 0.05 and k' was 2 for all parts and MSD and MPC converge smoothly.

This is a more realistic scenario but by no means comprehensive.

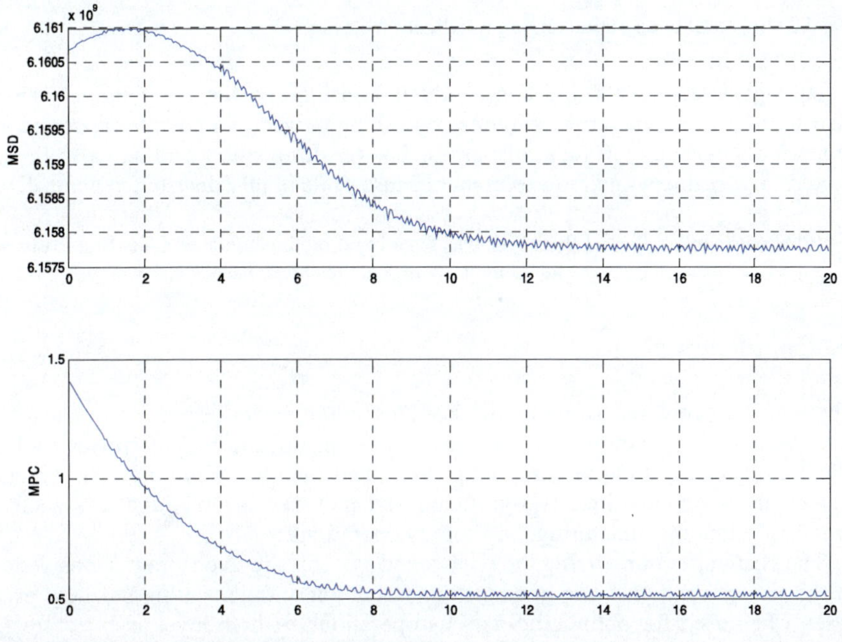

Fig. 9 Manufacturing performance cost (MSD) and mean power cost (MPC) for 50 parts

The concept of intelligent entities autonomously and cooperatively controlling manufacturing performance and power cost can be extended to intelligent factories as illustrated in Fig. 10.

The result would be *grid level cooperation* among intelligent and autonomous factories through their constituent distributed intelligences on the consumption side of the smart grid.

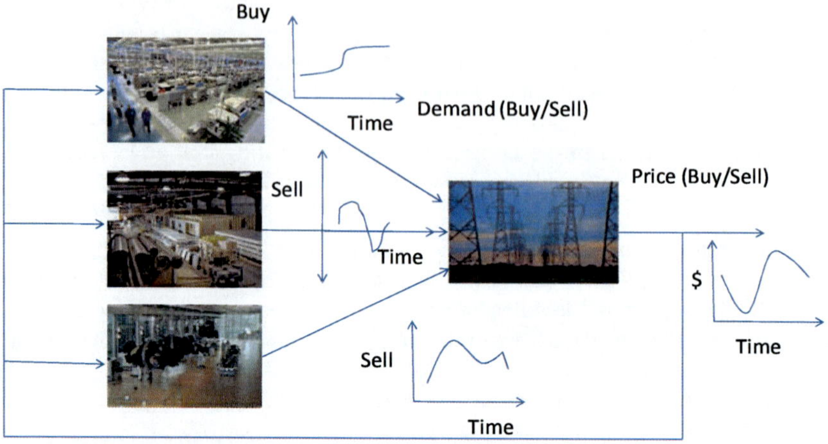

Fig. 10 Grid-level cooperation among intelligent factories

Intelligent factories would cooperatively consume power, cooperatively share distributed power generation resources by selling power, and potentially cooperate in storing power in their local resources. The resulting smart grid can also be expected to be resilient and fault tolerant because of its highly distributed control.

Acknowledgments. Part of this work was performed by the author as a Visiting Professor with the PSI Group, TEMPO Lab, University of Valenciennes, France.

5 Conclusions

This chapter presented some of the key opportunities for services to enable the next generation of competitive and sustainable manufacturing enterprises by leveraging advances in cloud computing, SOA and mobile computing. The breadth and depth of opportunities is significant and includes several functions such as product design, manufacturing, and supply operations.

The challenges in realizing the envisioned systems are to ensure responsiveness of the services and ensuring humanistic automation of the resulting business processes when used for optimizing various operations. A high-level architecture for simulation-based feedback control using SOA and cloud computing was proposed. This architecture extends distributed decision-making and autonomous control approaches developed for manufacturing to include additional functions of energy generation and storage.

A topology for combined control of manufacturing performance and power price was discussed along with preliminary insights into dynamics of such systems using simulation.

There are many interesting research issues that can be pursued to realize the approach presented in this chapter. The dynamic analysis using analytical and simulation techniques should be studied. This should be systematically linked to develop service level agreements to ensure proper operation and stability of the system, especially when the underlying business processes are automated.

Another rich topic of investigation would be to engineer the resulting processes to be more humanistic in terms of cognitive burden on the users. Finally the triple bottom line of such approaches should be quantified in realistic contexts.

References

1. Banerjee, U.: Cloud Economics – Amazon Microsoft Google Compared. Cloud Computing Journal (2010), http://cloudcomputing.sys-con.com/node/1257999 (accessed August 11, 2011)
2. Bennett, C., Highfill, D.: Networking AMI smart meters. In: Proceedings of IEEE Energy 2030 Atlanta, GA (2008)
3. Brown, R.E.: Impact of smart grid on distribution system design. In: Proceedings of IEEE Power and Energy Society General Meeting - Conversion and Delivery of Electrical Energy in the 21st Century, pp. 1–4 (2008)
4. Cheng, C.Y., Barton, D., Prabhu, V.V.: The servicisation of the cutting tool supply chain. International Journal of Production Research 48, 1–19 (2010)
5. Duffie, N.A., Prabhu, V.V.: Real-Time Distributed Scheduling of Heterarchical Manufacturing Systems. Journal of Manufacturing Systems 13, 94–107 (1994)
6. Esty, D.C., Winston, A.S.: Green to Gold: How Smart Companies Use Environmental Strategy to Innovate Create value and Build Competitive Advantage. Yale University Press, Yale (2006)
7. Fan, J., Borlase, S.: The Evolution of Distribution. IEEE Power and Energy Magazine, 63–68 (March/April 2009)
8. Ipakchi, A., Albuyeh, F.: Grid of the Future. IEEE Power and Energy Magazine, 52–62 (March/April 2009)
9. Mehroze, M., Dwolatzky, B.: Smart demand-side energy management based on cellular technology – a way towards smart grid technologies in Africa and low budget economies. In: Proceedings of IEEE Africon 2009, pp. 1–5 (2009)
10. Momoh, J.A.: Smart grid design for efficient and flexible power networks operation and control. In: Proceedings of IEEE PES 2009 Power Systems Conference and Exposition (2009)
11. NIST. Smart Grid Interoperability Panel Standards Catalog. NIST, Gaithersburg, MD (2009)
12. Pipattanasomporn, M., Feroze, H., Rahman, S.: Multi-agent systems in a distributed smart grid: design and implementation. In: Proceedings of IEEE PES 2009 Power Systems Conference and Exposition, pp. 1–8 (2009)
13. Prabhu, V.V.: Stability and Fault Adaptation in Distributed Control of Heterarchical Manufacturing Job Shops. IEEE Transactions on Robotics and Automation 19, 142–147 (2003)
14. Prabhu, V.V.: Performance of Real-Time Distributed Arrival Time Control in Heterarchical Manufacturing Systems. IIE Transactions 32, 323–331 (2003)
15. Prabhu, V.V.: Distributed Cooperative Control Approach for Smart Manufacturing in the Smart Grid. In: Proceedings of Mechatronics (2010)

16. Prabhu, V.V., Duffie, N.A.: Nonlinear Dynamics in Distributed Arrival Time Control of Heterarchical Manufacturing Systems. IEEE Transactions on Control Systems Technology 7, 724–730 (1999)
17. Rios, W., Crespo, J., Prabhu, V.V.: Financing and Mobile Technologies in Microfinance of Mexican Mango Exports. In: Proceedings of the Industrial Engineering Research Conference (2011)
18. U.S. Department of Energy. IMPACTS (December 2009)
19. Venayagmoorthy, G.K.: Potentials and Promises of Computational Intelligence for Smart Grids. In: Proceedings of IEEE PES 2009 Power Systems Conference and Exposition (2009)

Different Approaches Regarding the Operational Control of Production in a Flexible Manufacturing Cell

Nick Andrei Ivanescu, Mihai Parlea, and Andrei Rosu

University Politehnica of Bucharest, Dept. of Automation and Industrial Information
{nik,mihai.parlea,andrei}@cimr.pub.ro

Abstract. The chapter describes several modern solutions for controlling a complex manufacturing cell, consisting of conveyors, robots, stoppers and other devices, using one programmable logical controller (PLC). The objective was to design a control system for a Holonic Planned Manufacturing System running into an Intelligent Production Cell. In order to achieve this goal, each product processed in the manufacturing cell has to be identified and its own routing / workstation jobs schedule needs to be implemented without interfering with other products. Two solutions were implemented, one using an offline planner that calculates the necessary data for processing products on pallets and the other using Intelligent Electronic Devices (IED) attached to the pallets that can take decisions by themselves in real time. A special array of data structures was used to easily access and manage the pallets processed at the four robotized stations of the cell.

Keywords: operational control, HMES, order holon transposition, PLC, OPC server, failure management.

1 Introduction

Traditional networked assembly structures have either hybrid or heterarchical architecture. The first type allows data exchange and co-operation between lower-level (robot) controllers. In this architecture, a master initiates all the activities and then the subordinates cooperate to perform them. The second type of architecture is formed by a group of independent entities, called agents, which bid for orders based on their status and future workload [5]. The master-slave relationship is dismissed and due to this decentralized control architecture, the agents have a complete local autonomy and the system is able to react promptly to any event such as resource failure, rush customer's order etc [7].

In order to face resource breakdowns, job shop assembly structures use networked robot controllers with multiple-LAN communication facilities allowing production data saving and automatic re-planning of batch production using failure and recovery management [1].

T. Borangiu, A. Thomas, and D. Trentesaux (Eds.): SOHOMA 2011, SCI 402, pp. 241–254.
springerlink.com

2 System Architecture and Production Flow

The manufacturing cell in which this project was implemented consists of six work stations and a transporting system (Fig. 1). These six workstations are built around an equal number of Adept robots that can execute assembling / profiling operations and are organized in the following way:

- Two stations equipped with Cobra 600 robots achieving:
  - Mounting different type of parts
  - Video inspection in the intermediate and final stage
- Two stations equipped with vertical articulated Viper 650 robots and one CNC being able to:
  - Mount different types of parts
  - Manipulate raw parts in order to be processed on the CNC
  - Perform visual inspection of products in intermediate and final fabrication stages
- One feeding station equipped with a horizontally articulated Cobra 800 robot and two parts feeders that can achieve:
  - Identification of the raw parts from the feeder's recipients
  - Their positioning on the transport pallets.
- One station equipped with a Cartesian robot responsible with introducing the pallets in the cell.

Pallets entering the system have the possibility of being written with a unique magnetic code for identification during the production process. Each pallet, according to a previous planned production, stops at one or more working stations in order to have the operations done. Once all operations are completed, the pallet exits the system. The maximum number of pallets that can enter the system for a batch is 256, limited by the eight bits available for the magnetic code. For practical reasons, each pallet can support maximum 16 operations. An offline planner has to create an optimal schedule by maximizing the load of all available machines.

The existent stoppers in the system can block or allow the movement of the pallets on the conveyors, four of them being related to code-read sensor and other two to a read/write code sensor heads. Lifts have the task of transferring the pallets from one main conveyor belt to another or from the main conveyor loop to the robot conveyor belts that take the pallet to the working stations. The conveyor system is controlled by a PLC that enables all mechanical actions to be taken in the transport system [4].

When the products enter the system, their memory capsule is written with their unique code for identification throughout the production of the batch. The products are then always transported to the next working station by advancing on the main conveyor loop. At the five locations where they can be deviated there are sensors that identify their unique code, and according to the data received from the planner will take the necessary actions for the deviation, if the products should be brought to a robot-controlled workstation in order to be processed in one or more operations.

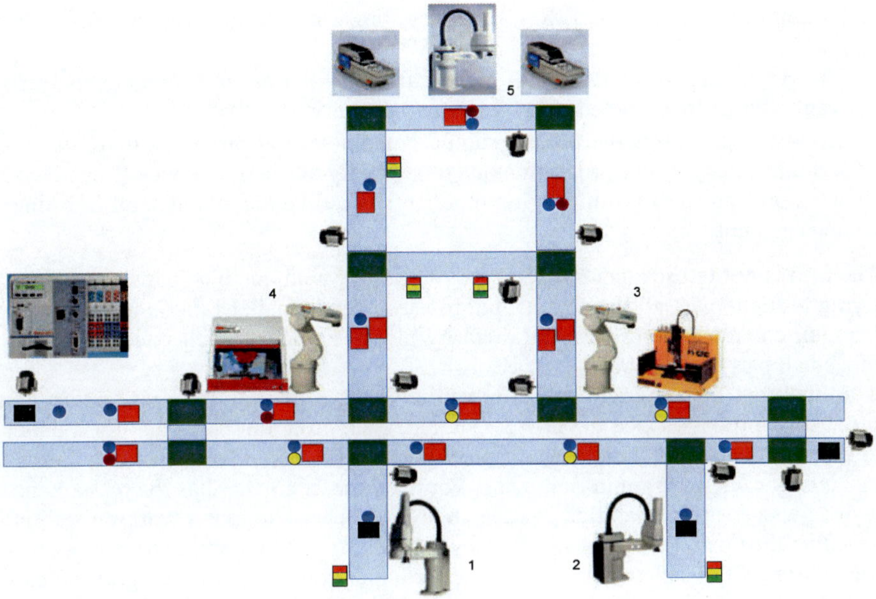

Fig. 1 The main configuration of the manufacturing cell

From the five robot stations, four are used for product execution; they support operations like assembling, mounting, cutting, fixing, gluing, polishing, painting. Two of the four product-processing robots (of SCARA type) can share a common workspace to cooperate for complex tasks. The other two robots, vertical articulate, are able to feed machine tools extending thus the services performed on parts.

Some constraints apply to the system:

- At any time in the system there can be no more than 4 products.
- The products coming out of a post have priority over the ones moving on the main conveyor.

3 First Control Solution and PLC-Based Implementation of Order Holons

3.1 Theoretical Backgrounds

When dealing with complex systems having a big number of input and output signals to be controlled by a single PLC, a modern solution for software design must take into consideration the following issues:

- The PLC must be chosen such as to support advanced software development kits, containing most of the standard programming languages for PLCs, like

Sequential Function Chart (SFC), Ladder Diagram (LD) or Structure Text (ST) [3, 5].

- The project should contain both graphical sequential programs and cyclic programs (for actions needed to be performed at every PLC cycle).
- In most of the cases the project should contain several programs that will run simultaneously, every program managing usually one action device (like relays, actuators and so on) or managing other activities like communication or timing management.

The idea is not to use a centralized type of control, with one big "parent" program trying to implement all the actions, but to use several smaller programs running in parallel, communicating between them by global variables, each one taking care of a small part of the process [10, 12, 14].

Initially all the programs are in a waiting state, expecting the user to start the production process. Once the process started, each program then waits for a pallet to enter the zone that it controls. Priority rules on the conveyor are locally implemented by each program in each zone. For example, if a pallet has to cross a zone, it must wait for this operation in case another pallet has to exit a working station, avoiding thus pallet collisions or conveyor blocking. The control program inhibits the motion of the pallet trying to cross a particular conveyor area and grants clearance for the one exiting by means of a global Boolean variable.

The holarchy for the resources and information counterparts of this production cell is implemented according to the PROSA reference architecture [8]. The offline batch planner and scheduler (providing an optimal product sequence, operation sequence for each product and resource allocation for each operation) is the Staff Holon and the workstations are the Resource Holons. The Order Holons are inserted into the Staff Holon which, using its knowledge about products (via the Product Holon) and resource capabilities - workstations and the transporting time on each conveyor segment (via Resource Holons), computes offline an optimal production plan that will be implemented then by the PLC to execute the manufacturing of each ordered product.

3.2 Project Structure

Following the guidelines previously described, a Bosch PLC together with Indralogic development software was used to accomplish the task. The software system has 3 working modes:

1. Manual: the human operator has the possibility to individually act upon the system's elements in order to check their functionality or to bring the system in a particular status.
2. Diagnose: this mode allows the movement of a pallet between the workstations, stopping it in some important points of the transport system such as the whole system can be verified by confirmation or error messages.
3. Automated: this is the normal mode. It receives the data from the global planner, transforms it in commands for the execution elements and detects abnormal states of the system's components, reporting them to the supervisor.

The programs system is a complex structure consisting of many programs that run simultaneously, as shown in Fig. 2.

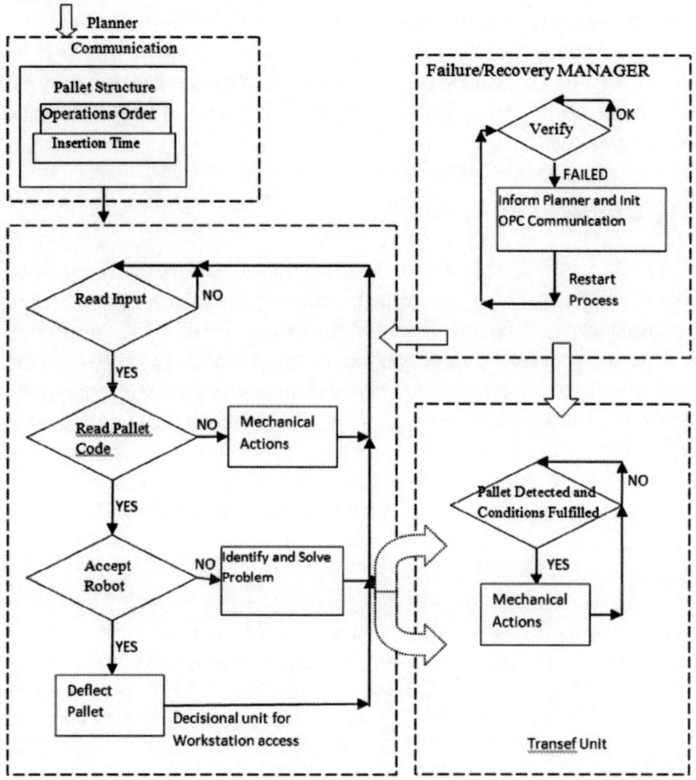

Fig. 2 General diagram of the program structure

One of the issues for solving the control problem of the transport system was the entry. Because products enter the system at different time moments, and other may need to be transferred from one main conveyor belt to the other, some kind of inter blocking must be implemented. The first solution one might think of is a simple TAS (Test And Set). The product that needs to enter the system checks for clear transfer, signals to the other products interested that the region has been occupied by setting a variable, and takes the necessary mechanical actions to make the product enter the system; upon finishing it resets the variable, clearing the transfer. If by any chance two products do the test in the same time then they are both going to occupy the transfer and a crash is possible. The solution is to check first the sensors that detect the products and give priority to the one that needs to be transferred, because it must reach another working post. When one product is exiting a working post and another product that is on the main conveyor would like to cross to the next point, then a priority management was implemented. It was decided that the exiting product has a higher priority than the products on the main conveyor loops.

The robots present in the manufacturing system should have a minimum of idle time, although in some particular cases some of the robots may not be involved. For achieving a maximum load, the conveyor system should never be blocked by any product that is waiting to be processed by a robot. If the transport system is not overloaded, the robot station can always be reached without waiting times and can always carry out a task.

3.3 The Routing Challenge

A programmable controller is responsible for issuing the necessary commands to the executing elements of the transport system in order to introduce the pallets to the planned workstations for processing the raw materials. In this first approach the PLC should be able to receive the necessary data from the global planner (the supervisor computer) and act consequently. The first step was to define the form in which the data from the planner will be stored. A pallet entering the cell should support different operations, every operation needing supplementary information such as the workstation that should perform it, the time needed for completing it and so on. Analyzing this, the following structure was settled to define an operation:

```
  TYPE datemasina :
  STRUCT
       post:BYTE;  (*number of the robot workstation*)
       operation:BYTE;(*a code representing the operation
done at this station*)
       timemin:WORD;(*the minimum amount of time necessary
for completion *)
       timemax:WORD;(*maximum amount of time for completion
of the operation*)
       report:BYTE;(*a small report about the result of the
operation*
  END_STRUCT
  END_TYPE
```

This structure is repeated for each operation that the product has to perform and these structures compose an array named "`sir_palete`".

It was decided that 16 such structures are sufficient to completely describe a product and due to limitations of the system there can be no more than 256 products in one batch, these being the reasons why "`sir_palete`" was defined as an array of 256 by 16 structures of type "`datemasina`" (Fig. 3).

In order to have access to the data in the array one needs two indexes: the first index is the product number (0-255) and the second is the number of operation (1-16). The array named "`sir_index`" has a length of 256 fields each of them being the operation index reached for the corresponding product. The data structure

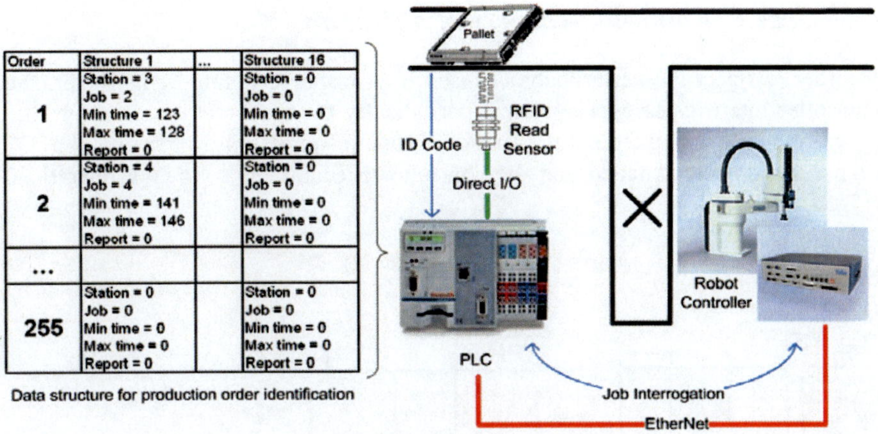

Fig. 3 Pallets identification system using RFID sensors

necessary for the information regarding the production is very large 256 (products) x 16 (operations) x 5 (elements in a structure) = 20480 items. A string that shows the moment of time the product should enter the system extends this data structure; such strings are kept in an array "`time_insertion`".

In order to have access to a particular element of this matrix two coordinates must be identified (column and line) together with the desired field from the selected structure. The line is given by the identifier of the pallet (bar code) while the column is the operations index. For an easy access to this index these values are stores in an index array with 256 elements, each element corresponding to a pallet identifier.

Using this array of structures the PLC has all the necessary data to route a pallet through the cell. The transport system was software divided in segments, each segment being managed by an independent program that, taking into account the status of the nearby elements sends, at a proper moment of time, the correct commands to transport the pallet to the next segment. When a product reaches a stopper on the conveyor, there are different decisions that can be taken. In this points the PLC can read its bar code and after verifying its current element of the data structure allows eventually the product to pass, deviated on a secondary conveyor branch or to the nearby workstation (if it matches the current workstation).

4 Failure and Perturbations Management

The nature of the perturbations in the frame of this system can be of two types:

- Failure or repair of a production resource (robot, CNC)
- Finish of the raw material stock at the location of a production resource

4.1 Case 1: Failure / Recovery of a Resource

In order to reject the perturbations caused by a resource failure or its repair, the controller interrogates cyclically and periodically the resources status, using the ping protocol detailed later in the chapter. If one resource fails to respond in a due time it is considered failed and the controller executes the following operations (Fig. 4):

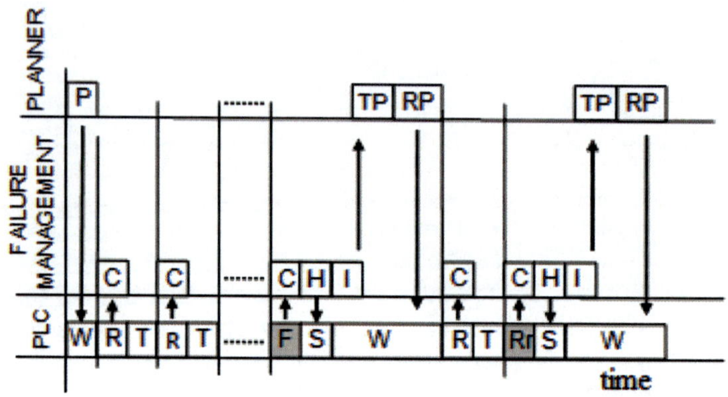

Fig. 4 Rejection of perturbations caused by a resource (failure / recovery): P – planned operation; W – wait for instructions; C – check the status; R – operational robot; T – transport operations; F – failure of the robot; H – stop signal; S – stop of mechanical actions; I – initializing the planner; Rr – repair & recovery of the robot; TP – finish of current production; RP – re-planning

- Stops all mechanical actions
- Refresh the resource holon regarding the new status of all the resources
- Reads the status of the products inside the cell
- Evaluates all the remaining orders if they can finished or not
- Initialize the global planner with the current status of the system
- Receives the new planning from the global planner
- Resumes all the mechanical actions

4.2 Case 2: Re-supplying Mechanism

The second type of disturbance that can affect the production is the shortage of raw materials necessary for processing the parts. In the case of a product arriving at a robot workstation that needs additional parts, it triggers the re-supply mechanism as shown in Fig.5.

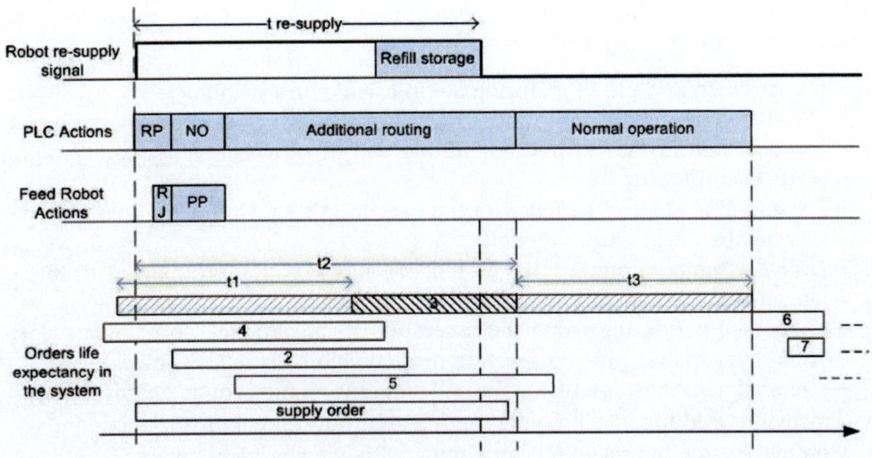

Fig. 5 Re-supply mechanism: t1 – initial planning time; t2 – waiting time for re-supplying; t3 – actual execution time; t re-supply – the time spent for supplying the robots with parts; RP – re-supply planning; NO – normal operation; RJ – receive re-supply job; PP – place parts

This means that the robot raises its re-supply signal from low level to high level; the PLC detects this event and starts planning the supply holon and then sends the request for a re-supply sequence to the part feeding robot SCARA, assisted by vision, with dual feeding unit [14]). This robot, after receiving the supply job command, starts to select the necessary parts and places them on the supply order pallet. During the time it takes the feed-robot to prepare the supply order, the PLC has a normal operation and deals with the other orders in current execution.

The moment the feed robot signals that the supply order is ready, the PLC initiates additional routing operations in which the supply order is executed with a lower transport priority then the rest of the products in the system, until the parts are placed in the empty storage and the supply pallet returns in its wait location.

After the robot recovers and the order reaches it, production is resumed. It takes a different time from the initial planning to make this product (t3), as the other orders have already been partially or totally finished. T2 is the time delay due to the feed-robot filling the supply pallet, the supply pallet travelling to the empty storage and the processing robot refilling its storage. The re-supply event triggers a system blocking; new orders cannot be inserted in the system as long as the ones in current execution are not finished. After order 3 is finished the production resumes as previously planned although there is a delay (approximately equal to t2-t1+t3).

5 Second Approach: Using Intelligent Products

Semi-heterarchical control assumes decentralization of both the effort to take the decision and the effort for executing the decision. In order to achieve this goal, a new approach was taken into consideration, adding intelligence to the product pallets [11, 13, 16]. Thus, the pallets travelling in the system were expanded with an electronic circuit composed from:

- On-board memory, used for storing the following information:
  - Product / assembling cell model [the timing model of the current assembled product and of the current assembling cell in which an active entity holon (HEA) is situated]
  - Assembling stages [the list with the execution stages of every operation described by the product's model]
  - Local assembly program [the current assembly program (or part of it) that is in course of execution]
  - Reports [reports regarding the assembly of the product, components positioning precision, video inspection results, reports from CNC machines]
  - Internal data [position inside the cell, data for running routing algorithms]
- Processor: used for executing the routing algorithms, tracking production
- Communication interface: for interfacing with other services entities

The solution chosen for this Intelligent Embedded Device (IED) was Gumstix OveroAIR - a very small device having an ARM processor, a Wi-Fi interface and Linux operating system. The hardware architecture of the new system is shown in Fig. 6. OverAir is mounted directly on the pallet carrying the product throughout the system and communicates with other devices using its Wi-Fi interface [9].

Resources (robots) and the PLC are connected in the same way, using digital inputs/outputs together with Ethernet. Communication between the PLC and OveroAir is achieved by dedicated software developed on the gateway PC directly connected to the PLC, as in Fig. 3. The communicating application has two functions:

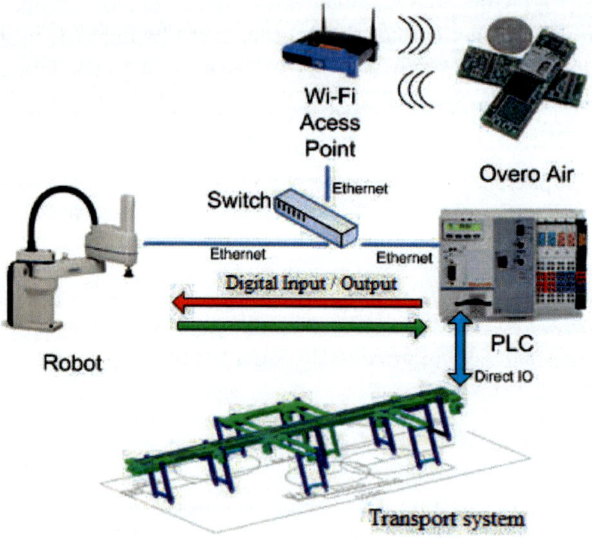

Fig. 6 New system architecture

- Communication of data through TCP/IP with OveroAir.
- Acting as an OPC client that can log in the existing server in order to facilitate the data transfer to the PLC

This application receives from OveroAir the standard data structure for the current operation and sends this information to the PLC. The PLC performs routing in the same way as in the first approach but this time it does not receive an offline planning from the supervisor computer but only the current operation from the product itself. Moreover it tracks the pallets through the cell and stores this information in several integer variables, one for every pallet existing in the cell, representing the segment of the conveyor where the product is located at that moment.

5.1 Routing a Pallet

In order to manage the route of a pallet, the PLC has now a slightly modified data structure containing the following information:

- *Pallet ID*: the ID of the pallet is written on the RFID memory on the pallet and is fixed during its production lifecycle
- *Current Position*: represents the conveyor segment where the pallet travels at the current moment (the conveyor can has maximum 256 motion segments)
- *Routing step*: is the identifier of the next destination workstation for that pallet
- *Routing Destinations*: is a list of destinations that the pallet should reach

For routing purpose, OveroAir can send a Routing destination and can read the Current Position anytime, in order to check the correct execution of the routing commands issued by the PLC. The routing algorithm works in the following way:

- Initially the Routing Step is 0 and the pallet is outside the cell; anyhow, before the entrance of the pallet in the cell the AveroAir IED will write at least the first value of the Routing Destinations.
- Once the pallet entered in the cell, the PLC will route the pallet on the main loop of the conveyor while the Routing Step will take the first value of the Routing Destination.
- At every stopper the PLC reads the Pallet ID, accesses the data corresponding to that pallet and checks if the Routing Step is equal to the robot's ID from that location; if so, the pallet will enter the workstation; else it will travel further on the next segment of the conveyor.
- After a pallet is completely processed by the robot, the Routing Step will be updated with the next address from the Routing Destinations list.
- Once the pallet is routed on segment 255, it is considered that the pallet was taken out from the cell so that the PLC will update the Routing Step with the value 0, the pallet being prepared to re enter the cell (for a new product) if necessary.

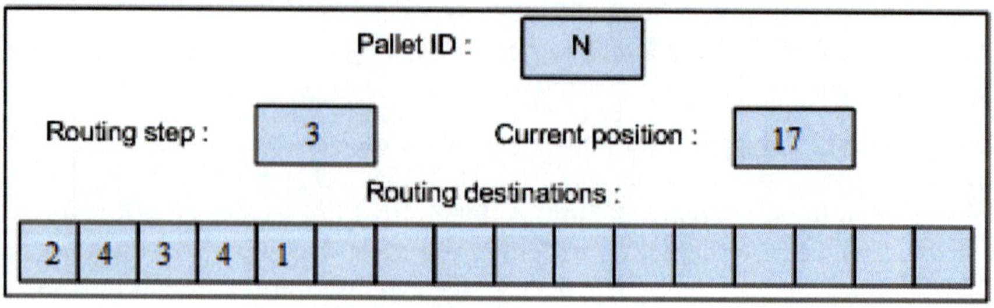

Fig. 7 New data structure for a pallet inside PLC

6 Communication Inside the System

The PLC is the key element of the cell because it is the central node that facilitates the communication between all involved system's components. It has also the task of controlling the execution of the transport operations involved. So, the communication between PLC and Robots is based on a dialog type model. According to this model, any communication protocol is initiated by the PLC; the Robot executes the orders and answers only in the mode depicted by the protocols.

Any communication protocol that was implemented is able to perform three main tasks:

- to monitor the robot's online / offline status
- to coordinate the robot task execution
- to transmit the codes corresponding to the requested job and complementary execution report.

Considering the existing requirements and the available communication interfaces, two protocols were developed:

- *Ping* – this protocol detects the online / offline status of the robots
- *Synchronization* – this protocol implements the robot task execution

Since it is of up most importance to know which robots are online and when a robot changes its online / offline status, a protocol was implemented in order to provide continuously this information [2]. Because of the high redundancy required by this protocol, it uses only the I/O lines. The protocol runs as follows:

- the PLC detects that Ready is 0 so it sets Request_Job to 1 and transmits the job code over TCP/IP.
- the Controller reads the TCP/IP code and evaluates if it can execute the job; in this case it will send the job acceptance code, if not it will send a job reject code; if the job is rejected then communication stops (we assume that the job is accepted).
- the PLC brings the pallet in the working position and sets Pallet_In_Position to the value 1.

- the Controller sets Ready to 1 (the PLC sets Request_Job to 0) and begins executing the job.
- upon job execution, the Controller sets Job_Done to 1, Ready to 0, and sends the job completion report over TCP/IP.
- the PLC takes the pallet, sets Pallet_In_Position to 0, the Controller sets Job_Done to 0 and is ready to restart the protocol.

7 Practical Results and Conclusions

After several tests performed with both planning solutions, the conclusion was that the control solution completely fulfils the initial objectives, meaning that products are routed in such a way that cell blocking is impossible and the all the products are processed exactly like it is planned.

Some evaluation was carried out regarding the insertion time and finish time for a batch of products. The results were rather satisfactory; the insertion time was almost always the same as the planned one for every product. Also, slight errors can be observed in achieving the production execution times defined by the planner; this is caused by the small speed variations generated by the friction conveyor and the small time variations of robot task execution (part locations detected by vision may vary).

It is worth mentioning that the errors for product execution time do not add up / multiply over time, which confirms the correct implementing of offline schedules.

The main contribution of this work was the proposal and implementation of a specific data structure together with distributed software architecture for PLC as conveyor (product routing) controller and for robots that control operations (services for products). Using this approach a maximum of four products can be inserted and processed simultaneously in the manufacturing system, ensuring that no transportation blocking happens.

Future work will consider ambient intelligence solutions gathering process, resource and environment data for on line heterarchical resource allocation [11, 16].

References

1. Babiceanu, R.F.: Framework for control of automated material-handling systems using holonic manufacturing approach. Int. J. Prod. Res. 42, 3551–3564 (2007)
2. Barata, J., Camarinha-Matos, L.M.: Shop floor reengineering to support agility in virtual enterprise environments. In: E-Business and Virtual Enterprises, pp. 287–291. Kluwer Academic Publishers, London (2000)
3. Borangiu, T., Gilbert, G., Ivanescu, N., Rosu, A.: Holonic Robot Control for Job Shop Assembly by Dynamic Simulation. In: Proc. of the 16th Mediterranean Conference on Control and Automation – MED 2008, Ajaccio (2008)
4. Lastra, J., Delamerm, I.: Semantic web services in factory automation: Fundamental insights and research roadmap. IEEE Trans. on Industrial Informatics 2, 1–11 (2006)

5. Morel, G., Panetto, H., Zaremba, M., Mayer, F.: Manufacturing enterprise control and management system engineering: Rationales and open issues. IFAC Annual Reviews in Control (2003)
6. Pétin, J.-F., Morel, G.: A product-driven reconfigurable control for shop floor systems. Studies in Informatics and Control 16 (2007)
7. Rahimifard, S.: Semi-heterarchical production planning structures in the support of team-based manufacturing. International Journal of Production Research 42 (2004)
8. Van Brussel, H., Wyns, J., Valckenaers, P., Bongaerts, L., Peeters, P.: Reference Architecture for Holonic Manufacturins Systems: PROSA, Computers in Industry. Special Issue on Intelligent Manufacturing (1998)
9. Overo Air documentation (2010), `https://www.gumstix.com/`
10. Flammini, A., Ferrari, P., Marioli, D., Sisinni, E., Taroni, A.: Wired and wireless sensor networks for industrial applications. Microelectronics Journal 40, 1322–1336 (2009)
11. Nett, E.: WLAN in Automation - More Than an Academic Exercise? In: Proceedings of Latin-American Symposium on Dependable Computing, pp. 4–8 (2005)
12. Neumann, P.: Communication in industrial automation—What is going on? Control Engineering Practice 15, 1332–1347 (2007)
13. Senol, S., Leblebicioglu, S., Schmidt, E.G.: An algorithmic approach on networked control system design. Journal of Network and Computer Applications, 1326–1341 (2011)
14. Rosu, A., Borangiu, T., Ivanescu, N.A., Raileanu, S.: Vision-Guided Part Feeding in a Holonic Manufacturing System with Networked Robots. In: Proc. of the 17th International Workshop on Robotics in Alpe-Adria-Danube Region RAAD 2008, Ancona, Italy, p. 97 (2008) ISBN: 978-88-903709-0-8
15. Tatara, E., Cinar, A., Teymour, F.: Control of complex distributed systems with distributed intelligent agents. Journal of Process Control 17, 415–427 (2007)
16. Khalgui, M., Mosbahi, O.: Intelligent distributed control systems. Information and Software Technology 52, 1259–1271 (2010)

Using Hybrid Petri Nets for Performance Analysis in Manufacturing Systems

Calin Munteanu, Simona Caramihai, Mihnea Alexandru Moisescu, and Ioan Stefan Sacala

Politehnica University of Bucharest, Control Systems and Computer Science Faculty, 313 Spl. Independentei, Sector 6, Bucharest, Romania
calin.munteanu@aii.pub.ro

Abstract. The chapter presents a relatively new way of modelling manufacturing systems using an extension of Petri nets formalism – Hybrid Petri nets. The formalism allows a macroscopic analysis of manufacturing systems evolution, presenting the evolution tendencies. For a better understanding, an analysis algorithm and a case study are presented.

Keywords: Hybrid Petri Nets, manufacturing, modelling, performance analysis.

1 Introduction

The manufacturing systems are, by their nature, one of the most well known, but also complex types of discrete event systems. From the tools one can use to model and analyze discrete event systems, the automata theory and Petri nets formalisms are of large use. But taking into account the modelling power of the two formalisms and the fact that today's system are more and more complex, the Petri net formalism is best suited for modelling such systems, and this includes flexible manufacturing processes.

In present chapter two classes of Petri nets – continuous and hybrid – will be briefly presented and a representative example will be given in order to better understand the way a model can be build and how the analysis is done using these formalisms.

The Hybrid Petri Net (HPN) formalism was introduced in [2]. Basically, a HPN consist from two parts: a discrete one (as presented in [5]) and a continuous one (presented in [1]). The discrete Petri Nets are very well known and will not be described in this chapter. The Continuous Petri Nets (CPN) are relatively new related to discrete ones and will be detailed in the next section.

2 Continuous Petri Nets

The CPN represent a limit case of T-timed Petri nets, with a large number of tokens and which have, accordingly, a huge marking graph. In order to prevent such a

T. Borangiu, A. Thomas, and D. Trentesaux (Eds.): SOHOMA 2011, SCI 402, pp. 255–264.
springerlink.com

marking graph, CPN are introduced. The evolution graph associated to a CPN is much simpler than the one associated to the corresponding Discrete Petri Net (DPN), because it consist from macro-markings - general states that indicate the evolution tendencies (either accumulation or loss of markings) and not each state / value.

To better understand the CPN, an analogy (places - tanks with fluids) on one side and (transitions – pipes) on the other side can be made.

There are many new issues when we talk about CPN. Some of them will be explained in the following.

The first aspect is related to the domain from which the numbers (tokens and arc weights) take their values. While DPN's have a natural number of tokens in places and a natural number as an arc weight, in CPN's the markings and arc weights can be positive rational numbers. We make a difference between the terms *token* and *marking*. The term token is used in DPN's and refers to the quantitative information related to places. Marking is the term used to refer to the quantitative information related to CPN's places and (as defined) is composed from an infinite number of tokens. Related to this delimitation, the minimum quantity that can be found in a place (representing one token) is denoted in CPN with 0^+.

The second aspect is related to the way a transition fires. In DPN one has a discrete firing (one at the time, with some time interval between two consecutive firings). In CPN we have a continuous firing (like liquid flowing) and one distinguishes two characteristics: a maximal firing speed (denoted with V_j) and an instantaneous firing speed (denoted with v_j). The maximal speed is a net/system characteristic (like the maximum flow that can pass through a pipe). The instantaneous speed is an evolution characteristic and depends on the actual state of the system. The instantaneous speed can range between zero and the maximal speed and represents the transition's firing speed.

If a place has at least one input transition with an instantaneous speed greater than 0, then the place is called fed.

There are also two types of transition enabling. If all the input places of a transition have a positive marking, then the transition is called *strongly enabled* and it will fire at its maximal speed.

A transition whose input places have either a positive marking or a zero marking but are fed by other transitions (at least one input place having a zero marking and being fed) is called *weakly enabled.* The weakly enabled transitions fire at a speed that is equal to the minimum between its maximal speed and the feeding speed of its input places as shown in the formula [3]:

$$v_j(t) = \min_{P_i \in Q_j(t)} \frac{1}{\mathrm{Pre}(P_i, T_j)} \sum_{T_k \in {}^{o}P_i} \mathrm{Post}(P_i, T_k) * v_k(t), V_j \tag{1}$$

where:

- $Q_j(t)$ represents the subset of input places for T_j transition such as $m_j(t) = 0$ (or 0^+);
- $Pre(P_i, T_j)$ – represents the weight of the arc from P_i to T_j;
- $Post(P_i, T_j)$ – represents the weight of the arc from T_j to P_i;
- ^oP_i – represents the set of input transitions for P_i place.

All the above mentioned information applies if there are no conflicts. The conflict situation and its resolution are a very important aspect when talking about CPN's.

First, let us define the conflict situation. In a CPN, there is an actual conflict if:

1. The input place has a zero (or 0^+) marking.
2. The input place is fed.
3. The feeding of the input place is not sufficient to satisfy all the enabled transitions demands.

With the help of the examples in Fig. 1 we shall point the different possibilities.

In the first situation, case a) there is not an actual conflict because there is a positive marking in the P_2 place.

In the second case b) the P_1 place is fed with the speed of 4, so it can give further with the speed of 4. The same applies to places P_2 and P_3. So T_4 (and the same applies to T_5) can fire at an instantaneous speed equal with minimum between (4 – the feeding speed from P_1, 4 – the feeding speed from P_2 and 2 – the maximum speed of T_4), that is at a speed of 2. So, because the input in the P_2 place is 4 and the output is 2+2=4, there is no problem in splitting the input. Also in this case there is no actual conflict.

In the third case, although the sum of maximal speeds of the output transitions (2+2=4) exceeds the speed of the input transition (3), the actual firing speeds for T_4 and T_5 are, according to (1), 1.2 and 1.7, and so, both transitions can fire without any problem at that speeds (the sum of firing speeds is 2.9, less than the input speed).

The fourth case is an example of actual conflict. The T_4 transition can fire with a speed equal with 1.2 and T_5 with 1.7. The output sum (2.9) is greater than the feeding speed from T_2 (2). That means that we have to decide how to split the incoming flow.

So far two conflict resolution policies have been defined: *priority* and *sharing*.

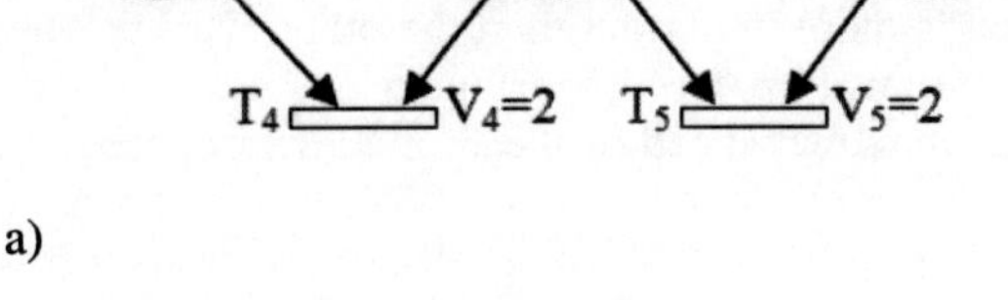

a)

Fig. 1 Structural conflicts in CPN a), b), c) – no actual conflict, d) actual conflict

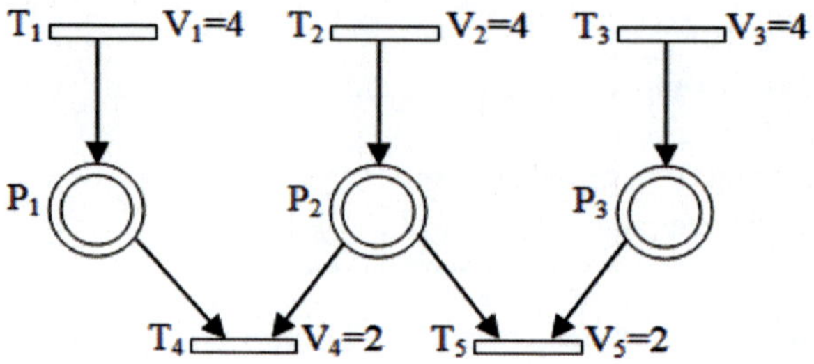

b)

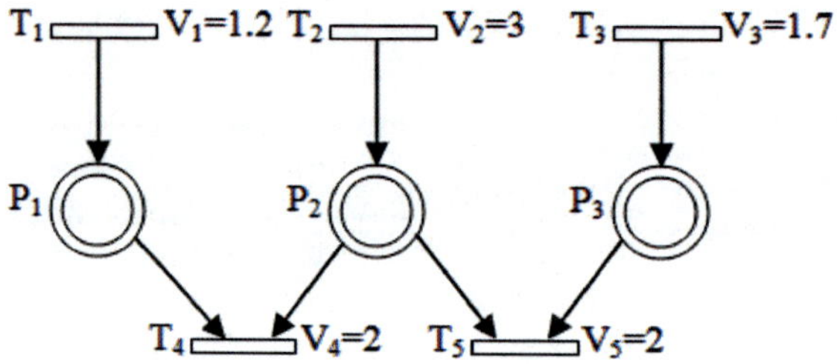

c)

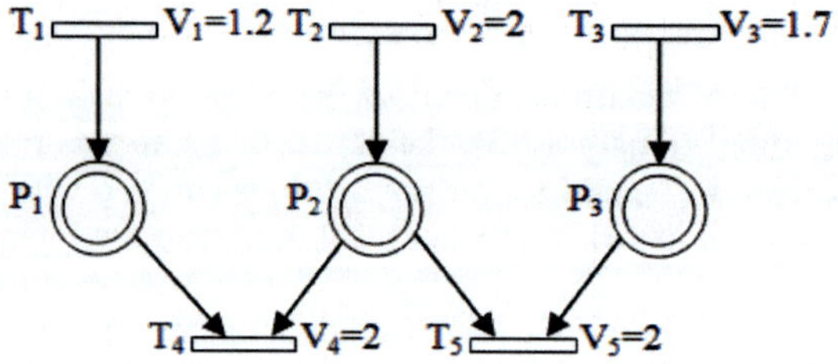

d)

Fig. 1 (*continued*)

We will use for explanation the case where there are only two transitions. The general case is presented in [3] where multiple combinations and multiple priority levels are introduced.

Priority means that one transition is more important than the other. This is why we will always give markings to that transition until its instantaneous speed cannot grow anymore (either it reaches the maximal speed or the limitations imposed by the other places in the net). This resolution rule represents a very strict rule which should be considered carefully.

For example, in the d) case, if we give priority to T_4 transition (over T_5), then it will fire at its maximal possibility (with a speed of 1.2). Then, the remaining flow from place P_2 (2-1.2=0.8) can be given to T_5. If T_5 takes priority over T_4, then T_5 will fire at its maximal possibility (1.7) and the remaining flow (2-1.7=0.3) will be given to T_4.

The second resolution rule is represented by *sharing*. This is a more relaxed rule. It associates with every transition a sharing weight and tries to give such instantaneous speeds so to equalize the instantaneous speeds multiplied by the corresponding weight. But if the speeds necessary to respect the equality cannot be obtained, then the rule is bypassed.

We will give examples using the CPN from Fig. 1 d). If the sharing will aim to obtain equal speeds between T_4 and T_5, then the incoming flow (2) will be split, and we will obtain for both T_4 and T_5 the same instantaneous speed of 1, speeds that respect the restrictions.

But if we aim to obtain such speeds that the speed of T_4 to be 4 times greater than the one of T_5, then, according to the sharing rule, we must have v_4=1.6 and v_5=0.4. But the T_4 transition can fire only with a maximum instantaneous speed of 1.2. The rest (0.8) will be given to T_5. So, in this case, we will not respect the sharing restriction (but we will be as closely possible of doing that).

Another very important issue related to CPN is related to its evolution. For DPN we build a marking graph in which every state corresponds to a state of the system.

For CPN an evolution graph is built, each state of it (called *macro-marking*) representing not a state of the system but infinity of states of the modelled system. The important feature of a macro-marking is that the instantaneous speeds for each transition in the net remain unchanged for the duration of the macro-marking. That means that for a macro-marking the marking evolution for each place is a linear one and can be characterized by its tendency – called *balance* that is equal to the difference between all the input and all the output. A change between two macro-markings appears when a continuous place is emptied.

3 Hybrid Petri Nets

The HPN represent a combination between DPN and CPN [4]. The CPN can be viewed as the process evolution and the discrete part as a commanding part. The discrete places that are connected with continuous transitions model the servers that can do that action / activity. The actual number of servers is given by the number of tokens presented in that place.

In HPN there can be four types of conflicts:

a) between two discrete transitions (having as input place either a discrete or continuous place) solved by firing the transition having the shortest delay,
b) between two continuous transitions having as input place a continuous place (solved as in CPN's case),

c) between one discrete and one continuous transitions (having as input place either a discrete or a continuous place) – in which the discrete transition has always priority,
d) between two continuous transitions having as an input place a discrete place – the number of servers represented by the number of tokens in that place is split in order to do both actions. As in continuous case, resolution rules are given (either priority or sharing).

In the case of HPN, the transitions have associated flow rates from which the maximal firing speed is computed [3].

The macro-markings corresponding to a HPN are characterized by two elements:

- the instantaneous speed for continuous transitions is unchanged;
- the marking for discrete places remains unchanged.

In the case of HPN there can be two reasons for a macro-marking change:

- a continuous place is emptied OR
- a discrete transition is fired and the effect of that firing is either a change of the discrete marking, or a discrete change of continuous marking (a continuous marking either gains or looses a positive amount of marking).

4 Modelling Manufacturing Systems with Hybrid Petri Nets – Case Study

We will present different aspects related to modelling a manufacturing system using HPN by the help of an example.

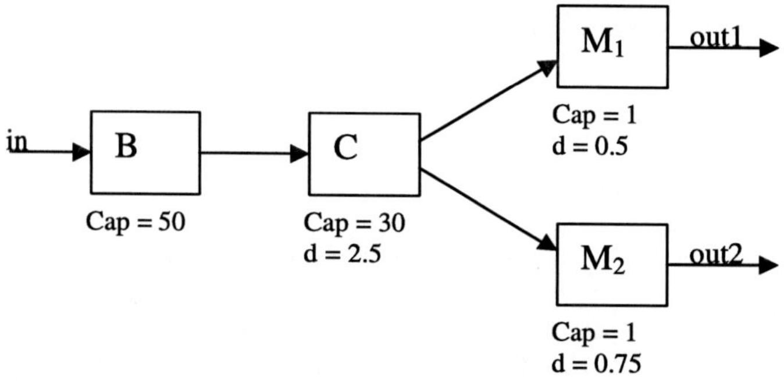

Fig. 2 Manufacturing system

Let's consider the manufacturing system in Fig.2. The system processes one type of parts. The parts are entering in the system and stored in a 50 place capacity buffer (B), from which a conveyer (C) takes them for manufacturing at two machine tools, that are performing the same operation, but at different speeds. After the processing on machines, the parts are leaving the system. When the B buffer is full, the feeding of B is stopped. The feeding will be resumed when the number of parts in B reaches 20. When B is fed, a part arrives each 0.1 Time Units (TU). On the conveyer, the time needed for transporting one part to a machine is 2 TU. The

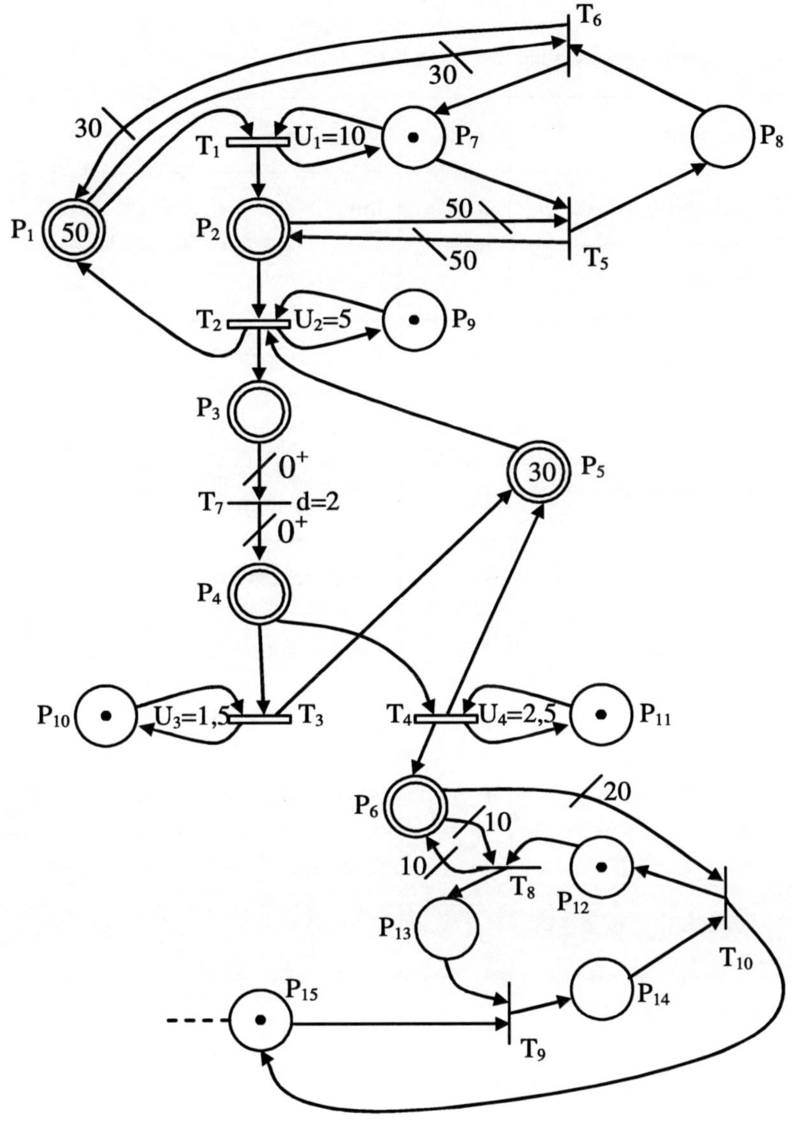

Fig. 3 HPN model of the system

processing time of M_1 machine is 0.4 TU and the one for M_2 is 0.67 TU. After processing on M_2, the parts leave the system without any other restriction. After processing on M_1, the parts are stored into pallets of 20. When a pallet has 10 parts, a request will be issued for an AGV to pick up the pallet. When the pallet is full and the AGV available, the pallet leaves the system.

The HPN model corresponding to the system in Fig.2 is represented in Fig. 3. The significance of each element is given in Table 1.

Table 1 Significance of nodes in Fig. 3

Node	Meaning	Node	Meaning
P_1	Availability of B	T_1	Input in B
P_2	B occupation	T_2	Input in C
P_3-P_4	C occupation	T_3	Processing on M_2
P_5	Availability of C	T_4	Processing on M_1
P_6	Number of parts in pallet	T_5	Blocking entrance on B
P_7	Allow entrance in B	T_6	Allowing entrance on B
P_8	Restrict entrance in B	T_7	The transportation on conveyer
P_9	C functional	T_8	Request for an AGV
P_{10}	M_2 functional	T_9	AGV reservation
P_{11}	M_1 functional	T_{10}	Sending out the pallet
P_{12}	Availability for AGV request		
P_{13}	AGV requested		
P_{14}	AGV reserved		
P_{15}	AGV available		

There are some important structures in the model that will be presented next.

First, the cycle P_7 -> T_5 -> P_8 -> T_6 -> P_7. It corresponds to enabling/disabling parts entrance on the B buffer. The condition for existence of 20 markings in P_2 is reversed checked: we are checking the complementary place (P_1) for complementary amount (50-20=30).

Next follows the P_3 -> T_7 -> P_4 structure that models the conveyer transportation. This structure is used to introduce the delay: each marking that arrives in P_3 will appear in P_4 after the delay corresponding to T_7 (2 TU).

Finally, the P_{12} -> T_8 -> P_{13} -> T_9 -> P_{14} -> T_{10} -> P_{12} models the availability, request and usage of AGV.

4 Analysis

Let us now examine how the net evolves. First some steps will be explained, followed by the whole evolution. Because the discrete marking of P_9, P_{10} and P_{11} will always be equal to 1, we will not further refer to them. Instead, the discrete marking will refer to (in order): $m_D=[m(P_7)\ m(P_8)\ m(P_{12})\ m(P_{13})\ m(P_{14})\ m(P_{15})]^T$. The continuous marking will have the following structure: mc=[m(P_1) m(P_2) m(P_3) m(P_4) m(P_5) m(P_6)]T.

The initial moment is t=0. The discrete marking is $m_D=[1\ 0\ 1\ 0\ 0\ 1]^T$ and the continuous one is: $m_C=[50\ 0\ 0\ 0\ 30\ 0]^T$. T_1 is strongly enabled and it will fire at its maximal speed (v_1=10). Because of it, P_2 will be fed and it will allow for its output transitions to fire at maximum speed of 10. Because the maximal speed of T_2 is only 5, $v_2 = 5$. This will feed P_3 and will allow a continuous firing of T_7, but only after 2 TU. So P_4 will not receive markings sooner than t=2. Having an input of 10 and an output of 5, the P_2 place will have a positive balance (5). That means that it can reach the amount needed for enabling T_5 in 10 TU.

To resume the speeds for the first macro-marking: $v=[v_1 = 10, v_2 = 5, v_3 = 0, v_4 = 0]^T$. The balances for will be $B(P_1) = v_2-v_1 = 5-10 = -5$, $B(P_2) = v_1-v_2=10 - 5 = 5$, $B(P_3) = v_2 = 5$, $B(P_4) = 0$, $B(P_5) = -v_2 = -5$, $B(P_6) = 0$.

The second macro-marking will start at t=2 with the firing of T_7. At the beginning of the macro-marking, the continuous marking will be: $m_C=[40\ 10\ 10\ 0\ 20\ 0]^T$. Because of the 0+ weight of the arcs, the firing will be a continuous one, so it will replicate the speed of the T_2. Accordingly P4 will be fed at the speed of 5 and it will enable the firing of both T_3 and T_4 at the speeds of 1.5 and 2.5. So the speed vector will be $v = [10\ 5\ 1.5\ 2.5]^T$ and the balances: $B(P_1) = v_2-v_1 = 5-10 = -5$, $B(P_2) = v_1-v_2= 10- 5 = 5$, $B(P_3) = 0$, $B(P_4) = v2-v3-v4 = 5-1.5-2.5 = 1$, $B(P_5) = v3+v4-v_2 = -1$, $B(P_6) = v4 = 2.5$.

The important events for the net at this stage are the moments when:

- P_2 reaches 50 (T_5 will become enabled) – denoted E1;
- P_5 reaches 0 (T_2 will fire at a speed lower than 5) – denoted E2;
- P_6 reaches 10 (T_8 will become enabled) – denoted E3.

According to the making at the beginning of the state and the computed balances, the above mentioned events will occur: E1 after 8 TU, E2 after 20 TU, and E3 after 4 TU.

So the third macro-marking will begin at t=6, when E3 occurs (the marking of P_6 reaches 10 and T_8 becomes enabled and fires). In order to simplify a little bit the analysis, although we introduced in the model the possibility that the AVG to be linked with other operations (the dotted line), we will assume that it is available and will not be used in other operation than the pallet transport, and that the transport duration is null. So, after firing of T_8, a token will be placed in P_{13} and the T_9 will become enabled and will fire. Accordingly, the discrete marking will be $m_D=[1\ 0\ 0\ 0\ 1\ 0]^T$. For this macro-marking, the continuous part evolution is not changed, so that $v = [10\ 5\ 1.5\ 2.5]^T$ and $B(P_1) = -5$, $B(P_2) = 5$, $B(P_3) = 0$, $B(P_4) = 1$, $B(P_5) = -1$, $B(P_6) = 2.5$.

As important events from the third macro-marking we will still have E1, E2 and in addition the event corresponding to the fact that P6 reaches 20 markings (and so T_{10} will become enabled and will fire) - denoted with E4. E1 and E4 will occur after 4 TU and E2 after 16 TU. So the next important moment will be $t=10$, when two discrete transitions (T_5 – stopping of parts input and T_{10} – pallet leaving the system) will fire.

The system's evolution is analyzed following the steps mentioned above until a cyclic behavior is reached (the system reaches a state in which it was before).

5 Conclusions

The Petri net is one of the well known and used formalisms for modelling manufacturing systems. Taking into account the fact that manufacturing systems become more and more complex, a technique was presented in this chapter of modelling them using HPN formalism, as well as the way to perform the related analysis. The tool is very useful if the system has some components with linear time-depending evolution, components having a great number of states or components which are processing a great number of parts. This is so because the analysis results are more compact (the obtained evolution graph has a much simpler structure than the corresponding discrete Petri net marking graph) and they still encompass all the information obtained from a discrete Petri net marking graph but at a higher level (a macroscopic one) related to the system's properties (liveness, blocking, conflicts, a.o.) and its evolution tendencies.

References

1. David, R., Alla, H.: Continuous Petri Nets. In: Saragosse (ed.) 8th European Workshop on Applications and Theory of Petri Nets, pp. 275–294 (1987)
2. David, R., Alla, H.: Du Grafcet aux Réseaux de Petri. Hermes Publications, Paris (1992)
3. David, R., Alla, H.: Discrete, Continuous, and Hybrid Petri Nets. Springer, Heidelberg (2010) ISBN: 978-3-540-22480-8
4. Le Bail, J., Alla, H., David, R.: Hybrid Petri Nets. In: Proc. European Control Conference, Grenoble (1991)
5. Petri, C.A.: Communication with Automata, Supplement 1 to Technical Report RADC-TR-65-337, N.Y (1965); Translation by C.F. Greene of Kommunikation mit Automaten, PhD Dissertation, University of Bonn (1962)

A JADE Environment for Product Driven Automation of Holonic Manufacturing

Silviu Raileanu[1], Mihai Parlea[1], Theodor Borangiu[1], and Octavian Stocklosa[2]

[1] University Politehnica of Bucharest, Dept. of Automation and Industrial Informatics, 313 Spl.independentei, Sector 6, 060042 Bucharest, Romania
{mihai.parlea,Silviu.raileanu,theodor.borangiu}@cimr.pub.ro
[2] East Electric, 159, Bd. Basarabia, Sector 6 Bucharest, Romania
octavian.stocklosa@eastelectric.ro

Abstract. This chapter presents a system control frame for implementing the switch between three different production strategies, each with its own planning objective (e.g. makespan, resource load) and perturbation avoidance capabilities. In order to implement such a system, Intelligent Products that possess computing and decision taking abilities are used, the system model and class diagram being presented for all the system agents that are implemented using the JADE Environment.

Keywords: multi-robot production cell, product driven automation, strategy switching, agent orientation, intelligent embedded device.

1 Introduction

The present economy trends demand that production cells equipped with robotized workstations offer the best possible performance and behave predictable over time. The only way that can guarantee the best possible performance is a hierarchical (or centralized) driving strategy, which executes an off-line generated schedule controlled by a central entity. Unfortunately, perturbations occur when driving such cells, such as failure or stock depletion of a robot workstation. This means that the entire cell needs to stop, at least until the planning is recalculated, with unfavourable consequences on the production schedule. In order to avoid cell downtime, efforts have been made to eliminate production schedule and distribute the driving scheme so that a perturbation will no longer stop the entire cell, thus resulting in a decentralized approach. To summarize, in the control of production cells two main approaches have emerged, both having complementary advantages and disadvantages:

- Decentralized
  - production starts straightaway with high perturbation rejection;
  - performances are bad as no elaborate planning is made [1, 2, 3, 4, 5].

T. Borangiu, A. Thomas, and D. Trentesaux (Eds.): SOHOMA 2011, SCI 402, pp. 265–277.
springerlink.com

- Centralized
  - shorter production times as high level planning algorithms are used;
  - high level planning algorithms have poor perturbation rejection and take considerable processing time before providing a production schedule [4, 5, 6, 7]

Intelligent production cells are usually highly oriented towards either the centralized or the decentralized approach, the internal product tracking mechanism usually rendering impossible the implementation of both approaches. But now, as we are entering a new era of Intelligent Products that bring processing power on the products to be manufactured (machined, assembled, ...), it is not only possible to implement multiple strategies with various degrees of centralization in batch planning and resource allocation, but also to switch between these strategies during production.

2 Using Intelligent Products for Decision Taking in an Industrial Environment

2.1 Control System Model

Implementing intelligent product control architecture requires implementing a Control System Model of the type represented in Fig.1. Production orders are generated by the Production Management and Control Application (PMCA) based on the requests issued by clients. The PMCA has two main roles, the Management and the Control of the automated production.

The **Management** part refers to production tracking, meaning that the production orders that are introduced in the system are tracked in real time so that one always has an updated real-time description about the orders waiting to be introduced in the intelligent production cell, the orders that have already been executed and the status of the orders that are currently in the production cell.

The state of the production cell is also tracked, which provides information about the current position of the *Product Agents* and the current state of the workstations (the status of workstations - online, failed, what is their job offer and part supply). All this information (except for the orders that have already been completed) is inserted into the Offline Planner at production start or when an internal (workstation breakdown) or external (occurrence of rush order) perturbation is encountered and it is decided to continue production using a hierarchical approach.

The **Control** part refers to changing the current production driving strategy. This is done by managing the production orders using two distinct buffers. Orders are entered from the Heterarhical Buffer without any prior planning while the Hierarchical Buffer contains the orders scheduled and with the job sequence planned / resources allocated as they were generated by the Offline Planner. The role of the control part is to make sure that both buffers are managed accordingly with the current strategy that is run on the production cell and that the **Product Agents** receive the orders from the correct buffer.

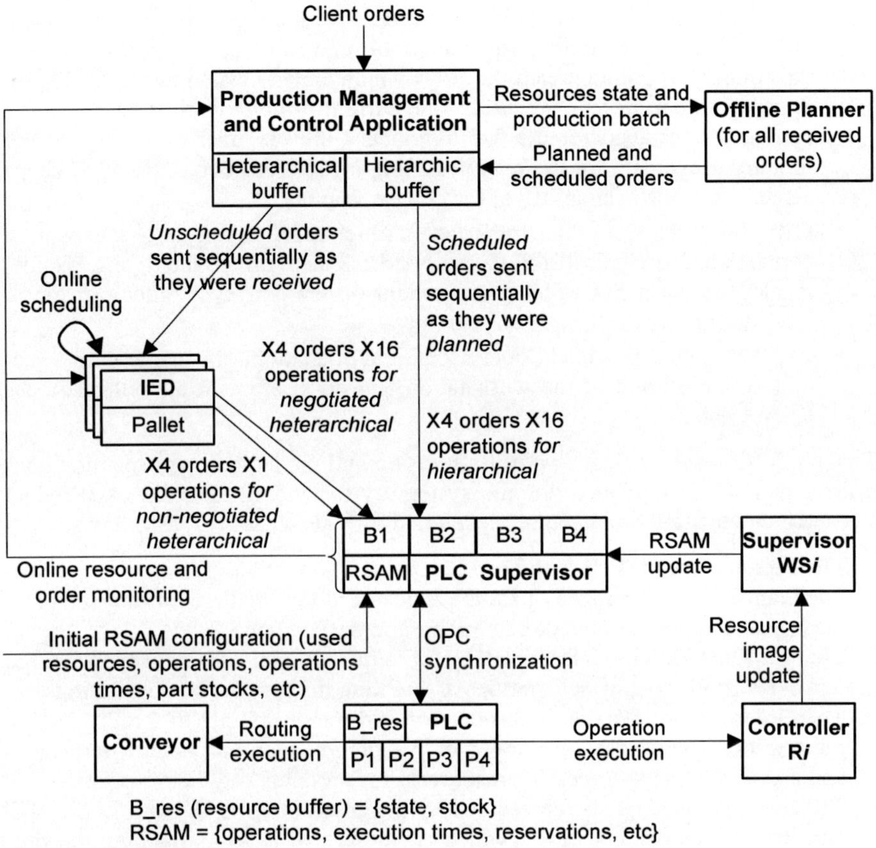

Fig. 1 Control System Model

The other entities included in Fig.1 have the following assignments:

- **IED / Pallet**: this is the Intelligent Product composed of an Intelligent Embedded Device (IED) mounted on the to the pallet carrier (the product will be created progressively on the pallet circulating in the production cell); using the assigned product model, the IED reads the Resource Service Access Model (RSAM) Agent to obtain information about the production cell architecture and current status of resources in order to take production-related decisions (for example which workstations will be used for which jobs).
- **PLC Supervisor**: this is the PC used to input the production orders in the cell's PLC; it contains a buffer with the partial or total planning of the product currently active in the production cell and the RSAM.
- **RSAM**: this is the Resource Service Access Model which keeps track of the reservations made on the conveyor segments and robot workstations.

- **PLC**: it is used for issuing transport orders for the conveyor and processing orders for the workstations (machine tool- and robot controllers) to execute the commands coming from the IEDs which are associated to the Intelligent Products under current execution (the current *product packet*) based on the commands received for the four products - the maximal dimension of the product packet is 4 for the manufacturing layout considered; the PLC keeps also track of the availability of the cell's resources.
- **CRi**: this is the Controller for Robot *i*, which receives application execution commands from the cell's PLC and performs them on the Intelligent Product located on the pallet which was brought on the conveyor branch of the ith workstation.
- **PC WSi**: this is the local Supervisor for Workstation *i*, in charge with a secondary monitoring of the workstation and duplication of all cell resources RSAM data.

This production scheduling, resource allocation and manufacturing execution control architecture is seen as a holonic system by expanding the PROSA (Product Resource Order Staff Architecture) developed by Valckenaers [9] as follows:

1. The Product Holon (PH) is the *product model*, which is transferred to the Offline Planner - in centralized operating mode or to the Intelligent Embedded Device (IED) - in decentralized (heterarchic) operating mode.
2. The Resource Holons (RHs) are the PLC - driving the cell's conveyor and the Robot Controllers - physical entities assembling the product in workstations.
3. The Order Holon (OH) is the production order that is input to the Production Supervisor; it comprises a number of *production plans* [operation sequences and association with processing resources] for products and their due times.
4. The Staff Holon (StH) is represented by the ensemble of PMCA, RSAM, PLC and Robot Supervisor entities, which are described as units tracking the production process and driving it through decisions and additional information.
5. For the *Product Agent* a new term will be used, that of an **Active Holon Entity** (AHE), in order to represent both its mobility and decisional taking abilities.

2.2 Production Driving Strategies

As was pointed out earlier, using the Intelligent Product it is possible to implement multiple control strategies in a production cell. The following strategies are proposed:

1. **Hierarchical**: a strategy based on *offline computed production plans* for all requested products (batch); it is assumed that: (a) an ordered list of production plans is the result of *product planning* at batch level, (b) all *operations were scheduled* for each planned product, and (c) for each operation scheduled *a resource was allocated*. Production plans are computed using a simulator which contains the sets of production rules, the topology of the transportation system (closed-loop cell conveyor) and the resources' features; the planning and alloca-

tions sequence with the shortest execution time is offered as the nearly optimal solution.

2. **Non-Negotiated Heterarchical**: when frequent perturbations occur (such as resource failures or storage depletion of components), offline batch product planning, operations scheduling and resource allocation is no more useful and instead, with no planning the operations order and resource allocation is done *stepwise for each product* entering the cell, so that when a perturbation occurs its impact will be minimal and it will be rejected with the least amount of effort possible. For this strategy only the next operation and resource are sought after finishing the current operation upon each product in execution.
3. **Negotiated Heterarchical**: products are still inserted in the cell with no a priori planning (they come from the Heterarhical Buffer) but the AHE will schedule its entire trajectory (complete sequence of operations and list of associated resources) through the production cell. This strategy is devised so that it will offer a better performance than the Non-negotiated Heterarhical one (quasi-optimal at the level of *product packet* - the *p* products in simultaneous execution in the cell) but still should support on line rescheduling when a perturbation affects one already allocated resource.

Based on initial configuration (decided by an expert) the system can run only according to one production strategy or switch between strategies as depicted in Fig. 2. The human expert will assure that there are no perturbations present when the system is brought online so the system will start with the Hierarchical strategy, then, if a perturbation occurs, the system will switch to the Negotiated Heterarhical strategy. If the situation continues to deteriorate then the control strategy will be switched again to the Non-negotiated Heterarhical strategy. If, on this strategy, perturbations will tend to reduce their frequency of occurrence or even disappear, then the system will switch back first to the Negotiated strategy and then to the Hierarchical one, if, in the absence of any perturbations the system attain the initial stable, predictable state.

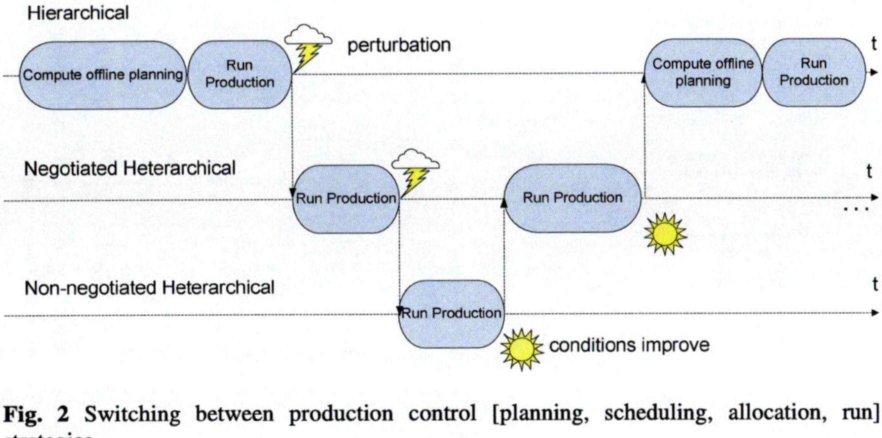

Fig. 2 Switching between production control [planning, scheduling, allocation, run] strategies

2.3 Using the CNP to Obtain Workstation Offers and Make Reservations

In order to obtain workstation offers, an improved version of CNP as described by Smith [10] will be used. The following algorithm implements the CNP (see Fig. 3 too):

1. When they are brought online, all resources will register to the RSAM Agent specifying all the jobs they can execute.
2. An AHE will request from the RSAM Agent all workstations that can execute the requested service.
3. The AHE will contact all designated resources in order to obtain data about their performances and component storage status.
4. The AHE will contact the RSAM Agent in order to obtain the reservations currently made on the conveyor segments.
5. The AHE will choose the resource with the best combination of offer and availability (as will be described in the following section).
6. The AHE will inform the RSAM about the new conveyor segments reservations that it will make.

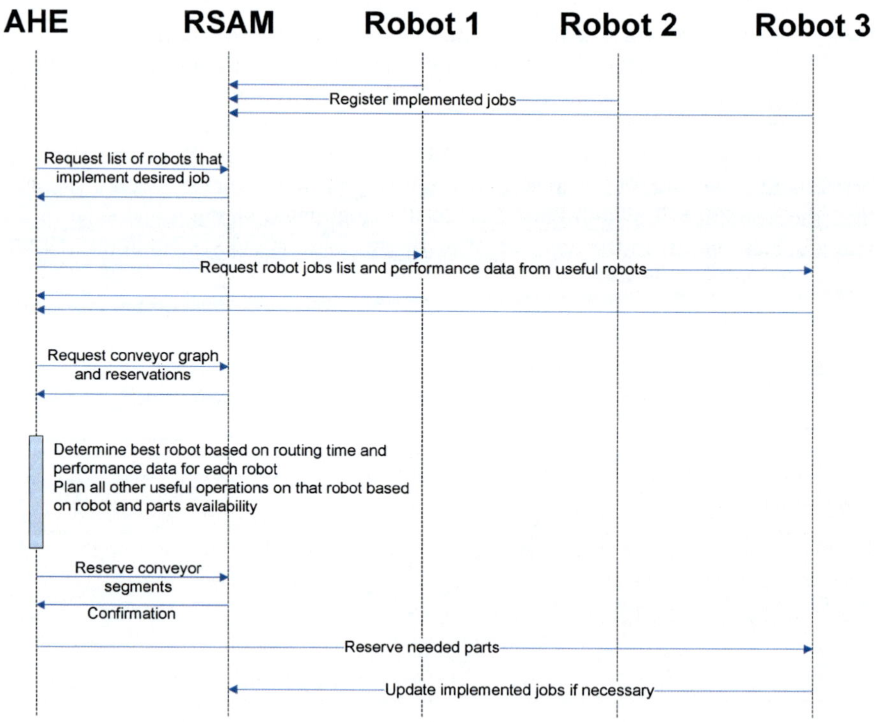

Fig. 3 Improved CNP

7. The AHE will reserve the needed parts on the selected Resource Agent (the Resource agent will then change its list of implemented jobs that it has previously send to the RSAM Agent).

The advantages over the initial CNP are that the AHE doesn't need to have to update a list with all the system resources, it just needs to have the address of the RSAM entity, furthermore, if a previously unknown resource is brought online all it needs to do is register itself with the RSAM and all the intelligent Products will take her into consideration.

2.4 Comparing Robot Offers

In order to take a Production Decision (which job to execute on which robot) first we need to decide which resource (here robot workstation) we want to execute the work order. This needs first the list of all robots which can execute that job (i .e. which can physically execute the job, have the necessary part / component or have the auxiliary needed equipment - video camera, CNC machine - available and online); there are three possible situations:

1. If no robot is found, then it is clear that there is no continuation and the particular product cannot be completed.
2. If only one robot that can execute the task is found, then it is clear that we will choose that robot.
3. If more than one robot is found, a list will be created containing all robots and one continues by evaluating each robot.

If the objective is minimal production time then we will only take into account the period of time needed for executing the job. Otherwise we can introduce other indicators when selecting a robot, such as: the number of previously failed executions of this job or of other jobs on that robot; power usage; precision; if the robot needs to be serviced or replaced; number of joints needed to do the job; auxiliary workstation equipment present in the workstation which is not needed but blocked by executing this task; overall human expert preference to use this robot, etc..

The problem with all these attributes is that they cannot be used as single robot selection criterion since the system will end up, for example, with all products using one specific robot just because its power consumption is 50W smaller than the others, or with the products generating a lot of conveyor usage just to access a robot that has a 99% success rate bypassing a robot that has only 98% success rate. In order to eliminate these "all-or-nothing" criteria, a system of "**virtual time penalizations**" is proposed that will be added to the operation time resulting in a "coefficient calculated with regard to the current cell objective (the robot with the smallest coefficient is declared the job winner):

- Shortest production time: $Coefficient = routing\_time + job\ duration$
- Equal working time among robots (load balance):

$$Coefficient = routing\_time + job\_duration + total\_robot\_work\_time$$

- Minimal power consumption:

  *Coefficient = routing_time * conveyor consumption + job_duration * robot_consumption*
- Consumption of allocated power supply:

  *Coefficient = routing_time + consumption_percent * 100 s*
- Not blocking other unused workstation equipment (e.g. robot vision system, CNC machine, etc):

  *Coefficient = routing_time + job_duration + no_unused equipment * 10 s*
- Avoid unnecessary robot wear and tear:

  *Coefficient = routing_time + job_duration no_robot_joints * 10 s*
- Encourage higher precision:

  Coefficient = routing_time + job_duration + precision * 100 *s / mm*
- Encourage better chance to execute the job:

  *Coefficient = routing_time + job_duration + robot_success_percent * 40s + job_succes_percent * 100s*
- Apply the human expert preference:

  Coefficient = routing_time + job_duration + robot_encouragement + job en couragement

2.5 Planning the Route

Planning the route is very important in order to take a Production Decision; for a good decision one needs to do more than just find the shortest way to the desired robot and add the transport time along all the segments in order to obtain the transport time. One needs first to consider the possible delays which are caused by other products using the same route; secondly one needs to make sure that delays are not generated for products that have already planned their route.

In order to enable these conditions, a *reservations system* is implemented, which will memorize a list of reservations for each conveyor subsystem. A reservation is described by two time moments:

- the moment a product enters the conveyor segment and
- the moment the product leaves the conveyor segment

The time difference between these two moments is equal or greater to the time needed for a product to traverse the conveyor segment; if we consider a conveyor that permits only one product on a particular conveyor segment, it will be not allowed that two reservations overlap or that a reservation includes another reservation.

In order to obtain such a plan, a path finding recursive algorithm is not sufficient; the algorithm needs to check if adding the next path generates a collision, case in which it must move backward to prolong the reservations it has made for the previous sections. This algorithm operates as follows:

1. IF the current solution takes more time than the current best calculated one or it makes too many loops around the central conveyor, or a robot we don't need to visit is blocked THEN **stop**

2. IF reserving the current segment causes a collision, then find the moment when the conveyor is free and **return** this value
3. IF we have reached the destination THEN IF the current solution is shorter than the current best calculated solution THEN save this solution as the best current solution and **stop**
4. FOR EACH neighbour of this conveyor, **recursively launch** the function
   a. WHILE the recursively launched function returns a collision delay, IF the delay can be added to this conveyor THEN add the delay to the current reservation and **recursively launch** the function on the same neighbour with the new start time, ELSE find the moment that this conveyor becomes free and **return** this value

Another problem that is solved by this algorithm is that if a robot is busy working on a product, that product occupies the conveyor in front of the robot, so, by waiting for this conveyor to become free, we will also be waiting for the robot to become free, so, if the robot is not available at the moment the reserving algorithm tries to reserve its conveyor, the routing time will also include the time needed for the robot to become free. This gives us the "Routing Time".

2.6 Taking the Production Decision

Having solved these problems, a Production Decision can be taken by running the following algorithm:

1. Make a list of robots that can execute the next available job on the product
2. Compute the "Coefficient" for each robot
3. Choose the robot with the smallest "Coefficient"
   Because we are in a production strategy that tries to make only simple short term planning, once we have managed to reserve a job on a robot, we will take another step in order to minimize the combinatorial explosion caused by taking the decision "*do we execute here or do we go to another robot*" for each product job that can be executed by that robot:
4. As long as the robot is free, continue by reserving all robot jobs that can be executed on that robot

3 Product Driven Automation

The dialogues needed to drive the production are presented in Fig.4. This figure is composed of three parts, each corresponding to a driving strategy. Each strategy has two main loops, the "Next Product" loop that activates when a new product needs to re enter the system and the "Perturbation" loop, which is triggered when a perturbation occurs.; the placement of those two loops is the main difference between the Hierarchical and the Heterarhical strategies.

The difference between the Negotiated and Non-negotiated Heterarhical strategies is given by the "Next station" loop which shows the difference between planning the whole product (its complete ensemble of operations) and planning only the next station to be visited (the next operation to be performed after terminating the current one). The left side of the figure shows the four "Strategy Change" loops; these loops are triggered when a signal issued to change the current production driving strategy is detected.

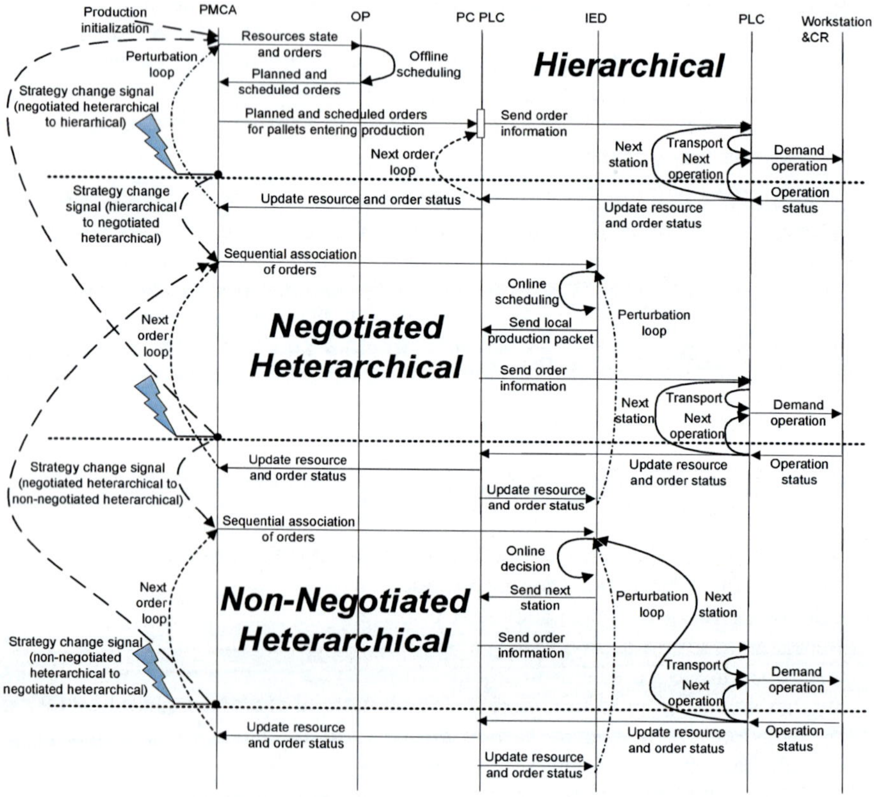

Fig. 4 Model for controlling the production cell with Intelligent Products

In order to avoid pallet collision, the conveyor system is equipped with stoppers so that a "one pallet per conveyor" system rule can be imposed.. In order to implement this idea, a two stage protection system is set in place:

- The first stage is maintained by the RSAM entity that has a list of reservations for each conveyor segment; when an Intelligent Product determines its route, it takes into account the reservations already made and when it finishes it will update the RSAM data with its own reservations so that they are visible to the next products planning their route (the RSAM software module will reject the reservations if they conflict with reservations already made);

- The second stage is coded into the low level PLC control; basically the PLC will not drop a stopper if the next segment is not free.

4 Software System for Implementation Using the JADE Environment

4.1 Message Exchange System

The JADE Multi-Agent Environment has been used for implementation mainly because of the communication capabilities that it offers. Basically, if a operating system can run Java JRE v5 or later then one can start a JADE Container on that system, this Container can then run Agents, and all Agents can communicate with each other regardless of the physical machine on which they are executed or their operating system [10, 12, 13].

This communication is based on a Message Exchange System, all one needs to know is the name of the recipient and if that recipient is present in the environment then it will receive the message; the system also offers a list of incoming messages, one can receive messages from any sender. This system does a lot to simplify the message sending problem as one does not need to concern himself with IP addresses, opening sockets, etc.. A message is composed of the following data:

- sender
- list of recipients
- message type (Inform, Request, etc. as described by FIPA)
- content

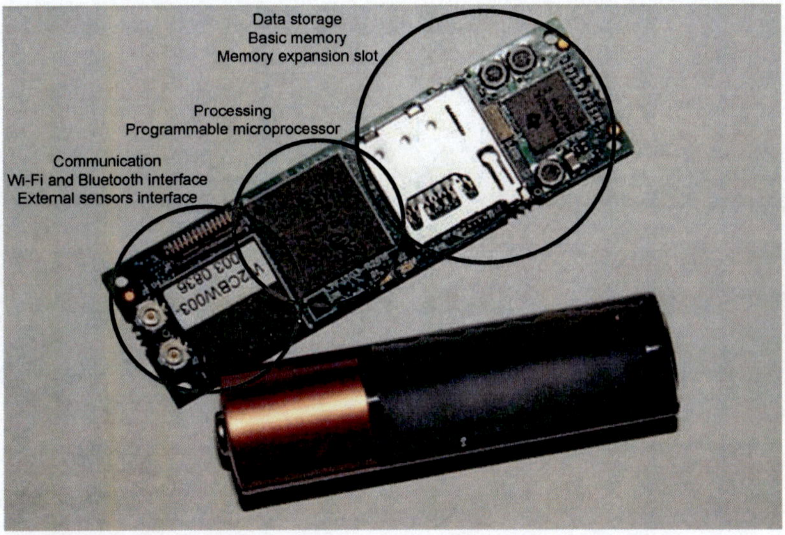

Fig. 5 Overo-Air Gumstix module

4.2 System Classes and Implementation

The Intelligent Product is run by the Overo-Air Gumstix modules (see Fig. 5); one such module will be placed on each pallet in the system, and is powered by batteries and will communicate via Wi-Fi. The Strategy Supervisors, Strategy Switch Mechanism and RSAM will run on a dedicated high availability server as they are critical applications. PLC and Robot Supervisors will run on the stations' PCs.

After having studied all the functionalities that need to be implemented, we came up with a class diagram presented in Fig. 6. The main communication architecture is offered by the JADE Environment with its Message Exchange System, other Agents such as the Offline Planner. The Robot and PLC Controller cannot run Java, so they will communicate using a TCP/IP connection [11, 13].

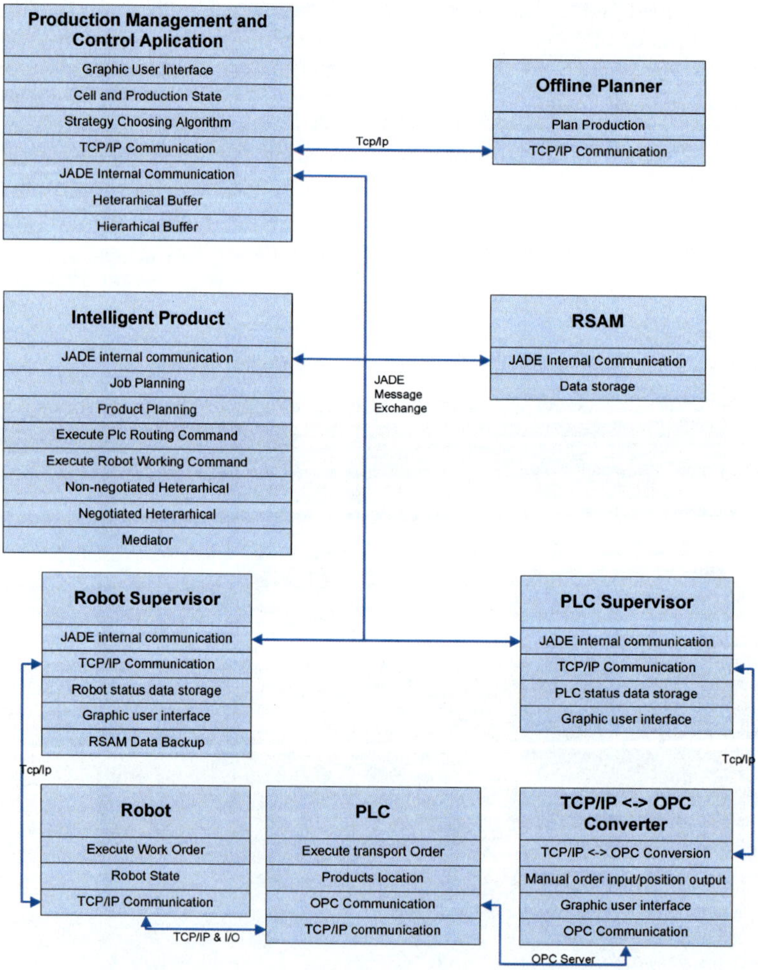

Fig. 6 Class diagram

5 Conclusions

Implementation on the intelligent manufacturing cell equipped with robotic workstations showed that the system, in a perturbation free state, has the same production performances (production time) as the previous centralized driving solution, this goes to demonstrate that using a distributed architecture implemented using the JADE Environment has the needed processing power and communication bandwidth to run the production cell without delays or hangs.

References

1. Gnonia, M.G., Iavagnilioa, R., Mossaa, G., Mummoloa, G., Di Levab, A.: Production planning of a multi-site manufacturing system by hybrid modelling: A case study from the automotive industry. International Journal of Production Economics 85(2) (August 2003)
2. Sahardis, D., Dallery, Y., Karaesmen, F.: Centralized versus Decentralized Production Planning –RAIRO. Operations Research 40(2) (April-June 2006)
3. Murillo, J., Muñoz, V., Busquets, D., López, B.: Schedule coordination through egalitarian recurrent multi-unit combinatorial auctions. Applied Intelligence Journal (April 10, 2009)
4. Kaminsky, P., Kaya, O.: Centralized Versus Decentralized Scheduling and Due Date Quotation in a Make-to-Order Supply Chain. In: Proceedings of the 2005 MSOM Conference. Northwestern University, Evanston (2005)
5. Barták, R.: Mixing Planning and Scheduling to Model Complex Process Environments. In: PACLP 2000, Manchester, UK (April 2000)
6. Borangiu, T., Gilbert, P., Ivanescu, N., Rosu, A.: An implementing framework for holonic manufacturing control with multiple robot-vision stations. Engineering Applications of Artificial Intelligence 22(4-5) (June 2009)
7. Heeks, R.: Centralized vs. decentralized management of public information systems: A core-periphery solution - Information systems for public sector, Working Paper, IDPM (1999)
8. Jay, B., Raymond, P.: Engineering Model For Enterprise Command and Control. In: Command and Control Research & Technology Symposium The Power of Information Age Concepts and Technologies – Storming Media (2004)
9. Van Brussel, H., Wyns, J., Valckenaers, P., Bongaerts, L., Peeters, P.: Reference Architecture for Holonic Manufacturing Systems: PROSA. Computers In Industry, Special Issue on Intelligent Manufacturing Systems 37(3), 255–276 (1998)
10. Smith, R.: The Contract Net Protocol: High-Level Communication and Control in a Distributed Problem Solver. IEEE Transactions on Computers - TC 29(12), 1104–1113 (1980), doi:10.1109/TC.1980.1675516
11. Meyer, G.G., Främling, K., Holmström, J.: Intelligent Products: A survey. Computers in Industry (2009), doi:10.1016/j.compind.2008.12.005
12. Guyon, D., David, M.: Towards the implementation of a product-driven control using infotronics technologies. Industrial Electronics Society
13. Verstraete, P., Valckenaers, P., Van Brussel, H., Saint Germain, B., Van Belle, J.: Towards robust and efficient planning execution. Engineering Applications of Artificial Intelligence 21(3), 225–274 (2008)

Physical Internet Enabled Open Hub Network Design for Distributed Networked Operations

Eric Ballot[1], Olivier Gobet[2], and Benoit Montreuil[3]

[1] Mines ParisTech – CGS, Paris, France
eric.ballot@mines-paristech.fr
[2] EPFL – TRACE, Lausanne, Switzerland
olivier.gobet@epfl.ch
[3] Université Laval – CIRRELT, Québec, Canada
benoit.montreuil@cirrelt.ulaval.ca

Abstract. Supply networks are still mainly based on organizations essentially centralized, dedicated and thus fragmented, whose sustainability becomes ever more problematic nowadays. The recently introduced Physical Internet tackles this problem by interconnecting all the logistics services through the encapsulation of the goods in smart modular containers. Within this framework, network adaptation with distributed routing problems take the lead over classical network design with flow assignment problems. Thanks to recent progresses made in the Digital Internet domain, decentralized approaches are foreseen to be applied for solving those problems on the large scale mandated by the Physical Internet. This leads us to propose here an evolutionist approach to solve the Physical Internet open hub network design problem. We model the problem, formally introduce the design approach, analyze empirical results and provide conclusion remarks and opportunities for further research.

Keywords: Physical Internet, Open hub network design, Holonic manufacturing, Holonic supply chain, Open Mobility Web, Open Supply Web, Supply network, Efficiency, Evolutionist algorithm.

1 Introduction

In the last decades, distributed, networked, holonic and multi-agent manufacturing and supply chains have been gaining a lot of attention both in industry and academia [1, 2, 3, 4]. Under this premise forthcoming performance requirements in terms of productivity, agility, robustness and resilience motivate the evolution of manufacturing and supply chains towards a much flatter decentralized organization that is networking, aligning and coordinating distributed, self-regulating manufacturing and logistics units belonging not only to a core company, but to a set of interlaced clients, suppliers and providers.

We assert that for such a targeted decentralization of manufacturing and supply chains to be efficient and sustainable, there is a need for an efficient and coherent physical logistics environment, beyond the current paradigm dominating the design, implementation and operations of logistics services, facilities and

T. Borangiu, A. Thomas, and D. Trentesaux (Eds.): SOHOMA 2011, SCI 402, pp. 279–292.
springerlink.com

infrastructures. The recently introduced Physical Internet [5] offers such a paradigm breaking environment that could both enable and take full advantage of autonomous and decentralized methods. In section two of the chapter, we briefly describe both the motivation for the proposed paradigm shift and describe the essence of the Physical Internet and the Open Mobility Web and Open Supply Web it enables.

As our core contribution in this chapter, we focus on describing the transformation in network design associated with the migration towards a Physical Internet. In section three, we emphasize the fundamental differences in the conception of the networks and webs, both structurally and geographically, and how they enable efficient and sustainable distributed, networked manufacturing, supply and logistics. Then, in section four, we use the case of consumer goods manufacturing and distribution in France, exploiting data on the operations of two of its largest distributors and their top 100 suppliers, to document and analyze the proposed transformation and its impacts. This leads us to provide conclusive insights and opportunities for further research.

2 A New Logistics Paradigm: The Physical Internet

2.1 *Motivations for a New Logistics Paradigm*

Research and practice in supply chain in general, and more specifically on distributed, networked and multi-agent manufacturing and supply chains, mainly focuses on relations and coordination between suppliers and customers at all levels from raw materials to finished goods delivered in the hands of the users. They emphasize the key roles played by information, benefits and risk sharing, by collaborative decision making and planning, to reach better solutions in terms of inventory, customer service and order-to-delivery time to name a few. While this is important, having significant value creation potentiality, it is also important to highlight what they do not emphasize, what they take mostly as granted, that is the overall logistics organization at local as well global levels. Yet the current logistics organization has severe shortcomings in terms of efficiency and sustainability, as reviewed in [5].

As illustration for this statement, consider the low level of use of logistics resources such as warehouses, trucks [1] and trains, even in advanced countries with intensive use of Third Party Logistics, or yet consider the difficulties faced by cities to manage their inbound, internal and outbound flow of goods. As a further highlight, consider the complete contradiction between on one side the growth of goods flow, between 30% and 70% over the last twenty years and the correlated CO_2 emissions and on the other side the environmental goals for CO_2 emission reduction set by numerous legislations, such as a factor 4 (-75%) reduction to be reached by 2050 in the European Union [2].

In fact, despite many works done in the field of transportation, to our best knowledge there is no evidence that transport-focused technological or organizational innovations will suffice to cope with the ambitious goal mentioned above [3, 4].

One can say that logistics is suffering form its own success. On one side, world container logistics, truck-based less-than-truckload logistics, parcel logistics, just-in-time and lean operations have raised cost, speed and reliability expectations. On the other side, the overall demand for logistics is ever higher, climbing faster than world trade. The past wide availability of affordable, fast and flexible transport and storage solutions around the world is currently fading. Furthermore, as large developing countries such China, Brazil and India get ever more to the core of world production and trade, with raised logistics expectations, the resource requirements from logistics as currently performed are to be explosive.

Hence, under the realistic assumption that the current development will not be sustainable in the future, we have to seek for a new logistics organization, aiming for an improvement in efficiency and sustainability by an order of magnitude, to cope with the main anticipated changes imposed notably by limited resources, the increasing transport prices, CO_2 taxes and the overall need for an environment footprint reduction.

2.2 The Physical Internet

In order to tackle the inefficiency and unsustainability issues raised above, a new logistics organization was proposed by Montreuil: the Physical Internet [5]. So as to steer innovation and break the current paradigms, this organization exploits the digital network-of-networks Internet metaphor, yet adapting it from dealing with digital data to dealing with physical goods.

As posted on the www.physicalinternetinitiative.org web site, the Physical Internet has been formally defined by B. Montreuil, E. Ballot and R. D. Meller as an open global logistics system leveraging interconnected supply networks through a standard set of modular containers, collaborative protocols and interfaces for increased efficiency and sustainability.

The required transformation is multidimensional, introduced by Montreuil [5] through thirteen Physical Internet characteristics. Without explicating them all here, let us emphasize its requirements for innovations in containerization, in freight transport and in material handling and storage. A set of smart, iso-modular and green containers is to be systematically used for encapsulating and securing goods during their sojourn through the Physical Internet (PI, π) [6]. Transportation, handling and storage means, systems and protocols have to be engineered so as to allow the efficient, seamless flow of these π-containers across the Physical Internet. The smart π-containers are to be connected to the Digital Internet through open platforms to ease their deployment, mobility, monitoring and securing.

The Physical Internet is to transform logistics towards seamless and efficient universal interconnection all logistics networks, enabling users to think and act in terms of open global mobility webs and supply webs.

A *mobility network* enables the transport, handling and buffering of objects (here embedded in containers) having to be moved from source nodes to destination nodes. A *mobility web* corresponds to a network of (potentially) interconnected mobility networks. An *open mobility web* is characterized by its inclusion of numerous transport, handling and buffering means, modes, actors and infrastructures that are accessible on demand, on a per-use basis, to numerous users, enabling a seamless interconnected flow of objects across its constituents. Ultimately, a modular container could be given by its shipper a destination, a target arrival window and a budget, and then launched by itself into the open mobility web, autonomously routing itself to destination and requesting help form its shipper only in exceptional situations. Many easier-to-implement less-ultimate alternatives are introduced in [5].

Whereas a *mobility web* is about moving objects, involving for example vehicles, conveyors, transits and hubs, a *supply web* is about realizing and deploying objects involving for example production, assembly, fulfilment and distribution centres. Supply Webs are networks of interrelated supply networks, each embedding interlaced supply chains, potentially involving multiple organizations with collaborative or competitive relationships. *Open supply webs* have the following three characteristics. First, their nodes are openly accessible to most actors, be they producers, distributors, logistics providers, retailers or users. Second, the service capacity of their nodes is available for contract on demand, on a per-use basis, be it for processing, storage or moving activities. Third, dynamic and interlaced virtual private networks are created by actors for realizing and deploying the products, services and solutions in anticipation of and response to stochastic demand from clients.

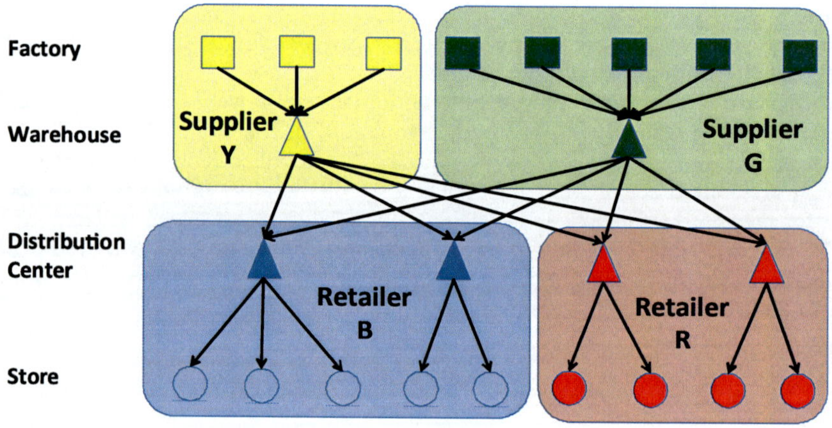

Fig. 1 Supply web composed of two private manufacturing supply networks interconnected to two private retail supply networks

Fig.1 provides a simplified typical structural illustration of a supply web composed by the union of four supply networks, involving four companies, two suppliers feeding two retailers. Each supplier has factories whose products flow into a centralized warehouse. From each supplier's warehouse, products flow to the distribution centres of the retailers. Quite contrasting, Fig. 2 depicts the same four companies now exploiting an open supply web. It shows four open logistics centres exploited by the four companies as well as many others. For simplicity of illustration, we show four open centres, yet in a Physical Internet enabled open supply web, there would potentially be many other open centres accessible to the four companies.

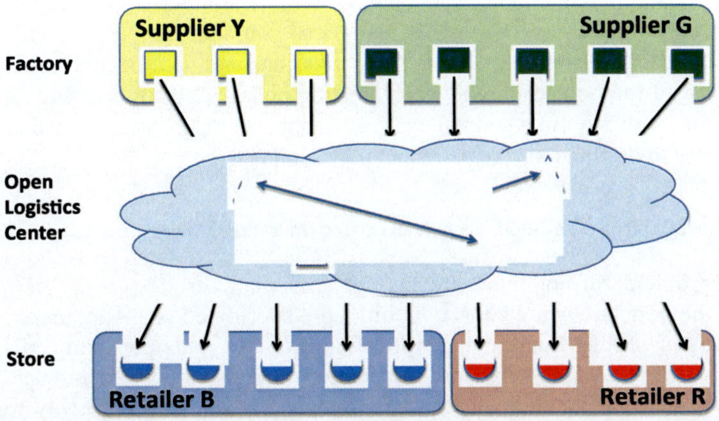

Fig. 2 Physical Internet enabled open supply web exploited by two suppliers and two retailers as well as by many others

As analytically assessed in [7], the Physical Internet can enable order-of-magnitude improvements on global logistics performance. The transformation from the current logistics organization to the Physical Internet is not to be a Big-Bang event [5]. The roadmap is to be decades long, with a gradation of adaptations, innovations and breakthroughs in the technological, infrastructural, business and regulatory domains.

3 Physical Internet Enabled Open Hub Network Design

3.1 Network Design and Assumptions

Network design is a universal and difficult problem, with applications as varied as, for example, irrigation networks, human body artery networks, road networks and logistics networks. It is generally formalized in terms of nodes, arcs and flows, and entails constraints and/or objectives in terms of node locations, network topology, arc direction, flow homogeneity and direction, cost function or combination of these factors. Fig. 1 shows location problems for two industrials serving each their only warehouse and the symmetric problem for the retail chains with the location of their two distribution centres independently from the other retail chain.

Network design is highly sensitive to modelled cost structures, especially if the cost function is sub additive [8], as it is usually the case in transportation, leading to an optimal network requiring extra nodes to consolidate flows. Indeed the sub additive cost function implies that it is most efficient (less costly) to join destinations during transportation for a given part of the transport. The optimal location of the extra node(s) depends on the flows as well as on the cost function. See, [9, 10] for a very case were the cost function is linked to CO_2 emissions and optimization relies only on the choice of the pooled hub between existing nodes. As shown in Fig. 2, the Physical Internet will require a general approach, as it is a supply web instead of a flow from sources to destinations the p median problem [11].

3.2 Typical Logistics Network Optimizations Problems

The p median problem formulations are less general than the design problem stated above as the p locations are chosen within a predetermined set. Another approach, closer to our problem, is the p hub location problem. This problem deals with optimal location of a set of p hubs to interconnect many sources to many destinations. Solving the hub location problem is even harder and the only way to solve it is to make strong assumptions on locations and connectivity allowed. A survey of the assumptions and related solving methodologies can be found in [12, 13] and [14] and examples of state of the art techniques in [15, 16]. We can conclude from the research done in this area that we have no MIP or quadratic tractable solutions for large problems and we must rely on heuristics [12]. These approaches are different from analytical formulations proposed in [8] with the Gilbert-Steiner problem or solutions proposed by the decomposition of the problem and are inspired from evolution methods [17, 18].

3.3 Open Hub Network Design for the Physical Internet

For purposes of proof-of-interest of the Physical Internet, we can also use current locations but the structure of the network is so different that there is absolutely no guarantee that a limited pre selected set of locations obtained for example from p median problems (warehouses or distribution centres) is the right one. Moreover the flows are not only directed from sources to destinations, irrigation like problems or transportation flow problems, but are from many sources and destinations, sent to several hubs and mixed in several hubs as in the hubs location problems. Therefore the hub network design problem transcends the classical independent supply chain design problems from suppliers to customers as illustrated by Fig. 3.

The network design was described in the previous paragraph from a functional point of view, which is different from the geographical point of view. Fig. 3 illustrates the principle of the current organizations where three plants from company A (left) serve a central warehouse and customers afterwards and three plants from company B (right) serve two warehouses and customers afterwards too. The stack of the two networks gives a more complicated but more realistic picture of supply networks for the two simple supply chains. It is rather impossible to have insights from the overlap of real supply networks extensive data.

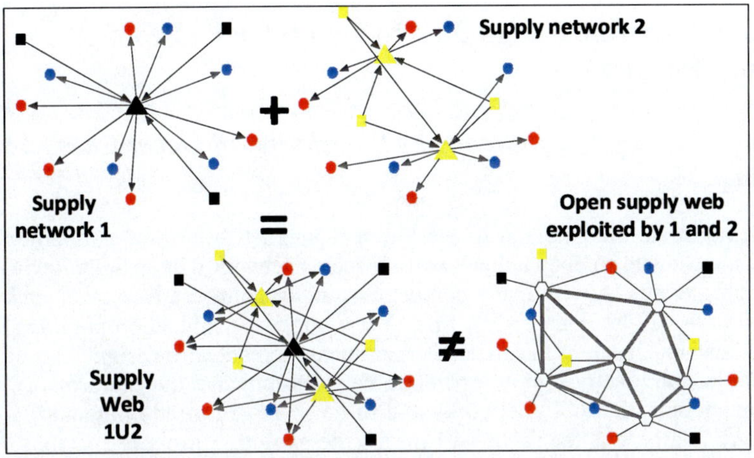

Fig. 3 Contrasting a network of independently optimized supply networks with a Physical Internet enabled open supply web embedding an open hub network

The supply networks depicted in the upper part of Fig. 3 stem from solving problems such as p-median or more complex supply chain design problems, and are justified by the fact that each supply network is dedicated to a single company. However these networks create indirect routes for the flows to their destinations thus creating extra t.km. See [7] for evaluation by continuous approximations of the detours, truckload and costs impacts. The stack up of the networks in Fig. 3 clearly indicates that it would be interesting to interconnect the line hauls to consider these networks as an integrated one. This is clearly the aim of the Physical Internet by interconnecting supply networks into an open supply web through the accessibility of an open hub network.

As a result, the design of this interconnecting hub network is a new and hard optimization problem. Fig. 3 (below right) shows one example of Physical Internet open hub network for the same sources and destinations. The hub's number and their locations have to be optimized according to network costs (hub transit cost and line haul cost). This total cost is a function of hub locations and number, but also a function of routing. Indeed, with a given network design the routing protocol will choose the best routes for the flows and consequently make a specific use of hubs. As flows are reassigned, hubs can be reorganized accordingly.

In fact network optimization by the p-median problem is already assigning flows to optimize the locations of facility within a given set but in a static manner. Here we propose an approach for complex and meshed networks such as hub networks for the Physical Internet, exploiting evolutionist and decentralized techniques. Different available techniques are presented in the next section before developing a specific evolutionist procedure.

4 Open Hub Network Design for Physical Internet Proof of Efficiency Purposes

4.1 A Need of Proof of Efficiency by Simulation of Decentralized Design

In order to demonstrate the potential of the Physical Internet in term of a mobility web, we need to postulate an open hub network. If such a network will be built, it will be done in a progressive way by many competitive and/or collaborative actors, and the resulting design will be improved by time. Yet a priori, for simulation purposes, we have to estimate the result of this dynamic multi-actor hub network design.

Having that in mind we design an open hub network to prove the performance of the interconnection of supply networks and to do so bio inspired computation methods is particularly adapted, as it will mimic the evolution process and these methods already worked successfully on design network problems.

The main interest for bio inspired algorithm is the parallel between, human or animal solutions to supply problems that leads to solutions honed by many cycles of evolution and are expected to produce "good" solutions to the previously mentioned combinatorial optimization problems. In addition, they do their task without centralized control and may represent a readily scalable solution for very large networks, such as supply networks. Many approaches are available: Ant Colony Optimization [19], Swarm intelligence with its "social" dimension and inertia [19, 20], or slime molds [21]. The core of these methods is a set of decentralized decision-making routines which is much more adequate with our Physical Internet proof-of-efficiency purpose.

4.2 Open Hub Networks Design by an Ad Hoc Evolutionist Algorithm

In our proposed algorithm, differing from many other methods, we do not start from scratch as we exploit an infrastructure network to start the evolution process in the initialization stage. Similarly to other bio inspired algorithms, the optimization loop is based on the evolution of flows throughout the network and the reconfiguration of the supply network itself. The flows are routed through the network and according to performance indicators a new collection of hubs and arcs is proposed and locations are adjusted. The evolutionist loop starts again until an exit criterion is reached.

The algorithm is described bellow in pseudo code and is implemented in Java.

```
load physical flows
load infrastructure map
initialize the starting hub set
  do{
    create and launch simulation
    try to improve hub set
  } while hub set changed
display report
```

The simulation starts with a given hub set initialization. This initialization can be set at the middle of flows or at highway intersections as they are likely the location of the hubs. Arcs are initialized not only with a single hub pattern but also with multiple hub assignments, except end node to end node links. See [12] for classification of hub networks topology. The main loop involves routing simulation of shipments throughout the updated network and could be seen as a more precise solution, than the one proposed by [15] but as a downside it requires more computation time.

```
init simulation log
  do{
    init or update routing
    update log with flows mapped on optimal paths
    evaluate objective function
  }while routing is not stable
```

The routing is not purely static with an obvious solution for the distance criterion, the shortest path, but implies the use of transportation means and evaluation of efficiency on one-way trip and round trip are done on a day-to-day basis. This evaluation process leads to more realistic routing and is repeated until a stable routing is found.

At this state of the optimization process, the objective function is currently mainly a cost or an energy minimization objective (total trucks*km or fuel consumption).

The next step of the optimization process is the adjustment of hub number and positions.

```
for each hub
  if it does not satisfy the required condition
    remove/merge hub
  else
    optimize the hub position
    if it does not satisfy the required condition
      create/split hub
```

A hub h_i is retained if it meets the following requirements:

- h_i is connected at least to a specific amount of supply sources and destinations;
- the quantity transiting through itself $f(h_i)$ is more than a given minimum $f(h_i)>f_{min}$

Removing a hub automatically balances its orders to the neighbours' hubs.

If a hub satisfies the previous conditions, the position of the hub is optimized. The location optimization is weighted according to the unit weight and the quantity of each linked sources, destinations and hubs. The new position is not the pure result of the location computation as it also takes into account the past with a balance between the new and the last position.

Symmetrically, a hub is split if the quantity that transited through itself is bigger than a given value $f(hi)>f max$. The split consists in adding one hub randomly on one of its arc, i.e. along one of the roads. The split can also occur to reduce the distance between two hubs to satisfy a social constraint.

At the end of the optimization process, when we reach a stable hub set, the network is recorded and the key performance indices (KPI) are computed. In addition, intermediary solutions are also available. The performance of the network is currently characterized by the cost of the network (number of hubs and hub sizes), transport efficiency (total distance travelled), transit cost (number of transit) and maximal length of route between hubs (social criterion) and robustness or fault tolerance (service completed with the removal of a hub even if not really achieved with actual organization).

4.3 An Open Hub Network for Food Distribution in France

In this section, an open hub network is designed to simulate the performance of food supply chains in the Physical Internet framework. The flows are given by a database of real flows from more 100 companies from the food supply chain in France from the industry as well as from the retail side of the supply chains and

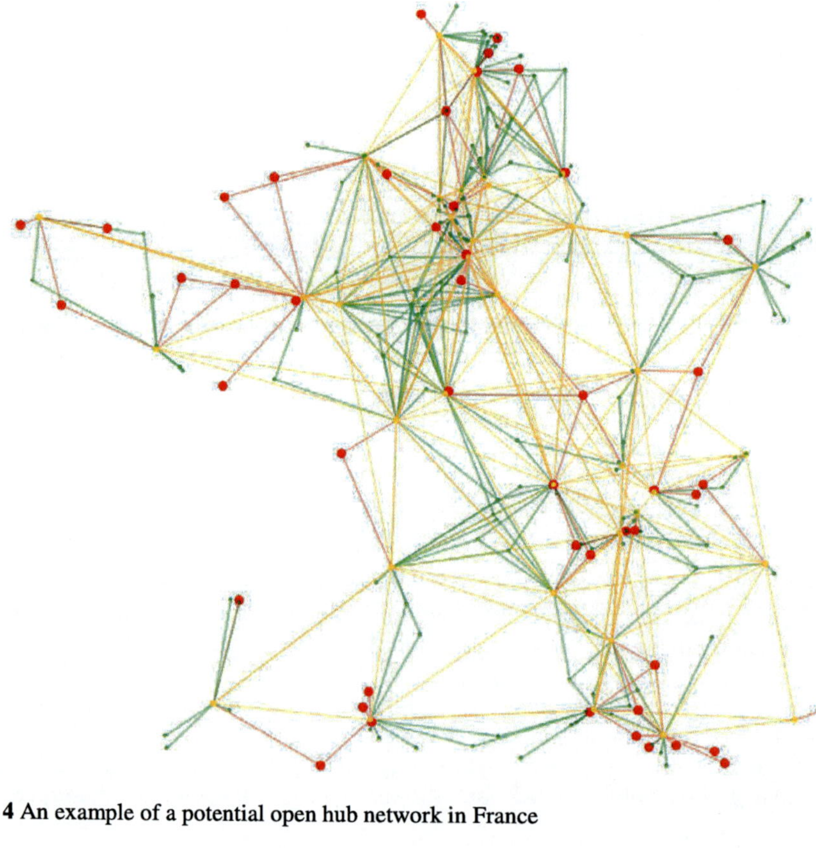

Fig. 4 An example of a potential open hub network in France

for a 12 weeks period of time. The infrastructural data comes from a shape file of France infrastructures of major highways, roads, interchanges, railroads and sorting yards and stations ©IGN 2011.

Fig 4. illustrates how an open hub network can be located and flows routed in order to connect several sources (plants) to destinations (distribution centres). This figure illustrates a step of the optimization process and therefore does not represent a final optimized solution.

Fig 5 illustrates the simulation tool window and a view of flows along roads. The purpose of the optimization tool is not to design the Physical Internet network but rather to explore the sensibility of different configurations of costs and constraints on the design.

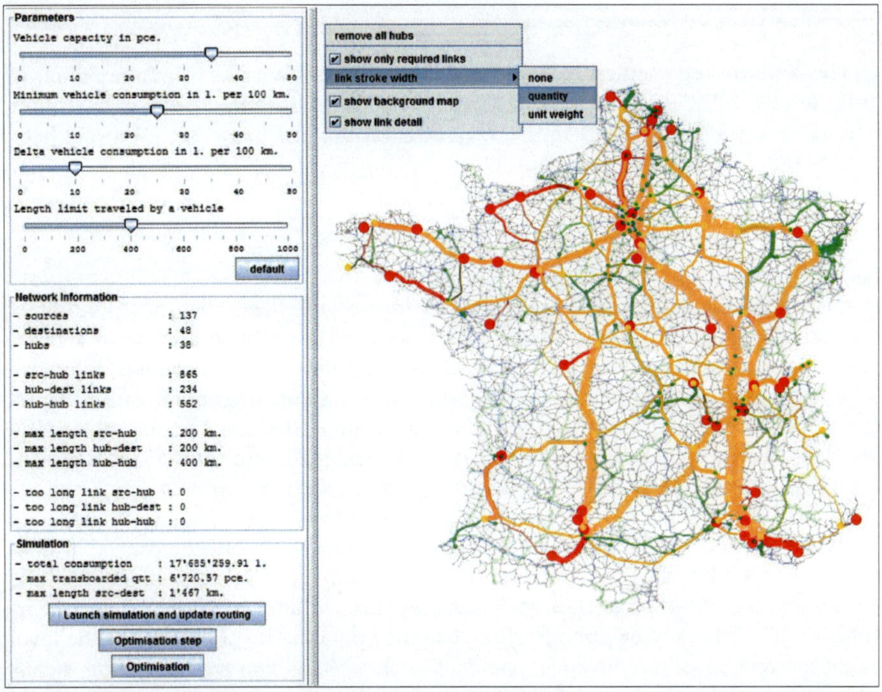

Fig. 5 Example of open hub network simulation depicting resulting flows

Table 1 provides numerical results from the open hub network optimization tool. The Physical Internet oriented key performance indicators, here reduced to four, are compared to the baseline, which represents the actual operations. The indicators are a from the sustainability point of view (fuel for environment and trip length for social,...).

Table 1 Illustrative numerical results for food distribution in France

Performance indicator	Baseline	PI open hub network
Hub #	0	37
Fuel consumption	21,380,650	17,104,524 l
Max trip length (with return)	1399 km	400 km
Transferred containers/hub	NA	<7000

The solution represented in Fig. 4 and 5 involves only trucks' line hauls but in order to take full benefit of the consolidation of flows operated here, we will take into account train operations in the very near future. For that case, rail–road hubs proposed in [22] will be exploited.

5 Conclusion and Future Work

In this contribution, we made the point that for decentralized, distributed, holonic, networked operations to be efficient and sustainable they have to rely on an enabling logistics organization. We emphasized that the newly introduced Physical Internet with its open supply network, offers such an enabling organization. In order to demonstrate the potential efficiency of this combination, an underlying Physical Internet open hub network has to be designed. We introduced a bio inspired evolutionary algorithm for designing such an open hub network. Then we illustrated its application on the large-scale case of food distribution in France.

Even if this chapter doesn't focus on holons, the highly decentralized framework proposed by the Physical Internet draws many research avenues to include design and study the properties of holonic agents within the physical Internet organization. This could be done obviously at the container level but also at the level of means related to the physical activities such as handlers, transportation means and at more global level with host (shippers) or hubs… thus enabling a really decentralized logistics and supply operations.

A forthcoming research work from our team will address the routing of the containers throughout the designed Physical Internet network in order to demonstrate the potential of the Physical Internet in terms of efficiency, environmental footprint and economy.

Yet, in the lines of [23], a lot of research is to be done in order to assess and enable efficient, economic, robust and sustainable performance of decentralized, distributed, holonic, networked operations, here enabled by the Physical Internet.

References

1. Ballot, E., Fontane, F.: Rendement et efficience du transport: un nouvel indicateur de performance. Revue Française de Gestion Industrielle 27, 41–55 (2008)
2. International Energy Agency. Transport, energy and CO2. OECD, Paris (2009)
3. Leonardi, J., Baumgartner, M.: CO2 efficiency in road freight transportation: Status quo, measures and potential. Transportation Research Part D: Transport and Environment 9, 451–464 (2004)
4. McKinnon, A.: Sustainable distribution: opportunities to improve vehicle loading. UNEP Industry and Environment (2000)
5. Montreuil, B.: Towards a Physical Internet: Meeting the Global Logistics Sustainability Grand Challenge. Logistics Research 3 (2011)
6. Ballot, E., Montreuil, B., Thémans, M.: OPENFRET: contribution à la conceptualisation et à la réalisation d'un hub rail-route de l'Internet Physique. In: Predit (ed.) MEDDAT, Paris (2010)
7. Ballot, E., Montreuil, B., Fontane, F.: Topology of Logistic Networks and the Potential of a Physical Internet. In: International Conference on Industrial Engineering and Systems Management IESM 2011, Metz (2011)
8. Bernot, M., Caselles, V., Morel, J.M.: Optimal Transportation Networks: Models and Theory. Springer, Heidelberg (2009)
9. Pan, S.: Contribution à la définition et à l'évaluation de la mutualisation de chaînes logistiques pour réduire les émissions de CO2 du transport: application au cas de la grande distribution. Doctorat Mines ParisTech., Paris (2010)
10. Pan, S., Ballot, E., Fontane, F.: The reduction of greenhouse gas emissions from freight transport by pooling supply chains. International Journal of Production Economics (2010) (in press, accepted manuscript)
11. Ahuja, R.K., Magnanti, T.L., Orlin, J.B.: Network flows: theory, algorithms and applications. Prentice Hall (1993)
12. O'Kelly, M.E., Miller, H.J.: The hub network design problem: a review and synthesis. Journal of Transport Geography 2, 31–40 (1990)
13. Campbell, J.F., Ernst, A.T., Krishnamoorthy, M.: Hub Arc Location Problems: Part I-Introduction and Results. Management Science 51, 1540–1555 (2005)
14. Daskin, M.S.: Network and discrete location. Wiley & Sons, Inc., NJ (2005)
15. O'Kelly, M.E., Bryan, D.L.: Hub location with flow economies of scale. Transportation Research Part B 32, 605–616 (1998)
16. Campbell, J.F., Ernst, A.T., Krishnamoorthy, M.: Hub Arc Location Problems: Part II-Formulations and Optimal Algorithms. Management Science 51, 1556–1571 (2005)
17. Dressler, F., Akan, O.B.: A survey on bio-inspired networking. Computer Networks 54, 881–900 (2010)
18. Tero, A., Takagi, S., Saigusa, T., Ito, K., Bebber, D., Fricker, M., Yumiki, K., Kobayashi, R., Nakagaki, T.: Rules for Biologically Inspired Adaptive Network Design. Science 327, 439–442 (2010)
19. Blum, C., Li, X.: Swarm Intelligence in Optimization. In: Blum, C., Merkle, D. (eds.) Swarm Intelligence. Springer, Heidelberg (2008)
20. Seidel, T., Hartwig, J., Sanders, R.L., Helbing, D.: An Agent-Based Approach to Self-organized Production. In: Blum, C., Merkle, D. (eds.) Swarm Intelligence. Springer, Heidelberg (2008)

21. Adamatzky, A., Jones, J.: Road planning with slime mould: If Physarum built motorways it would route M6/M74 through Newcastle. Journal of Bifurcation and Chaos (2009)
22. White, T., Pagurek, B., Deugo, D.: Collective Intelligence and Priority Routing in Networks. In: Hendtlass, T., Ali, M. (eds.) IEA/AIE 2002. LNCS (LNAI), vol. 2358, pp. 790–800. Springer, Heidelberg (2002)
23. Trentesaux, D.: Distributed control of production systems. Engineering Applications of Artificial Intelligence 22 (2009)

Volunteer Based Search Engine for Holonic Manufacturing Services

Cristina Morariu, Octavian Morariu, and Theodor Borangiu

University Politehnica of Bucharest,
Dept. of Automation and Applied Informatics,
Centre of Research CIMR,
Bucharest, RO 060042, Romania
{cristina.morariu,octavian.morariu,
theodor.borangiu}@cimr.pub.ro

Abstract. This chapter presents a framework for intelligent search of web services that expose the offer request management functionality for intelligent manufacturing systems. Currently research is being done in the area of Semantic Web Service discovery, which involves a search engine that indexes WSDL files, and UDDI repositories. A novel concept of volunteer based search is introduced, in which the search criteria is passed to the manufacturing system, to allow a self-assessment against the search criteria. This volunteer based approach is specifically designed for the cases in which the search engine cannot compare the search criteria against the target service, so it cannot determine the matching. The chapter presents the architecture and design of the web service provided by the manufacturing system and of a web service search engine implementing the volunteer based search concept. Some specific issues arising from this search approach are discussed, such as time available to decide and respond to an offer and offer invalidation. A locking mechanism helping to overcome this issue is given. The chapter also presents the advantages of integration of the manufacturing system with SOA and BPEL processes.

Keywords: Web services, semantic web service discovery, volunteer based search, SOA, manufacturing systems.

1 Introduction

Intelligent manufacturing systems have evolved in the last years, being able to expand and optimise production lines. Also, the manufacturing systems can respond to run-time issues that might arise, like unexpected failure of a component or resource, and can reconfigure themselves to react to the change. This flexibility enables the manufacturing of custom products that might need only a subset of the operations offered by the intelligent manufacturing system. Also, there is a quest for enterprise level control and planning for manufacturing processes, which will integrate the shop floor processes with enterprise business processes. The goal is

T. Borangiu, A. Thomas, and D. Trentesaux (Eds.): SOHOMA 2011, SCI 402, pp. 293–306.
springerlink.com

to be able to optimise enterprise level processes with minimum effort, in terms of work and investments. The standard for enterprise wide business process implementation is based on *Service Oriented Architecture* (SOA) platforms, having *Web Services* as an underlying technology [1]. Web Service interactions and invocations are described by a business process execution language. Business Process Execution Language (BPEL) provides a formal language for defining executable business processes. The BPEL language is focused on the *orchestration* concept, meaning that the BPEL process is executing the business process and has the central execution control [2].

Research is being done on how these manufacturing systems can be exposed as services, in order to be consumed by a BPEL business processes in a SOA environment [3-5]. The mentioned papers focus on a Holonic Manufacturing System (HMS) and state that a SOA system for production management control should provide means to implement:

1. **Offer Request Management** – manages the request for offers by evaluating the internal state, the capability to complete a product and the estimated time of arrival (ETA) for a product.
2. **Management of Client Orders** – once an order is being placed, it manages the internal actions required to manufacture the product.
3. **Order and Supply Holon Management** – manages the input of order- and supply holons in the HMES (Holonic Manufacturing Execution System).
4. **Order Holon Execution and Tracking** – execution of the order holons and tracking the status of their execution through services provided by shop-floor resources (machine vision inspection for quality control, resource availability and power consumption, feedback from the environment).

Fig. 1 illustrates the block architecture of the HMS system considered.

The Offer Request Management module is responsible for processing incoming requests and responds with appropriate offers. For this it is using Product Holons for product type identification, Resource Holons for resource availability and capabilities, supply constraints and current system state, based on already planned activities. The Customer Order Management is generating an Aggregate List of Product Orders (APO) based on the offer placed by the client. The APO is sent to the Job Shop Scheduler which is responsible for generating Supply Holons (SH) and Order Holons (OH). OH and SH are then executed and mapped to specific instruction sequences or programs for the device (robot, machine tool) controllers.

In this chapter we will focus on the Offer Request Management. Once the Holonic Manufacturing Systems are exposed as Web Services, a new challenge arises, specifically how these services can be discovered and used by business users across the enterprise.

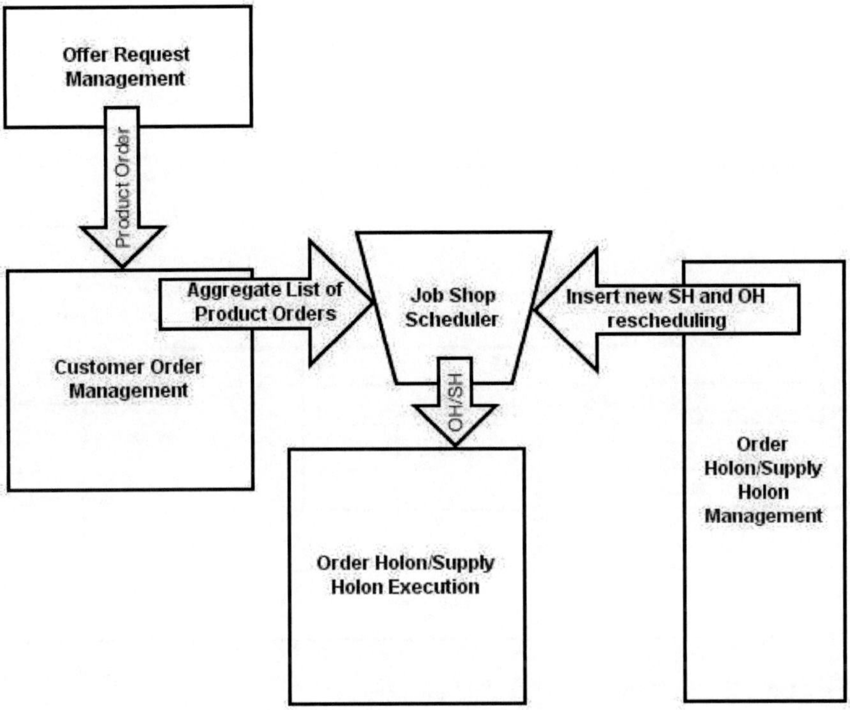

Fig. 1 Block architecture of the Holonic Manufacturing System: activity processing, data- and information flow centred around order- and supply holons

The study focuses on an active search mechanism that creates a bridge between the enterprise use cases and intelligent manufacturing systems. Our contribution to the Offer Request Management module consists in designing a Web Service search engine and a Web Service endpoint that would consider the internal state of the HMS system by implementing a volunteer based search concept. The search engine has a two phase search algorithm:

- A semantic search of web services and
- A volunteer challenge based search.

2 System Architecture

The architecture of the Offer Request Management module is presented in Fig.2 below.

The Web Service endpoint is implemented using Apache Axis [6], and is the interface of the module with the external system (SOA, Business Users). The communication protocol is Simple Object Access Protocol (SOAP) [7]. The web service endpoint module does the processing at SOAP level by parsing the incoming messages (requests) and mapping them to the corresponding Java objects. Also, it does the reversed process for sending the response back to the client, by creating SOAP messages from Java objects.

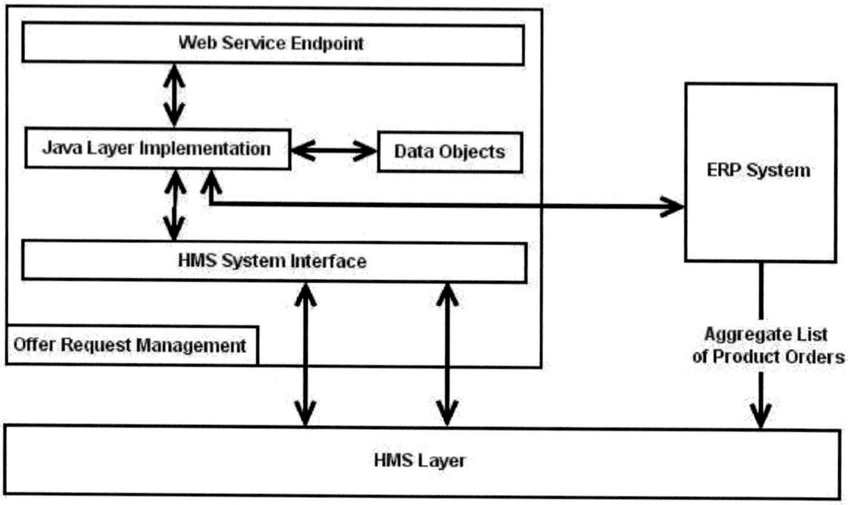

Fig. 2 Offer Request Management module architecture

The Java Layer implementation has the role of realize the logic for the web service functionality. Specifically, it implements the algorithm to determine if the system can respond to a volunteer challenge for a specific request. To do this, it will have to interrogate the HMS Layer and consult the Product Holons (PH), Resource Holons (RH) and Supply Holon (SH) and also check the Global Production Scheduler to determine the current state of the HMS system. The state refers to current Rush Orders or resource breakdowns that might affect the response to the volunteer challenge.

Once the order is placed, the ERP system is generating the Aggregate List of Product Orders, optimizing the execution, based on the ETA required for each product and current resource and supply constraints.

The interface between the Java layer web service and the native HMS implementation is done based on a JNI (Java Native Interface) wrapper, which allows method calls from Java to the native code [8].

If an enterprise is having multiple HMS systems that offer similar services, a specific search engine must be designed, in order to implement the volunteer based search concept.

The search engine operating principle is illustrated in the diagram represented in Fig. 3; a number of steps are carried out as follows:

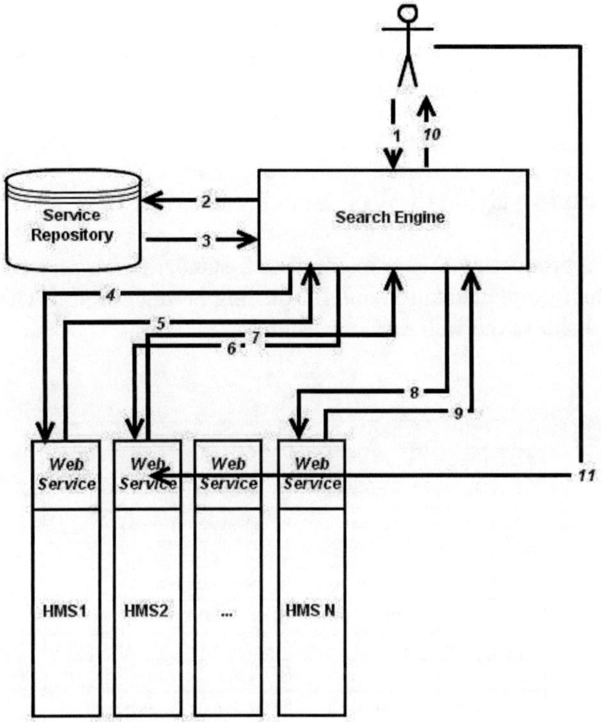

Fig. 3 Volunteer based search engine operating principle

- **Step 1** – User enters a query criteria
- **Step 2** – Search engine performs a semantic search in the web services repository
- **Step 3** – Service repository returns the result list of the semantic search
- **Step 4** – Search engine invokes the HMS1 web service with the volunteer challenge
- **Step 5** – HMS1 system performs a self-assessment against the query criteria and returns a response to the search engine
- **Step 6 → 9** is a repetition of the process for all HMS services returned by the semantic search
- **Step 10** – Search Engine consolidates the results and responds to the user's search
- **Step 11** – User chooses the HMS system to use, and invokes the web service to place the product order

An additional step is done by the search engine, by invoking the refuse order method on all the other HMS systems that were challenged but were not chosen by the user to place the order. This allows unlocking of the HMS systems, for future challenges.

3 Holonic Manufacturing System Web Service Design

In our approach each HMS production line is exposed as a standard web service, for the volunteer based search implementation and for placing product orders. The following diagram in Fig. 4 shows the web service design:

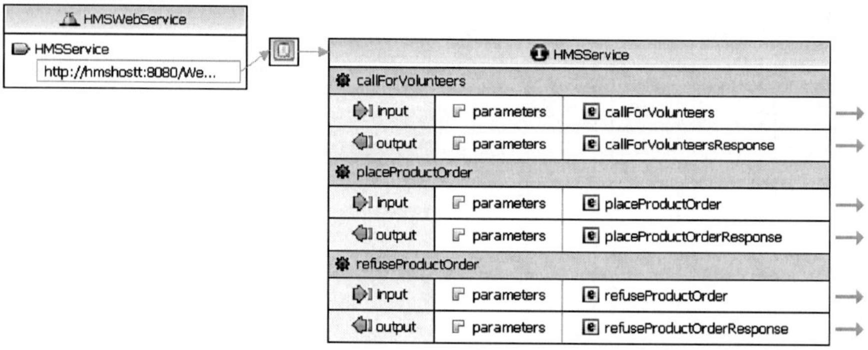

Fig. 4 HMS Web Service Design

The web service implements three methods:

1. **callForVolunteers** – invoked by the search engine for each HMS. This method takes a **OfferSearchCriteria** object as a parameter and responds with a **ProductOffer** object. The structure of the input parameter (OfferSearchCriteria object) is presented in Table 1 below:

Table 1 The OfferSearchCriteria data structure description

Type	Name	Description
Long	productHolonId	The holon ID of the product type for which an offer is requested
Date	requiredDeadLine	Required deadline for the product implementation (ETA requested)
Boolean	rushOrder	Boolean value that flags if the offer should be treated as a rush order

The structure of the output parameter (ProductOffer object) is presented in the Table 2 below:

Table 2 ProductOffer data structure description

Type	Name	Description
Boolean	accepted	Boolean flag signifying the acceptance of the order
Date	closestETA	The closest ETA as computed by HMS
Boolean	rushOrderPossible	Boolean value that flags if the offer can be treated as a rush order
Long	OfferID	The ID of the offer made by HMS (used to tack the outcome of the order)

2. placeProductOrder – invoked by the service consumer once it decided to place the order. This method takes an input parameter of type ProductOrder. The input parameter (ProductOrder) has the structure given in Table 3:

Table 3 ProductOrder data structure description

Type	Name	Description
Long	offerID	The ID of the offer returned by the callForVolunteers invocation
String	Details	An optional member data containing specific details related to the product order

3. refuseProductOrder – this method is invoked by the Search Engine for each HMS system which was not chosen to provide the service. The role of this method is to release the HMS, so that its resources are available for a future volunteer search.

4 Volunteer Based Search Engine

The search engine is an enhanced semantic search engine, based on the volunteer concept. The search operation is performed in two phases: first it will do a semantic search based on the user entered criteria and then it performs the volunteer challenge to the top services discovered.

Phase I – Semantic Search

The semantic search module proposed for this search engine is implemented as a *Vector Space Model* [9]. This algebraic model assumes creating a vector for each document that the search engine is indexing. This vector will then represent the document. Similarly a vector is created for the search criteria entered by the user who is performing the search operation. Then the vector criteria are compared with the document vectors in order to determine if the document vector should be included in the query results. Also, this approach allows ranking of the matched documents, so that the search engine can return the meaningful results first.

In our case, the documents are represented by the WSDL files provided by each web service. We will use for this the approach described by Platzer and Dustdar [10]. The method proposed there involves using both WSDL files and UDDI repository information to create the vector space. Then, by computing the cosine coefficient, the angle between the query vector and the document vector is computed, thus providing a relevance measure.

One important note on this is that from phase I of the search one is interested in a fixed number of results, specifically, the N most relevant services, which will be the subjects in phase II of the search operation.

Phase II – Volunteer Challenge

In this phase, the search engine will first construct the **OfferSearchCriteria** object from the data entered by the user. Once this object is constructed, the search engine will invoke each web service resulted from phase I. Specifically it will invoke the **callForVolunteers()** method, and pass the **OfferSearchCriteria** instance to each call. Then the service provider (HMS system) will execute the method and will respond with the product offer. The method implementation will evaluate the local HMS status and, based on the current resource holons data and rush orders that are currently in execution, computes the output.

Our proposed approach is superior compared to a simple semantic Web Service search for this type of application because the Search Engine cannot compare in this case the search criteria with the HMS system. This is because the internal state of the HMS system is different in time and has a big influence on the feasibility of the offer. The idea here is that the HMS system evaluates itself against the search criteria (the request for offer).

The diagram in Fig.5 depicts the concept of the two-phase search operation described above.

Another important aspect of the above mentioned approach is that the search results (Web Services) are invoked before they are presented to the user. This allows the HMS system on one hand to evaluate itself against the search criteria, implementing this volunteer concept and providing thus an accurate and documented response, and on the other hand raises some questions about how the answer on the volunteer challenge is relevant in time.

In essence, if two users are performing a similar search, and the same Web Service is present in both result sets returned by the search engine, there would be two invocations on the **callForVolunteers()** method of the same service. In this scenario, if the second user places the order, by calling **placeProductOffer** before the first user, then the **ProductOffer** returned to the first user will become invalid. This is because the HMS system state will change, due to the offer made to the second user. To better understand this scenario, one should refer to the sequence diagram in Fig.6:

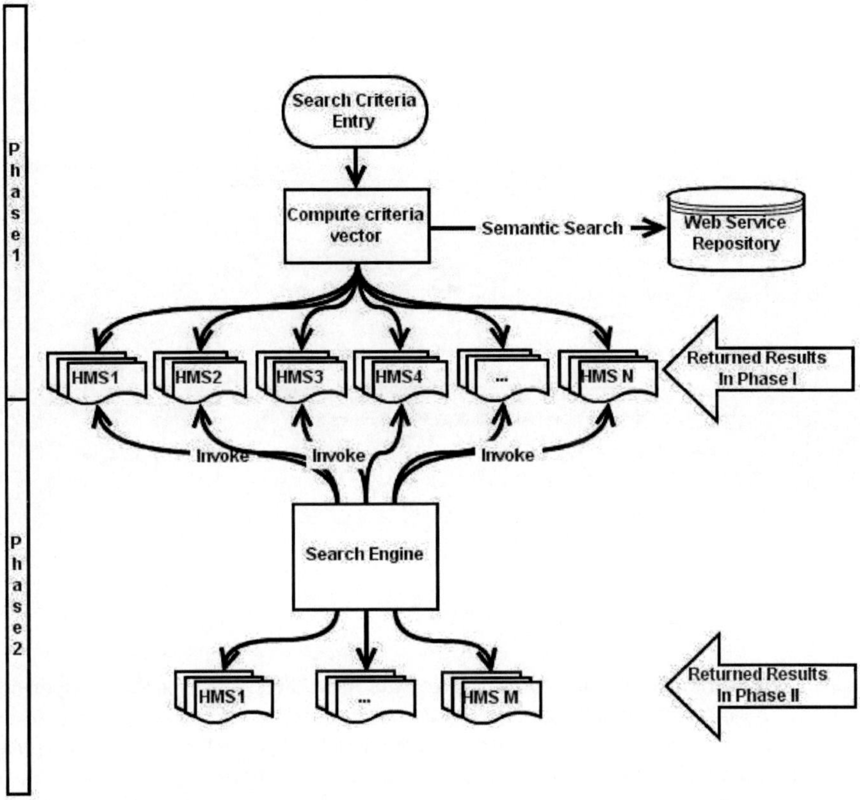

Fig. 5 Two-phase search operation concept

In this scenario, the following assumptions were made in order to better demonstrate the problematic scenario:

1. There are two concurrent users performing searches
2. The searches will return the same HMS service to both users
3. Both users chose to place the product order on the same HMS service
4. The second user "decides" faster then the first user

Note: only phase II of the search algorithm is discussed, as phase I is not relevant in this context.

Having the above assumptions, the scenario further considered is that User 1 performs a search (Search 1). The search engine (in phase II), invokes **callForVolunteers()** for two HMS Web Services (HMS 1 and HMS 2). Both services respond with a product offer (**ProductOffer1_1** and **ProductOffer1_2**) to User 1.

So, at this point, User1 has two results, and needs to decide if an order will be placed.

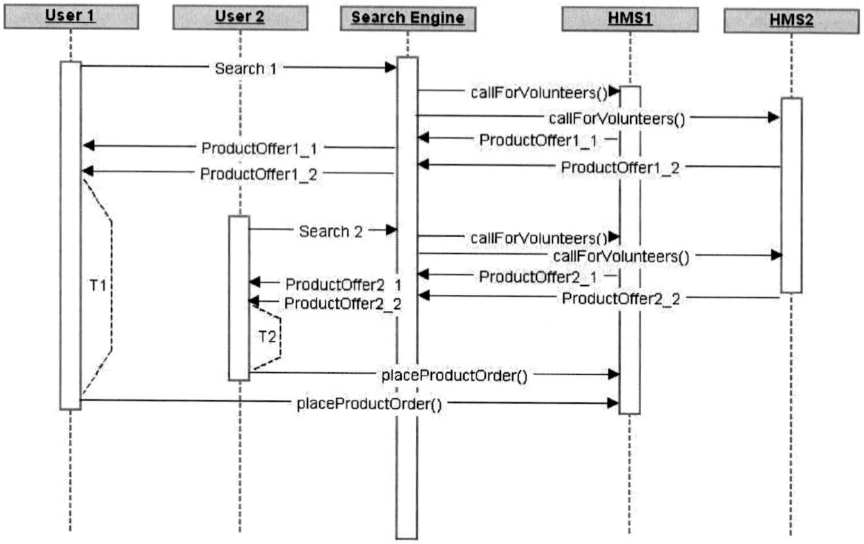

Fig. 6 Sequence diagram of two concurrent users

We will define T1 as time to decide for the user. This time-span is the period between the moment the ProductOffer has been received from the search engine until the user decides to place the order for the product to a selected HMS system. T2 is the same time interval for User 2.

Now, the next step is that User 2 performs a search (Search 2). The search engine invokes **callForVolunteers()** for the two HMS Web Services (HMS 1 and HMS 2). Both HMS services respond with a product offer (**ProductOffer2_1** and **ProductOffer2_2**) to User 2. At this point, User 2 accepts the **ProductOffer2_1** and places the product order for the corresponding offer. When this happens, the internal state of the HMS 1 production line will be changed, by the scheduling of the order from User 2, and the **ProductOffer1_1**, which was returned to the User 1 will not be accurate anymore, and consequently will become invalid.

The above problem is introduced by the fact that the web service is invoked, or in other words the target system (HMS production line) is interrogated in advance, letting the user to decide which offer he wants to take. In this decision time, the state of the HMS production line can change, making the offer invalid by the time the user decides.

To solve this problem, we will introduce a locking mechanism for the HMS system, to assure that the state of the system does not change until the user decides whether an order will be placed or not. The **callForVolunteers** method implementation on the HMS side will implement this locking mechanism, in conjunction with the **refuseProductOrder** method in the web service interface. Even if this locking approach is not feasible for general purpose web services search, it is appropriate for HMS systems search. This is because the number of HMS systems is generally limited for an enterprise, and also the number of users performing the searches. Also, the focus here is not on the speed of returned results, but rather on the quality of the results, which should reflect with high accuracy the capability of the HMS line to complete the product order in the given time frame.

The locking mechanism logic is presented in Fig.7. The HMS system will unlock itself if the order is placed and completed. The term *completed* means that the product is in a stage in which it does not influence the decision to accept a future challenge. The time when the system can unlock itself is an intrinsic property of the HMS system, in conjunction with the specific product being manufactured.

This locking approach will prevent the HMS system to be available in a search result set if it is committed with a product offer to another user and so, it will solve the concurrency problem introduced by the self-assessment of the HMS system against the search criteria.

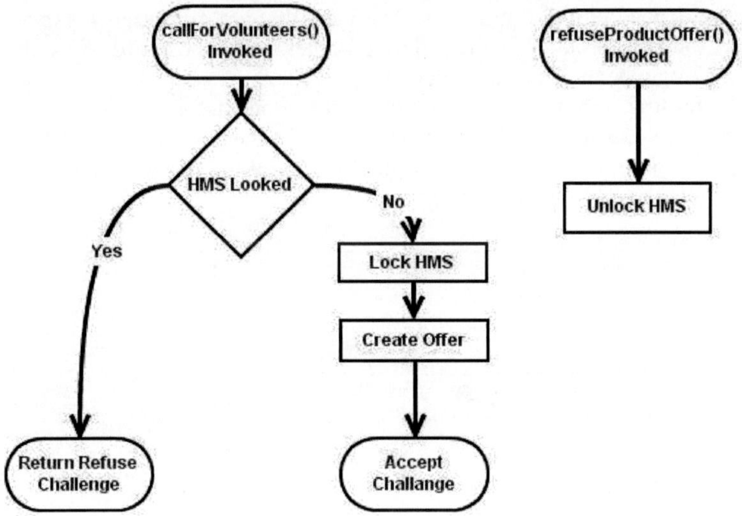

Fig. 7 Locking mechanism logic

5 Conclusions and Future Work

The integration of HMS systems in SOA/BPEL processes presents a great advantage to enterprises, allowing simple process modification and reconfiguration using standard tools. Also, integrated enterprise architecture allows a better tracking and auditing of business process executions, providing valuable information based on which the processes can be optimised and improved. However, as we have seen in this chapter, the more complex the service providers (HMS) become, the more difficult becomes finding the best provider. A semantic search is only useful for limiting the candidates, but is not enough. This is because complex systems evolve in time, and so the initial description of the service might not reflect the current state of the service provider.

The volunteer based search engine (Phase II) and locking algorithm proposed for the HMS system was implemented and tested in a simulation environment. The implementation was based on Axis framework for the Web Service implementation using Eclipse IDE. The Application Server was Apache Tomcat 6.0, running on Red Hat Enterprise Linux 5.0 x64, with 8 GB RAM. The test was done with the following assumptions: 10 concurrent users are searching for a similar product offer, one available HMS line; all 10 users place an order simultaneously. The following assumptions were made in regards to the HMS system: the product takes a fixed amount of time to be manufactured (20 seconds considered), the first stage of the product manufacturing takes 5 seconds, after which the system can respond to the next volunteer challenge, the time to respond to the challenge (self-assessment time) depends on the number of products that are currently in execution (2 seconds for each product in line). These values were chosen to show the impact of the locking mechanism implemented.

Fig.8 shows the time required to manufacture the product for each of the 10 concurrent users considered. The graph shows that the time for the product N is always higher than the time taken for the product N-1, and the increase in time is almost linear, due to the locking algorithm.

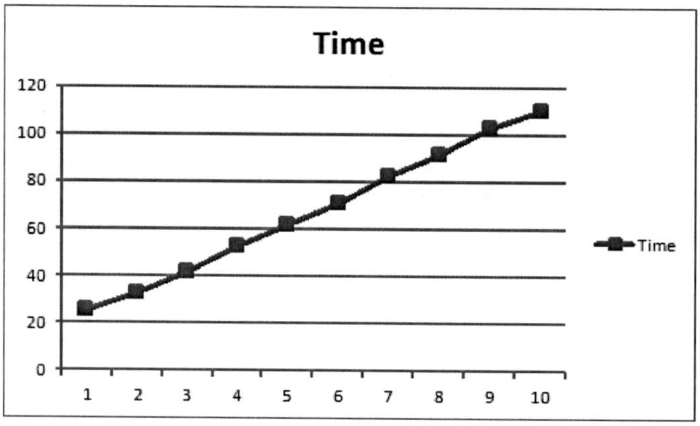

Fig. 8 Production time for 10 concurrent users

The following figure shows the variation of the time for product N and N-1 in relation with the number of products that are in execution phase.

Fig. 9 illustrates that the self-assessment time of the system increases with the number of products in the production line. This is caused by the fact that the more products are in execution, the more internal operations (resource holons interrogations) are required.

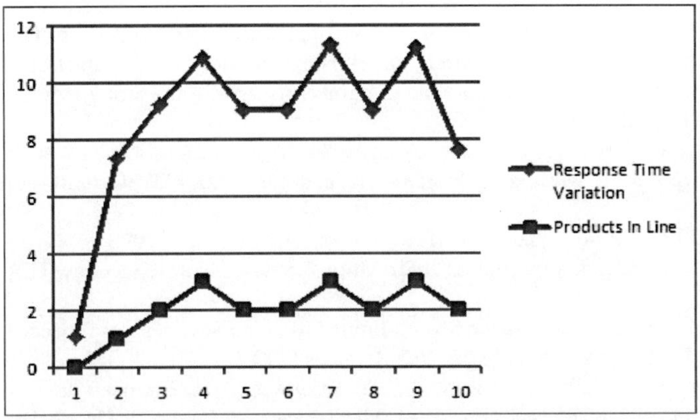

Fig. 9 Response Time Variation in regards to Products in Execution

The Offer Request Management module having the architecture described in Section 2 and using the Volunteer-based Search Engine defined in Section 4 of the paper is in course to be integrated, according to the block diagram in Fig. 1, in the Holonic Manufacturing Pilot Platform of the CIMR Centre within the University Politehnica of Bucharest (http://www.cimr.pub.ro).

Future work is focused on two aspects of the solution. First on the locking mechanism designed to prevent offer invalidation if the system state is changed during the time it takes for the user to decide if the offer is accepted or refused. An intelligent locking mechanism that takes into account the possible impact of a new offer in regards to a previously created offer, would greatly improve the system availability in a concurrent environment. Then some improvements can be made in the self assessment algorithm used by the HMS system, so that the algorithm should be externalised as a *plugin*. This extension of the framework would allow greater flexibility to the solution, as it will allow, with a small programming effort, to create algorithms that are specifically designed according to each HMS particular system configurations. This will permit better quality in the self assessment of the system against the query criteria object.

References

1. Real-World SOA: SOA Platform Case Studies, Forrester Research, Inc. (2005)
2. SOA Best Practices: The BPEL Cookbook - BPEL howto's from Oracle (2005), `http://www.oracle.com/technetwork/articles/soa/index-095969.html`
3. Borangiu, T.: A Service-orientated Arhitecture for Holonic Manufacturing Control. In: Rudas, I.J., Fodor, J., Kacprzyk, J. (eds.) Towards Intelligent Engineering and Information Technology. SCI, vol. 243, pp. 489–503. Springer, Heidelberg (2009)
4. Borangiu, T., Pascal, G., Ivanescu, N., Rosu, A.: Holonic Robot Control for Job Shop Assembly by Dynamic Simulation. In: 16th Mediterranean Conference on Control and Automation – MED 2008, Ajaccio (2008)

5. Borangiu, T., Pascal, G., Ivanescu, N., Rosu, A.: An Implementing framework for holonic manufacturing control with multiple robot-vision stations. The International Journal of Intelligent Real-Time Automation - Engineering Applications of Artificial Intelligence (2009) 0952-1976
6. Apache Axis Project (2006), `http://axis.apache.org/axis/`
7. SOAP Version 1.2 Part 1: Messaging Framework, 2nd edn (2007) (W3C retrieved, 2011)
8. Gordon, R.: Essential Jni: Java Native Interface, 1st edn. Prentice Hall (1998)
9. Salton, M.J.M.G.: Introduction to Modern Information Retrieval, vol. 1. McGraw-Hill (1983)
10. Platzer, C., Dustdar, S.: A Vector Space Search Engine for Web Services. In: Proceedings of the Third European Conference on Web Services ECOWS 2005 (2005)
11. Raileanu, S., Berger, T., Sallez, Y., Borangiu, T., Trentesaux, D.: The open control concept for holonic manufacturing. Int. J. of Mechanics and Control 11(1) (2010) 1590-8844
12. Pétin, J.-F., Guyon, D., Morel, G.: Supervisory synthesis for product-driven automation and its application to a flexible assembly cell. Control Engineering Practice 15, 595–614 (2007)
13. Demirkan, H., Kauffman, R.J., Vayghan, J.A., Fill, H.-G., Karagiannis, D., Maglio, P.P.: Service-oriented technology and management: Perspectives on research and practice for the coming decade. Electronic Commerce Research and Applications 7, 356–376 (2008), doi:10.1016/j.elerap.2008.07.002, 1567-4223
14. Vitharana, P., Bhaskaran, K., Jain, H., Wang, H., Zhao, L.: Service-oriented enterprises and architectures: state of the art and research opportunities. In: Hoxmeier, J., Hayne, S. (eds.) Proceedings of the Americas Conference on Information Systems, Keystone, CO (2007)
15. Zdun, U., Dustdar, S.: Model-driven and pattern-based integration of process-driven SOA models. Int. J. of Business Process Integration and Management 2(2), 109–119 (2007)
16. Zhang, Z., Liu, R., Yang, H.: Service identification and packaging in service-oriented reengineering. In: Chu, W.C., Juzgado, N.J., Wong, W.W. (eds.) Proceedings of the 17th Int. Conf. on Software Engineering and Knowledge Engineering, pp. 219–226. IEEE Computer Society Press, Taipei (2005)
17. Zhao, J.L., Tanniru, M., Zhang, L.J.: Services computing as the foundation of enterprise agility: Overview of recent advances and introduction to the special issue. Information Systems Frontiers 9(1), 1–8 (2007)
18. Peltz, C.: Web services orchestration and choreography. Computer 36(10), 46–52 (2003)
19. Singh, M., Huhns, M.: Service-Oriented Computing: Semantics, Processes, Agents. John Wiley and Sons, New York (2004)
20. Leymann, F., Roller, D., Schmidt, M.T.: Web services and business process management. IBM Systems Journal 41(2), 198–226 (2002)
21. Krafzig, D., Banke, K., Slama, D.: Enterprise SOA: Service-Oriented Architecture Best Practices. Prentice Hall, Englewood Cliffs (2005)
22. Overby, E., Bharadwaj, A., Sambamurthy, V.: Enterprise agility and the enabling role of information technology. European Journal of Information Systems 15(1), 20–131 (2006)
23. Arsanjani, A.: Service-oriented modelling and architecture: how to identify, specify, and realize services for your SOA. In: Developer Works. IBM Corp., Armonk (2004)
24. Erl, T.: Service-Oriented Architecture: A Field Guide to Integrating XML and Web Services. Prentice Hall, Upper Saddle River (2004)

Impact of Information Technology on the Quality of Health Services

Radu Dobrescu and Victor Purcarea

[1] Faculty of Automatic Control and Computers, POLITEHNICA University of Bucharest, 313 Spl. Independentei, Bucharest, Romania
rd_dobrescu@yahoo.com
[2] University of Medicine and Pharmacy "Carol Davila" of Bucharest
victor.purcarea@gmail.com

Abstract. The chapter offers a report on the impact of health information technology on quality of services, work efficiency and related costs of healthcare. The major benefits on quality of health services are considered to be the increased adequacy to standard based care procedures, improved surveillance and decreased medication errors. The key solution to improve the quality of the health services is considered to be the implementation of the Service-oriented Interoperability Paradigm for specifying and standardizing medical services. The chapter demonstrates that Health Level Seven (HL7) specification fits in the context of Service-Oriented Architectures (SOA) that support enterprise distributed processing. The dynamic model of SOA Interoperability Paradigm and its compatibility with the Reference Model for Open Distributed Processing (RM-ODP) are then analyzed, in order to ensure interoperability between applications based on the HL7 Version 3 Standard. Finally the architecture of a system that implements the principles of the HSSP/HL7 SOA methodology and its specific objectives are presented.

Keywords: Service-Oriented Architectures (SOA), healthcare, medical services, Healthcare Services Specification Project (HSSP), Health Level Seven (HL7).

1 Introduction

The literature of the last years is very interested in presenting articles oriented towards the identification of a meta-model and the unification of the IT concepts which are specific to a certain enterprise, institution or domain of interest.

The rate at which all new concepts and principles of building Information Technology Systems (ITS) are utilized by healthcare providers has far outpaced the most optimistic forecasts. Due to the evidence of clinical and operational benefits associated with such systems, information management is fundamental to healthcare delivery. The modern integrated eHealth systems allow improving access to quality health services, for better health of citizens. The efforts of the IT specialists focus now on the deployment of appropriate, complete, consistent and interoperable health information systems, that integrate public health and clinical requirements for overall health systems management ([1], [2], [3]). Information

T. Borangiu, A. Thomas, and D. Trentesaux (Eds.): SOHOMA 2011, SCI 402, pp. 307–319.
springerlink.com

management is essential to healthcare delivery. While the advantages of Health Information Technology (HIT) applications are theoretically evident, adapting outstanding information systems to healthcare objectives has proven difficult [4]. The existing applications have centred mostly on administrative and financial issues rather than on clinical care services delivery [5]. In this chapter we focus first on the effects of health information technology in three main domains: the quality of medical services, the efficiency and the costs of implementation, and then on the solutions to implement HIT using service-oriented architectures as infrastructure.

2 Major Effects of Health Information Technology Implementation

2.1 Effects on Quality of Medical Care

The most relevant benefit is the increasing adherence to protocol-based medical care, which provides decision support. Usually the decision support functions are embedded in *electronic health records* (EHR). EHR has been targeted as an essential tool for ensuring the secure availability of patient health records across healthcare entities. The utilization of EHRs can improve documentation, minimize medication errors, and enhance the mechanisms of patient surveillance and the efficiency of the clinicians' visits [6], [7].

Major advances in expanding HIT systems are offered by ubiquitous or pervasive computing which has also entered in the field of healthcare. For a u- (ubiquitous-) healthcare system to be a complete system, it should ensure scalability, reusability, interoperability and availability. Moreover, there are security issues in using a mobile u-healthcare system [8].

Another theme showing the capacity of HIT to improve quality of medical care is the large-scale screening based clinical monitoring and the data aggregation. One important issue is the use of the automated laboratory measurement. Among the first aspects to be automated were the computerized display of laboratory results and the digitization of radiological images [9].

The third HIT mediated effect on quality is the reduction in medication errors, large decreases in adverse drug events and in mistakes of medication [10]. An important role in this direction is played by electronic prescribing (e-prescribing) systems which offer one method of avoiding misinterpretation of terms which can appear when using the classical hand written formulary.

2.2 Effects on HIT Efficiency

One can consider two types of HIT-related effects on efficiency: the utilization of care and the provider time. The studies that have examined the effect of HIT systems on utilization of care showed decreased rates of health services utilization [11] due especially to the computerized decision support. The types of decision support include automated pre-calculation of the diagnostic tests, the display of the results and of the costs of the tests and a computerized schedule of the results delivery, leading to decreases in utilization rates around 20 - 25 percentage points [12]. The effect of HIT on provider time was mixed, because some studies showed increases in the time a physician spends related to computer use, while others showed a little clinically increase in clinic visit time [13].

2.3 Effects on Costs

Most of the costs related to the new medical procedures and services due to the introduction of health information technology are related to system implementation or maintenance. Other significant costs are due the computer storage costs. But because HIT systems were implemented and evaluated incrementally over time, it is very difficult to evaluate accurately the total development and implementation costs. There is still debate about the cost of HIT investment, especially when the infrastructure is a Healthcare Information Network (HIN). If we would be constrained to introduce a new point-to-point interface every time a new request appears, the cost to build a sequence of such as interfaces would be unsustainable.

To avoid this drawback, the HIN infrastructure must accomplish the following:

- Create data interoperability in the network
- Reduce the number of interface points
- Address the hardware and software architecture and related functions as a cohesive unit
- Ensure scalability in order to allow large scale applications
- Support current and evolving data standards

Fortunately there is a solution to overpass all the difficulties and its name is Service-Oriented Architecture (SOA). SOA is an evolutionary step in the direction of providing true interoperability in applications and use cases that reflect real-world. Healthcare, which as a domain has can achieve great benefits from the proposed methods to ensure interoperability, is behind others in achieving it.

3 The Impact of Service Oriented Architecture on Health Information Systems

Service Oriented Architecture (SOA) is one of the recent achievements in the information systems development. The goal of SOA is to select and encapsulate certain elements of functionality as services that can be made available across the system organization in order to develop service-oriented applications which create a truly interoperable environment. In particular these new service-oriented applications can more accurately reflect the actual business functions in a healthcare environment. SOA allows the different interconnected HIS to interact when changes are made. SOA is different from classical systems integration in that it requires new principles for the system design and the system management, in order to support reuse and sharing of system resources across the organization. With

SOA, IT processing is represented as a collection of services, organized in a library. Each service is made available to the entire organization through a standard protocol and communicates with the systems to which it is related. Applications supporting a specific workflow will reference the same service, while data is synchronized between systems [14].

In evolving healthcare technology, data integration and interoperability are key requirements. By adopting this technology one can expect in the following benefits: reduced medical errors, more accurate diagnosis, and increased access for patients, greater claim accuracy. SOA takes on greater significance with the emergence of Healthcare Information Networks (HINs). The cost of integration can be reduced significantly when using SOA for HIN infrastructure. But the success of large scale network utilization depends on the facility of communication between the network entities.

As the number of systems increases, standard interface formats are adopted by many healthcare organizations. The recommended standard for HIT applications is Health Level Seven (HL7). The HL7 communication standard was developed especially for the healthcare environment and enables communication between meanwhile almost all institutions and fields of healthcare. In the communication model of ISO for interconnection of open systems (Open System Interconnection, OSI), this layer is the seventh, which led to the name HL7.

4 Service-Oriented Solutions for Healthcare

There are two main approaches to implement the service-oriented IT concepts and ideas in a cohesive framework: the European model – HCIS (Healthcare Information System Architecture) and the USA developed model - HSSP (Healthcare Services Specification Project).

The European standard establishes the general principles for the Architecture of the Information System for Health and the characteristics of a basic series of services requested in the frame of information systems for health, which are necessary for sustaining most part of the demands concerning the treatment and the health care. This European standard is independent from any technological specific environment and does not imply the direct adoption of specific organizational solutions or related to projective and operational issues.

The HCIS architectural model has been adopted in Romania according to the European standards, being translated from the English version of the European pre-standard ENV 00251003/1996 HCIS-Arch - Healthcare Information System Architecture. According to this model, the organizational structure of the health system consists in a network of centres distributed in the territory, characterized by a high degree of heterogeneity and diversity, from organizational, logistical, clinical and even cultural perspective. The structure of the centres, taken one by one, develops vertically, from aggregate organization towards the integration of a series of specialized functional zones with specific characteristics and needs. The architecture is projected as comparative basis for the evolution and integration of the existing system and for the planning and, as well, is meant to envision, on superior level, new open and modular systems able to provide a consistent and integrate support for the clinical expectations, for the organizational and managerial demands of the health' organizations.

In comparison, HSSP is more specific and adequate to implement service-oriented technology. HSSP is a standards development effort to create service-oriented architecture (SOA) standards supportive of the health industry. HSSP is a jointly sponsored activity operating within the Health Level Seven (HL7) and the Object Management Group (OMG) standards group. The main goal of HSSP is to develop useful healthcare middleware standards addressing interoperability challenges and to create common service interface specifications that can be used in a HIT context. The stated objective of the HSSP project is to create usable healthcare. To the extent possible, HSSP specifications complement existing work and standards.

Service-oriented solutions have three components: *behavioural interfaces*, *semantic signifiers*, and *business-focused interactions of services*. For any given solution, these three components are modelled and implemented in support of a particular business process. The resulting models and their realizations take different shapes as they process through a methodology, but they adhere to certain common characteristics. HSSP has introduced a number of mechanisms, including semantic profiles and semantic signifiers to allow a loose coupling to be achieved between functional constructs and the payloads that are involved in the support of the business process. Semantic Signifiers are those sets of semantic structures expressed using some formalism that supports the operations in a behavioural interface. The Semantic Profile is the mechanism whereby semantic signifiers are constrained and decomposed to fit the business purpose of the service and subsequently bound to the individual functions. These allow functional consistency to be associated with specific local information models, which increase interoperability when coupled with explicit service descriptions.

SOA provides an architectural framework that allows incremental model-driven methodologies to play their role in HIT. In particular, it provides the desired level of abstraction that allows capturing, modelling and implementing the dynamics of healthcare interactions. Moreover, it allows specific interfaces to maintain a level of granularity that is appropriate to the business context.

5 The HSSP/HL7 SOA Interoperability Paradigm

For realizing semantic interoperability, all models have to be derived from reference models. After the definition of an application, conformance statements as well as implementation guidelines need to be deployed to guarantee that the described interoperability becomes technically real. The methodology for the SOA Interoperability Paradigm described below can be embedded in the framework proposed by the Reference Model of Open Distributed Processing (RM-ODP). RM-ODP provides an established ISO standard (ISO/IEC IS 10746) as a technology framework to organize the relevant work streams and artefacts, as well as a pattern for conformance and development.

RM-ODP defines five Viewpoints [15] that collectively define a given architecture specification. These viewpoints are not hierarchical, i.e. they are not derived from each other. Each viewpoint provides a particular perspective of a complete system including its business context. Each viewpoint specifies several conformance points that form quality tests for the other viewpoints. The Five Viewpoints (see table 1) are as follows:

Table 1 RM-ODP Viewpoints

Viewpoint	Characteristics (Concerned with)
Enterprise View	The purpose and policies governing the activities of the specified system, describing underlying business model
Information View	The information handled by the system and the constraints on the use and interpretation of that information
Computational View	The functional decomposition of the system into a set of objects that interact at interfaces, enabling the aggregation of the basic components to functional units
Engineering View	The infrastructure required to support system distribution and technical details of the components
Technology View	The choice of technology to support system distribution, covering protocols, deployment strategies, maintenance and training

The first three viewpoints describe platform-independent aspects which characterize the platform independent model (PIM), while the last two describe platform-specific aspects which characterize the platform specific model (PSM) [16]. Summarizing the afore mentioned principles, a real system must be modelled considering the aspects of a specific domain and have to be developed using model-driven and service-oriented architectures according to a clearly defined unified process [17]. All different aspects have to be formulated defining the basic pieces or building blocks (which cannot be subdivided any further without losing their functionality) and appropriate composition rules for both business and information models.

There are several benefits if applying this approach to a SOA based HIS:

- It allows specialists in each viewpoint to specify the behaviour of the distributed healthcare system using differing methodologies
- Conformance statements emerging from each Viewpoint form the Quality of Service (QoS) statements in the technical realization of the system
- It allows for conceptual models to be part of a congruent architectural process and to drive solution architectures
- It emphasizes the concurrent nature of the SOA methodology and processes
- It provides a framework for the creation of service (behavioural interface) standards

When RM-ODP is profiled using the SOA Interoperability Paradigm, a clear framework emerges for both specification and standardization of services, and the relationship between the two. Viewed in the context of RM-ODP, the SOA Interoperability Paradigm's process flows can be seen to cross several different Viewpoints (see Fig.1). The Business Analysis (Enterprise Viewpoint) gives definition and scope to the Information-centric components (Information Viewpoint) and the Functional components (Computational Viewpoint). Additionally, a number of different artefacts emerge that both enable the architecture and provide conformance points for quality testing.

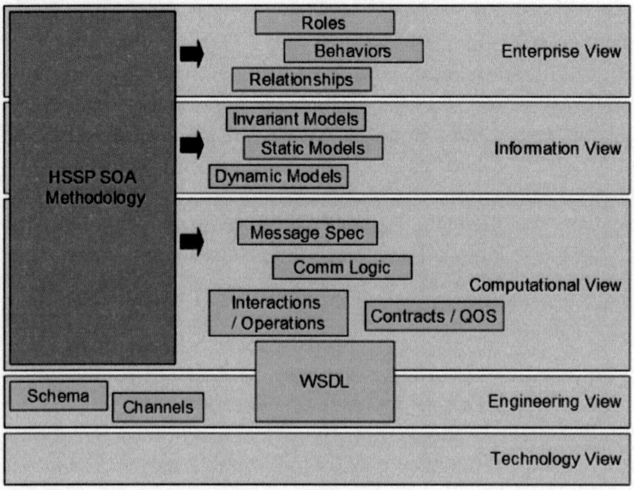

Fig. 1 Sample Artefacts from the SOA Methodology in the RM-ODP viewpoints

The HSSP Service Specification Framework (SSF) provides the standardization of behavioural interfaces using functional and informational profiling. These standards pass through two stages, first as a Service Functional Model (SFM), which at its turn forms the foundation for an RFP (Request for Proposal). The response to this RFP becomes the technical specification for the service, i.e. SSF is used to develop Functional and Technical Specifications.

6 State of the Art and Trends of HL7 Implementation

The HL7 version 2 standard was originally created in 1989, two years after the launch of the HL7 concept as the first operational version, which defines a series of electronic messages to support administrative, logistical, financial as well as clinical processes. Since 1989 the standard has been updated regularly, resulting in several versions from 2.1 to 2.6.

The HL7 version 3 standard aims at supporting any and all healthcare workflows. It consists of an elaborate set of ready-to-implement models (for messages, documents, or services) created using the HL7 Development Framework (HDF). These models are based on the object-oriented Reference Model (RIM), which is part of the standard. The RIM expresses the data content needed in a specific clinical or administrative context and provides an explicit representation of the semantic and lexical connections that exist between the information carried in HL7 messages, documents or services. The HDF, which is based on the UML, documents the processes, rules and artefacts relevant to development of all HL7 version 3 standard specifications.

HL7 Version 3 means much more than being just a new version in the course of developing the standard, because it follows a new paradigm. This change was a long-term process, demonstrated by the frequent changes of direction and the numerous series of versions of its basic elements.

The HL7 Version 3 communication standard is based on a new and comprehensive development methodology, which has been called the Version 3 Message Development modelling approach. Problems bound to that approach became obvious in performed extensions, frequently leading to a new arrangement of attributes or even classes. Thus, the model was hardly maintainable and extendable. As a consequence, in the second phase the RIM has been changed towards a stepwise abstraction of the RIM reducing it to only a few generic core classes by introducing the Unified Service Action Model (USAM).

The elaboration of such models has the role to transpose into abstract terms, as much as possible, the solution proposed for a certain domain, applicable in our case, to "healthcare". The Model HL7-RIM (see Fig. 2) belongs to the class "Model-Driven Development" and has a high level of generality, along with the Model HISA, offering the premise for the modelling of the solution of collaboration between the actors supposed to integrate a compatible healthcare information system HL7.

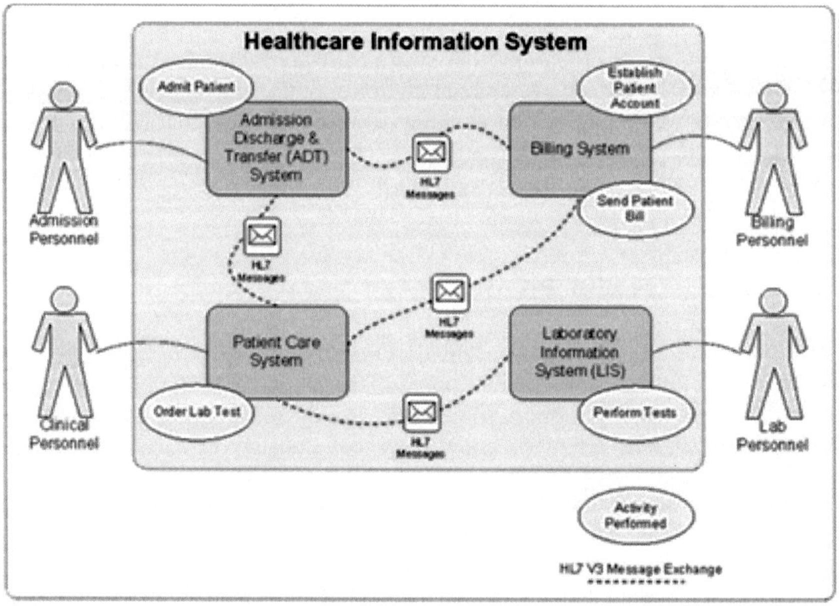

Fig. 2 HL7 Framework Model of collaboration between partners, based on messages

The model HL7 RIM consists of a great number of classes, grouped in a number of six blocks that integrate the so called "backbone" of the model (Fig. 3).

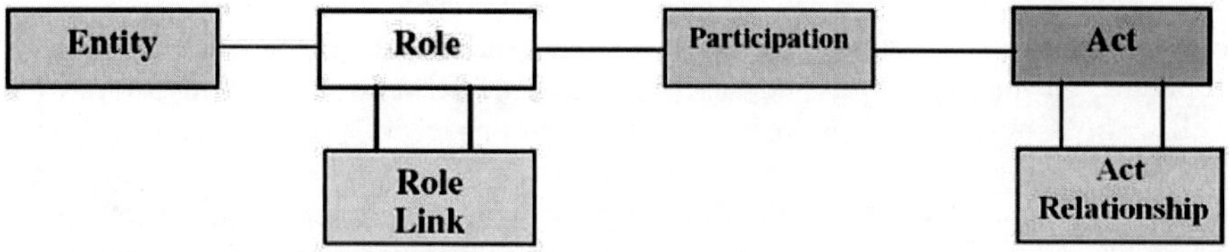

Fig. 3 The model HL7-RIM – main groups of classes HL7-RIM

These blocks are:

Entity: represents physical entities (things and beings) involved in the processes of healthcare;

Role: establish the roles allocated to these entities according to the way they participate to the processes of healthcare

Role Link: represents the relationship between the individual roles;

Participation: describes the context of a medical act in terms of: who executed it, to whom is destined, when and where the medical act happened etc.

Act: represents actions (medical acts) executed in purpose of healthcare that could be, for instance, documents and administrative data;

Act Relationship: represents the connection between the medical acts.

The "backbone" represented in fig. 7 could be read from left to right in the following way:

An Entity (ex. a person) plays a Role (ex. Patient) and takes part to some "Medical acts" (ex. Consultation). But it could also be read from right to left: to a "Medical act" (ex. Operation of appendix) an Entity (ex. a person) had participated, having a certain role (ex. Nurse).

The core classes contain some basic attributes such as Type_CD (Class_CD), Concept_ Descriptor, Time, Mood (determiner), Status, and ID. It is obvious that the core classes for *roles* and *participations* are specializations of the corresponding entities, whereby *roles* represent competence-related specializations and *participations* represent action-related specializations.

HL7's version 3 strategy of model-based message definition reduces optionality by modelling and defining every message according to its specific requirements and conditions. Thus, all specified components are required and being served, resulting in a set of similar but specific messages. Therefore, the interoperability striven for may be taken into question. The solution should be provided by the following principles:

- reference to a globally acknowledged Reference Information Model;
- specification of an accepted and binding vocabulary for all reference components as well as all domain concepts;

- development of application roles for characterizing the participation in message interchange;
- definition of requirements profiles, which lead to conformance statements.

For assuring interoperability between applications based on the HL7 Version 3 Standard, all messages must be based on the HL7 RIM, on agreed data types as well as on a binding vocabulary. At the domain-specific level, the temporal and procedural conditions expressed by Interaction Diagrams or State Diagrams as well as Application Roles, from which trigger events and interactions result, must be standardized. The Version 3 Publication is an automated process provided on the basis of the artefacts from HL7 Technical Committees (TCs) and Special Interest Groups (SIGs) collected in HL7 databases (repositories). For assuring the consistency of the standard, all specifications are verified with existing specification stored in such a repository.

According to Heitmann [18], the unified process to be followed has been defined within the HL7 Development Framework (HDF), which succeeded the Message Development Framework (MDF) clearly demonstrating the movement from a message orientation to an architecture paradigm (see Fig. 4).

The HL7 Development Framework describing the HL7 Unified Process is currently in its final specification phase. HL7 Version 3 meets all requirements and characteristics mentioned above. Service-orientation and process models within the business model are using clinical statements for the clinical domain. Moving from a message paradigm to an architectural one, HL7 V3 now meets the stated requirements for developing advanced health information systems.

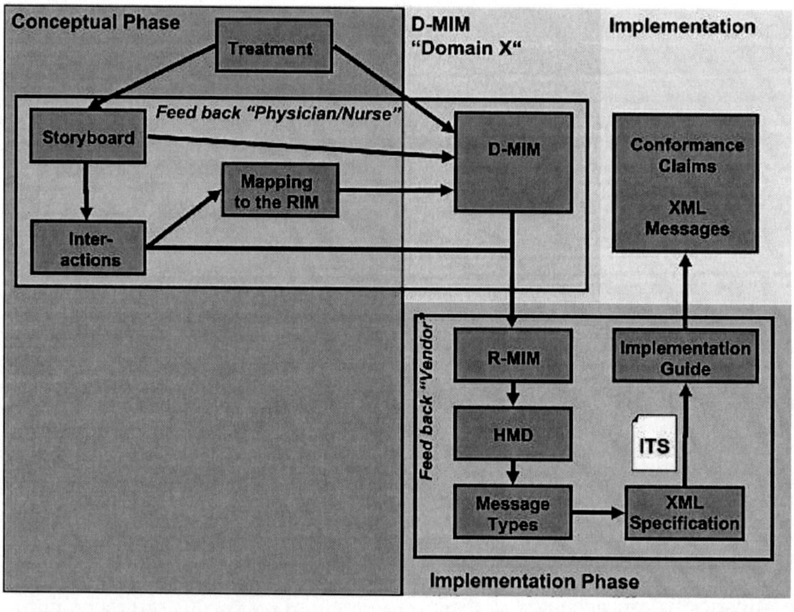

Fig. 4 HL7 modelling overview

7 HL7 Integration

To be efficient, the HL7 methodology should be integrated in a large framework, at the national level. For example, in USA was developed the Medicaid Information Technology Architecture (MITA) as an initiative of the Centre for Medicaid & State Operations (CMSO) intended to foster integrated business and IT transformation across the Medicaid enterprise to improve the administration of the Medicaid program. MITA aims at incorporating many of the relevant standards into an architecture model that will be used to build and enhance Medicaid healthcare systems as well as the data exchange between system components, and drive system design from a business and beneficiary-centric perspective.

The specific objectives of a system of this kind are:

1. Adopt data and industry standards
2. Promote secure data exchange
3. Promote reusable component through standard interfaces and modularity
4. Promote efficient and effective data sharing to meet, stakeholder needs
5. Support interoperability and integration using open architecture and data standards
6. Support integration of clinical and administrative data to enable better decision making

Fig.5 illustrates the block scheme of a system that implements the principles of the HSSP/HL7 SOA methodology.

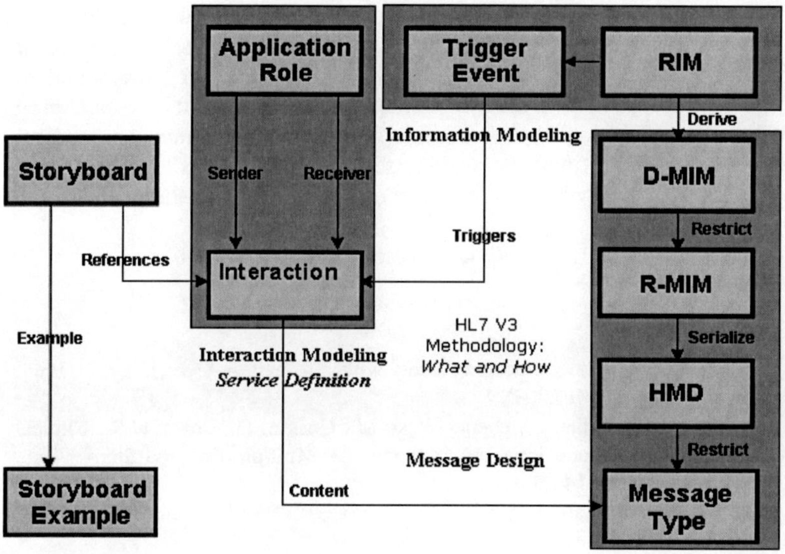

Fig. 5 HL7 methodology for hierarchy modelling

One can observe that the HSSP specification work is divided into two foundational components: *functional specifications* (manifested as Service Functional Models) and *technical specifications* (resulting in technology adoptions). The functional models enumerate *what* capabilities are to be supported but not *how* they are supported. The technical specification work details exactly how those functions are manifested and ultimately realized in a technology platform. During implementation of these technical specifications, errors and shortcomings of adopted Service Functional Models are captured, logged, and ultimately incorporated into subsequent versions of the Service Functional Models as part of a continuous process improvement approach.

9 Conclusions

Service-oriented architecture has moved into the mainstream of information technologies development strategies, and the public healthcare domain has not been left behind. This chapter has tried to demonstrate that though not appropriate for every situation, SOA paradigm can be a useful tool to enable rich functionality within existing public health systems, while minimizing the cost and time to deploy these new features. SOA also provides the necessary flexibility to replace system modules at some future date when superior qualitative modules become available. SOA can be considered the most appropriate architecture for a HL7 compatible health information system. Although the case studies analyzed in this chapter come from public health, the expertise and conclusions are applicable to other parts of the healthcare system as well. Hospitals and provider practices are also seeing an increasing need for interoperability with other systems outside of their organizations. SOA provides a useful strategy for either an incremental or more comprehensive approach to system redesign. One can conclude, considering the next evolution of HIS, that in a foreseeable future the managers of the medical organizations will be able to improve medical services, patients will become more empowered and healthcare will be delivered more appropriate to the needs of the people.

References

1. Bates, D.W., Gawande, A.A.: Improving safety with information technology. N. Engl. J. Med. 348(25), 2526–2534 (2003)
2. Amarasingham, R., Plantinga, L., Diener-West, M., Gaskin, D., Powe, N.R.: Clinical Information Technologies and Inpatient Outcomes. A Multiple Hospital Study. Arch. Intern. Med. 169(2), 108–114 (2009)
3. Ball, M.J.: Hospital Information Systems: Perspectives on Problems and Prospect. Int. J. Med. Inf. 69(2-3), 83–89 (2003)
4. Gambo, I., Oluwagbemi, O., Achimugu, P.: Lack of Interoperable Health Information Systems in Developing Countries: An Impact Analysis. Journal of Health Informatics in Developing Countries 1, 185–196 (2011)

5. Ash, J.S., Sittig, D.F., Dykstra, R.H., Guappone, K., Carpenter, J.D., Seshadri, V.: Categorizing the unintended sociotechnical consequences of computerized provider order entry. Int. J. Med. Inform. 76(suppl. 1), s.21–s.27 (2007)
6. Pizziferri, L., Kittler, A.F., Volk, L.A., Honour, M.M., Gupta, S., Wang, S.: Primary care physician time utilization before and after implementation of an electronic health record: a time-motion study. J. Biomed Inform. 38, 176–188 (2005)
7. Garrido, T., Jamieson, L., Zhou, Y., Wiesenthal, A., Liang, L.: Effect of electronic health records in ambulatory care: retrospective, serial, cross sectional study. BMJ 330, 581 (2005)
8. Varshney, U.: Pervasive Healthcare and Wireless Health Monitoring. Mobile Networks and Applications 12(2-3), 113–127 (2007)
9. Koppel, R., Metlay, J.P., Cohen, A.: Role of computerized physician order entry systems in facilitating medication errors. JAMA 293(10), 1197–1203 (2005)
10. Campbell, E.M., Sittig, D.F., Ash, J.S., Guappone, K.P., Dykstra, R.H.: Types of unintended consequences related to computerized provider order entry. J. Am. Med. Inform. Assoc. 13(5), 547–556 (2007)
11. Kaushal, R., Shojania, K.G., Bates, D.W.: Effects of computerized physician order entry and clinical decision support systems on medication safety: a systematic review. Arch. Intern. Med. 163(12), 1409–1416 (2003)
12. Chaudhry, B., Wang, J., Wu, S.: Systematic review: impact of health information technology on quality, efficiency, and costs of medical care. Ann. Intern. Med. 144(10), 742–752 (2006)
13. Epstein, A.M., Lee, T.H., Hamel, M.B.: Paying physicians for high-quality care. N. Engl. J. Med. 350, 406–410 (2004)
14. Marks, E.A., Bell, M.: Service-Oriented Architecture: a Planning and Implementation Guide for Business and Technology. John Wiley & Sons, Hoboken (2006)
15. Koisch, J., Hamilton, B.A., Honey, A., Permanente, K.: Healthcare Interoperability: A Service-Based Paradigm, Conformance Model, and HL7 Dynamic Model. In: Proceedings of SOA in Healthcare Conference, Chicago (2008)
16. Blobel, B.G., Engel, K., Pharow, P.: Semantic Interoperability – HL 7 Version 3 Compared to Advanced Architecture Standards. Methods of Information in Medicine 45(4), 343–353 (2006)
17. Blobel, B.G.: Concepts and Solutions for Future-Proof Health Information Systems and Health Networks. In: Blobel, B.G., Pharow, P. (eds.) Advanced Health Telematics and Telemedicine. The Magdeburg Expert Summit Textbook. Studies in Health Technology and Informatics, vol. 96, pp. 21–28. IOS Press (2003)
18. Heitmann, K.: HL7 V2 and V3 - A Family of Standards. In: Workshop Health Information and Quality Authority (2011), http://www.hiqa.ie/system

Competency Management System for IT Project-Oriented Organizations

Constanta-Nicoleta Bodea[1] and Robert Buchmann[2]

[1] Bucharest Academy of Economic Studies,
6, Piata Romana Street, District 1,
Bucharest, Romania
bodea@ase.ro
[2] Babes-Bolyai University,
1, Mihail Kogălniceanu Street,
Cluj-Napoca, Romania
robert.buchmann@econ.ubbcluj.ro

Abstract. The chapter presents a competency based system, developed for an IT project-oriented organization, during the implementation of a conventional Enterprise Resources Planning (ERP) system. The proposed solution is an ontology-based extension of the Human Resources module. Two competence ontologies, one for each relevant profession in the organization are parts of the ontological extension. Even if the proposed solution is based on a loose coupling between the knowledge base and the ERP system, it provides added value for ERP systems, expanding their ability to support all levels of management decision, but especially the executive ones. It is a practical and inexpensive approach for integration the semantic or knowledge oriented applications with the data-oriented or legacy systems.

Keywords: Competency management, competence ontologies, ERP, project management, semantic technology, knowledge-oriented applications.

1 Introduction

The human resources (HR) management is constantly changing due to the business dynamics. The financial crisis and the recession accelerate these changes and, therefore, the increasing of the business risks. According to the CBI's report [5], the organizations are reacting different as before to the business changes, especially to the social and demographic ones. The CBI group considers that the organizations are switching to more collaborative and flexible business models. The following four operational areas of the organizations are strongly affected by these new trends: capital and investment, HR, governance and sustainability as well as organization and location. On short term, the expected impact will be a slower, but a more sustainable economic growth. Considering the HR area, the main challenge is to overcome the actual shortages of critical skills. One way of doing that is to

T. Borangiu, A. Thomas, and D. Trentesaux (Eds.): SOHOMA 2011, SCI 402, pp. 321–334.
springerlink.com

replace the traditional human resource management processes with competence-based ones.

The competence-based human resource management processes have the potential to meet many business needs, such as: facilitate organizational change, enhance competitive advantage and increase financial performance and productivity. They offer the means of converting the traditional focus on work activities into an approach based on individual competencies and high performance [9]. Human resources management could take the lead in assuring the competitive advantage of the organizations.

According to Dubois, the main competence-based human resource management processes are the following: competency-based human resources planning, competency-based employee recruitment and selection, competency-based employee training, competency-based performance management, competency-based employee rewards and competency-based employee development. All these processes are person- and result-driven and not work- or job-driven as the traditional human resources management processes.

In a project-oriented organization, the project management (PM) competencies of the project personnel represent very important assets. The Global Alliance for Project Performance Standards (GAPPS) (http://www.globalpmstandards.org) provides a global platform to facilitate, develop, approve, publish, promote, maintain and review usage guidelines for performance project management standards for project and program management and governance roles. GAPPS differentiate between:

- Attribute based competence documents, such as: PM Bodies of Knowledge like PMBOK© (www.pmi.org), APMBoK (www.apm.org.uk), ICB (www.ipma.ch), Curriculum Vitae, assessment centre documents, and interview scripts. These documents are showing knowledge, qualifications, experience and underlying enabling attitudes and behaviours;
- Performance-based competency standards, such us: GAPPS, UK ECITB (www.ecitb.org.uk), Australian National Competency Standards for Project Management (ANCSPM) (www.aipm.com/globalstandards). These standards are showing demonstrable performance.

According International Competence Baseline (ICB), the competence baseline of the International Project Management Association (IPMA) (International Project Management Association [14]), the project management competencies can be grouped in three components and three categories. The PM competences *components* are: *knowledge* (the generally accepted practices of project management applied to specific technical disciplines), *skills* (the capabilities of applying knowledge in an efficient, effective, professional, and successful manner), *personal attitude* (the commitment to perform in an appropriate and acceptable professional and ethical manner) and *experience* (knowledge or skills gained from performing professional activities). The PM *competence categories* are: (a) *technical competencies* of delivering projects in a structured way, including the project management process, (b) *contextual competencies* in managing relations with projects within organizations, programs and portfolios, based on the knowledge of project characteristics, projects in the organizational context and project environment, and (c) *behavioural competencies* for a positive, collective, and dynamic thrust in nurturing project management professionalism such as leadership, communication, results-orientation, ethics, negotiation and so forth. IPMA defines the following four levels of competence:

- *IPMA Level A;* At this level, the individual has to demonstrate: successful use of the competence elements in the coordination of programs and/or portfolios; guided program and/or project managers in their development and in the use of the competence elements; been involved in implementing the competence elements or relevant methodology, techniques or tools in projects or programs; and having contributed to the development of the project manager's profession by publishing articles or presenting papers on his experiences or by outlining new concepts.
- *IPMA Level B;* At this level, the individual has to demonstrate successful use of the competence elements in complex project situations. She/he has also have guided (sub) project managers in their application and implementation of the competence.
- *IPMA Level C;* At this level, the individual has to demonstrate successful use of the competence element in project situations with limited complexity. He/she might need to be guided in the further development of the competence element.
- *IPMA Level D;* At this level, only knowledge related to the competence element is assessed by written examination.

The levels provide a suitable framework for developing career paths and organizational maturity models as well as for personnel development programs of individuals, companies and other organizations. The project personnel should continuously improve their project management competencies to meet the growth in demands and competition. Each project has its own specific needs which include the necessary competency needs of the project. Any gap between required and available competencies (in both internal and external resources) is a risk factor for the project's success.

For IT projects execution, a project-oriented organization relies on the IT competences of the technical experts. The most popular competence standard for software engineering is the Software Engineering Body of Knowledge (SWEBOK) (www.swebok.org). It describes the sum of knowledge within the profession of software engineering. The SWEBOK identifies and describes that body of knowledge that is generally accepted, even though software engineers must be knowledgeable not only in software engineering, but also, of course, in other related disciplines. The following ten knowledge areas are included in the guide: software requirements, software design, software construction, software testing, software maintenance, software configuration management, software engineering management, software engineering process, software engineering tools and methods and software quality. The SWEBOK also defines disciplines related to software engineering: computer engineering, computer science, management, mathematics, quality management, software ergonomics (cognitive ergonomics) and systems engineering.

The competence-based human resource management is often achieved without software support, or with stand-alone software, poorly integrated with other systems in place. The current trend of the research in competency management system (CMS) development shows that it is increasing very fast.

The next paragraph of the chapter presents relevant results in the CMS field. The following paragraph is dedicated to the proposed solution, defined for a Romanian IT project-oriented organization. The final part of the chapter presents the main conclusions and future work.

2 Literature Review

The HR-XML consortium has defined XML-schemes (HR-XML Consortium [12]) to facilitate the exchange of competencies-related information between organizations. The following definition of competence is used as basis for the XML-schema design: "A specific, identifiable and measurable knowledge, skill, ability and/or other deployment-related characteristic (e.g. attitude, behaviour, physical ability) which a human resource may possess and which is necessary for, or material to, the performance of an activity within a specific business context".

The description of competence has several components and the context of work is one of the most relevant. The context of work is "a particular working situation in which an individual is required to operate" [6]. Considering the project management key competences defined by the IPMA Competence Baseline, their description refers to specific situations in projects, for example the description of key competences for *1.05 Quality*, at level B is the following: *"Has successful managed the project quality situations of a complex project"* (International Project Management Association [14]).

Instead of the work context, in [20] "the learning and performing situation" is used in order to keep an adequate level of abstraction and to cover not only the work context, but also the competence development situations. The following components for the competence description are proposed: identifier, title, general description, definition, learning-performance situation, complexity factor, proficiency level, performance indicator, mapping function and metadata.

Taxonomies are traditionally used to model different competence categories [4, 13], but more sophisticated competence modelling approaches were also proposed [2, 7, 11, 19, 22, 25]. Vasconcelos [25] defined a competence model having the following two parts: taxonomy of primitive competencies with different competence categories (technical, cognitive, physical, project and business competencies) and taxonomy of application areas (process, activity, task, document, product, abstract device, language and customer). The elements of these taxonomies are connected by relationships. There are two kinds of relationships: work experiences relationships (has-experience-of, has-worked-on, participation-in and project-leader-of) and general competencies (author-of, background-of, practice-of, has-role-of, has-skill-of).

The ontology–based competency modelling approach allows the improvement of the definition of desirable competence profile [8, 15, 16, 26]. Ontology is a shared formal specification of a domain [10]. It defines and specifies the different domain concepts and the properties (relationships) of the concepts [17]. The benefit of developing ontology is that a machine-processable model can be directly used to make applications more aware of the domain semantics. Different software applications can use ontology to perform different tasks, including those involving reasoning based on the different relationships between concepts.

Ontology gives a better definition of the types of relations and hierarchies among competencies. It extends the measure scales to be used by the competence assessment process [21]. Schmidt and Kunzmann [22] proposed an ontology-based competence modelling approach with: competence ontology containing well-defined competences (structured by types and levels) and competence relationships (composition, generalization, subsumption). Competence ontology has three reference levels: upper-level competence ontology (specifying the notion of competence and main competence relationships), competence framework (using the upper-level ontology and defining reference competencies, mostly branch-specific, but company independent) and competence model/catalog (which is company specific and refines and contextualizes the framework). The core part of the upper-level ontology contains: competences, competence types, competence scales and competence levels. Bodea et al. [3] and Niculescu and Trausan-Matu [18] present competence ontologies developed for project management and IT, according to the existing professional competence standards, ICB (International Project Management Association 2006 [14]) and *Guide to the Software Engineering Body of Knowledge* (www.swebok.org).

The main functionality of a Competency Management System (CMS) is the identification; modelling and assessment of competencies represent the foundation of the competency-based human resources management. The CommOn [24] represents a framework for competence-based system development. CommonOn environment assists developers to build competence reference systems for particular domains, to represent competence profiles and to match different competence profiles. In Biesalski & Abecker [1] a comprehensive framework for integrated human resource and skill management is presented. The software solution is based upon an HR-Data Warehouse that integrates data from legacy systems, an ontology-based employee skill database, as well as three application modules.

3 The Proposed Solution for the Competency Management System

The solution was developed for a Romanian IT company that decided to implement a Microsoft Navision solution. This decision was based on the default interoperability of the ERP with Microsoft products (Office, especially Excel and Access), by the results of queries. For the HR module, the IT company asked for extended functionalities, such as: skill gap analysis, and the evaluation of competence profile of the project personal. The implementation team concludes that a CMS has to be developed and implemented, instead of the classical HR module.

The starting point for developing a CMS is to define the development approach. The authors propose to develop the CMS as an ontology-based extension of the HR module of the ERP system. The usage of two separate competence ontologies, one for each relevant profession in the organization is considered as more appropriate for structuring the CMS knowledge base instead to use only one competence ontology, combining the components of two different professions. Software engineering and project management are two different professions even if some project management activities are still considered in the knowledge of body of the software engineering profession. The IT organization practices make a clear differentiation between technical and managerial roles of the project team members, so it is not difficult to decide what competence ontology is applicable in which case.

The ontologies used to develop the CMS will be presented after the proposed system architecture.

In order to develop an ontology-based CMS, an ontology-centric approach in developing its IT architecture should be adopted, meaning to shift the focus from a function-oriented, tool-centric view, towards a semantic-oriented, ontology-centric view. This approach assures the semantic unification of the enterprise IT architecture. Traditional Enterprise Resource Planning (ERP) systems, even if capable to offer an integrated environment are often not integrated with other isolated applications (Biesalski and Abecker [1]). The mentioned ontologies were used to define the Business Terms and Glossary, extensively used during the ERP implementation by the entire team.

The following two approaches for integrating ontologies into ERP were considered: integration at the interrogation layer and integration at the report level. The first approach is based on the integration of the ontology as a separate module of the ERP system. In this way, the ERP report builder will receive different reasoning results like any other data input. The second approach is feasible only if the ERP report generator produces decomposable reports, being thus possible to separate the report content from the format, via XML-XSLT or XHTML-CSS coupling.

Because most of the report generators are user oriented, meaning that the focus is on the visibility and the printing properties of the report and not on the data structures and report decomposability, the best way to integrate the ontologies into the ERP is at the presentation layer instead of the report level. The report builder should use those data structures, which are provided by the ERP environment, and not to be changed every time when the semantic context is different.

3.1 The Architecture of the Competency Management System

Fig.1 presents, on the right side, the CMS architecture, as a component of an extended architecture of the ERP system. On the left side of the figure, there are shown the initial components of ERP architecture and on the right side, there are presented the new elements, included into the system in order to convert the HR module into a CBS. The figure shows also how the ontological extension was integrated into the initial architecture.

The extended architecture is based on the following layers of classical three-tier architecture:

- The presentation layer, incorporating the user interface components and the client module which provides the incoming flows for these components;
- The control layer (or the business logic), with the programs assuring the communication between interfaces as well as the processing of data passing through different levels;
- The models layer, with the initial database and the knowledge base, included in the extended architecture.

Between the control and models layers there is the access layer.

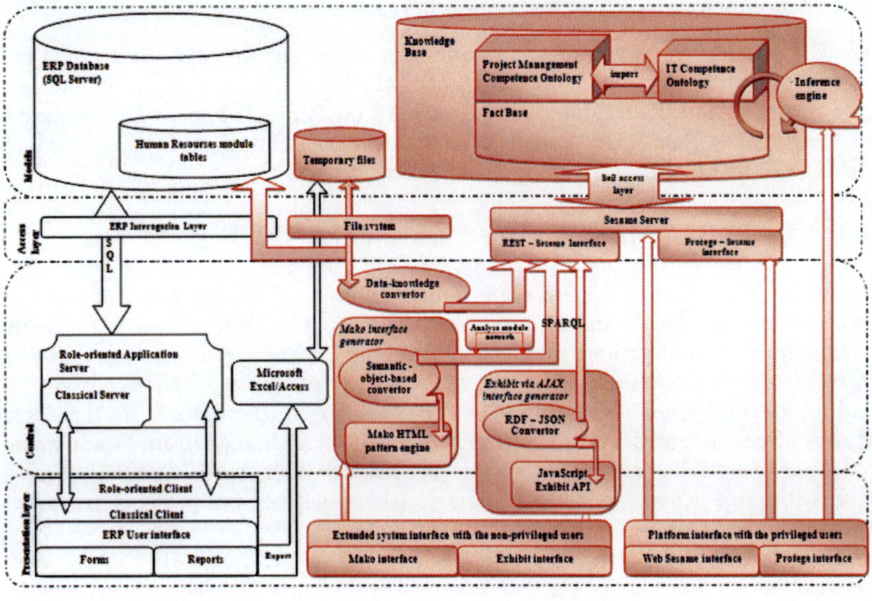

Fig. 1 The CMS architecture, as part of the extended ERP system

Protege and Sesame are the platform elements supporting the proposed models. They interface with the privileged users, when the reasoning processes should be executed by the inference engine. MySQL database system is used as platform for the knowledge base of the ERP. This platform assures a better control than those manipulating text and/or XML files. It is also possible to assure a relational view on the knowledge base. MySQL is one of solutions which Sesame server uses to store knowledge, as Postgre is doing. And finally, even if the MySQL performance is not as good as In-memory or Java graphs, it assures a high degree of scalability. Protege environment interfaces with the Sesame repository via MySQL. The *Sail access layer* is implemented on the Sesame platform in order to assure the independence of the knowledge base interrogation from data storage solution.

Considering the control and presentation layers, Sesame exposes two different interfaces, namely:

- *Web interface*, based on Java servlet, and named Workbench. This interface assures the knowledge upload from different sources, such as: web, text and XML files. The interface is used for the knowledge update and interrogations, having a direct access to ERP ontologies.
- *REST interface*, making Sesame to be exposed as a web service. In this way, all the programs implementing REST can send SQL interrogations to Sesame, receiving the results in JSON format, easy to be converted into locale data structures, such as Pyton repositories.

The REST interface is a common solution for accessing web services, which are using XML vocabulary for the communication between service requester and service provider. The REST interface relies only on the well known *http* requests, such as: GET, POST, PUT and DELETE. These requests are considered as READ, CREATE, UPDATE and DELETE operations.

3.2 The Project Management Competence Ontology

As shown in the Fig. 2, the PM competence ontology, also named as *PMCatalog*, has four abstract classes and 11 concrete classes [3]. The abstract classes are set in a taxonomic hierarchy: the class *Catalog* from Fig.2 is the super-class for the other three classes – *Technical_Competence_Range*, *Behavioural_Competence_Range* and *Contextual_Competence_Range*. The sub-classes correspond to the three groups of competences from ICB (International Project Management Association, [14]): 20 technical competences for project management, 15 behavioural competences for project personnel and 11 contextual competences of projects, programs and portfolios.

The *Catalog* class has only one slot, named *description*. All four abstract classes, among them is the *Catalog* class, have the class *MyMetaclass* as a metaclass. This metaclass was created to modify the characteristics of those four classes so that the class editor could display a supplementary property called *Explanation*.

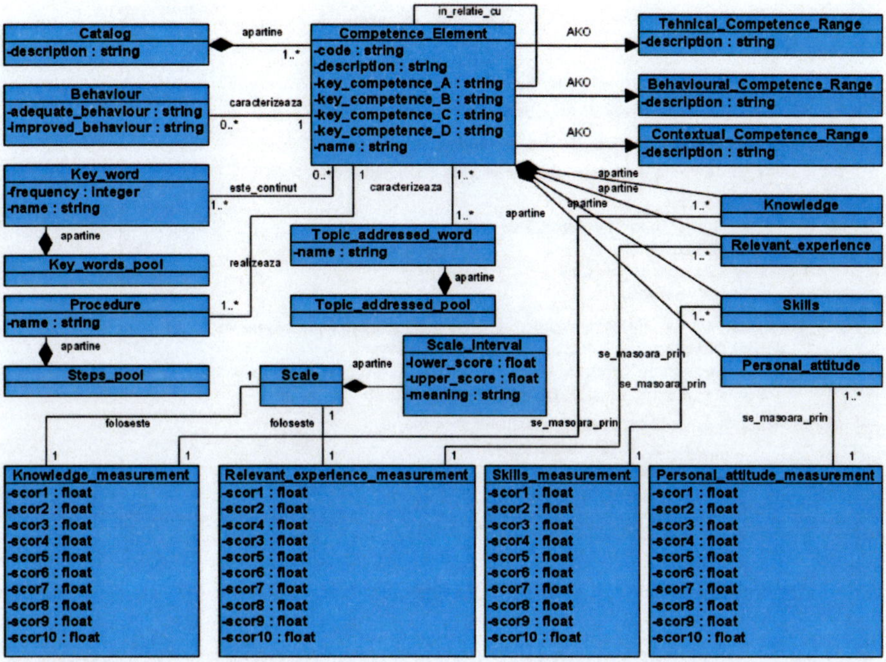

Fig. 2 The *PMCatalog* class diagram

3.3 The IT Competence Ontology

The IT competence ontology contains the basic concepts, such as:

- company-job,
- competence,
- domain,
- group,
- person, etc.

and their relationships with other concepts, instances and properties (Fig.3).

This ontology was built using the *Guide to the Software Engineering Body of Knowledge* (www.swebok.org).

Additionally to the representation of concepts, relationships and properties, the ontology contains also rules for integrity validation and inference [18].

The structure of the ontology is conceived so that description logics (http://dl.kr.org) can be used to represent the concept definitions of the application domain in a structured and formally well-understood way.

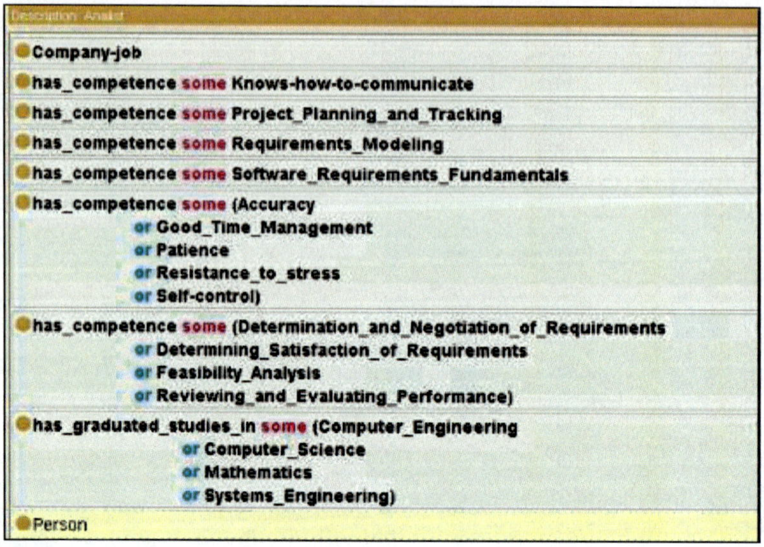

Fig. 3 Basic concepts of the IT competence ontology

4 The Experimentation of Competency Management System

A set of experiments was defined, in order to evaluate the proposed solution. The starting point in the experiment design is the identification of the use cases for the extended ERP system, as shown in Fig.4.

Starting with these use cases, the several experiment scenarios were defined (see the Fig.5) and executed (see the Fig.6).

	Scenario 1	Scenario 2
Query	Identify the required competence level of a project manager in order to successfully manage the project *A*.	Identify all the projects having an assigned project manager who does not fit from the PM competence point of view and display all the project manages candidates, having the required level of PM competences.
Results	Being a complex project, project A (id 4) needs a B-level project manager.	The project C (id 6) needs a B-level manager and does not have a proper project manager assigned (the assigned project manager is *Popescu Vasile*, a D-level certified associate. The compatible project managers for this project are *Lupu Andreea* (B level) and *Ionescu Ion* (B level). The project CONTO (id 3) needs a B-level manager and the assigned project manager is *Popescu Ion*, a C-level certified project manager. For this project, the compatible project managers are *Lupu Andreea* (B level) and *Ionescu Ion* (B level)

Fig. 5. Examples of scenario design

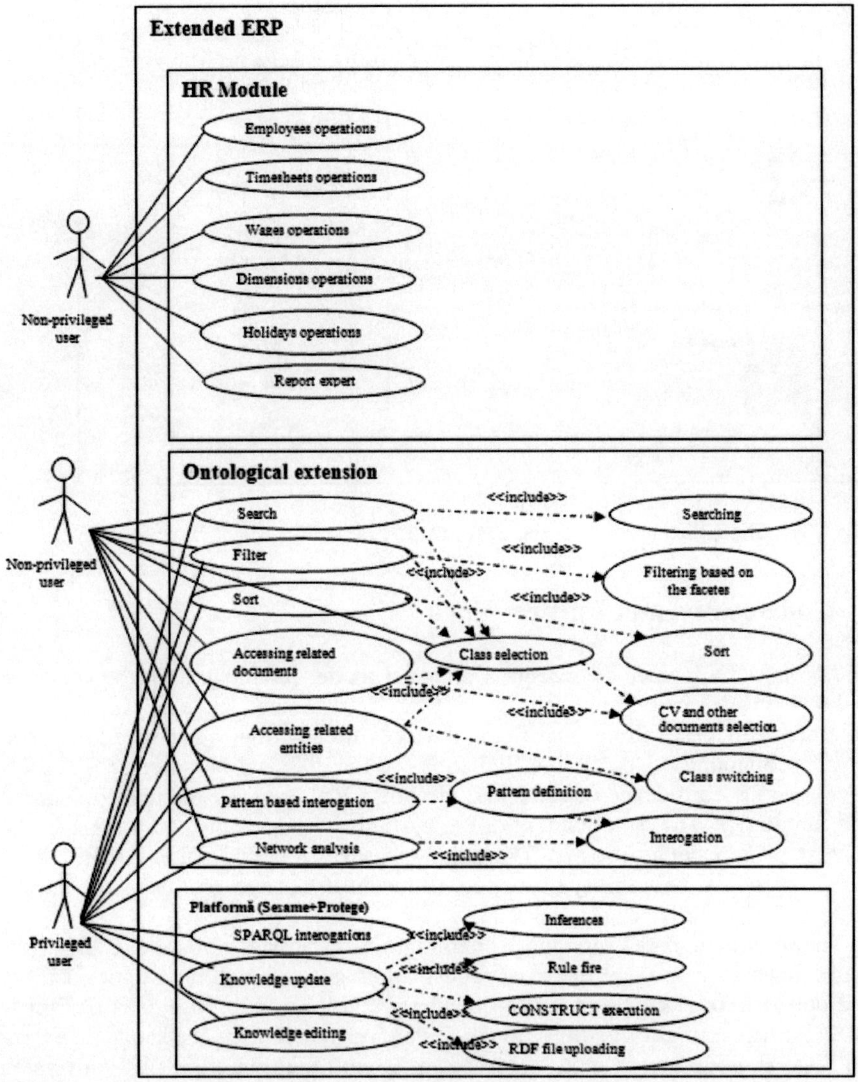

Fig. 4 The use cases for the extended ERP system

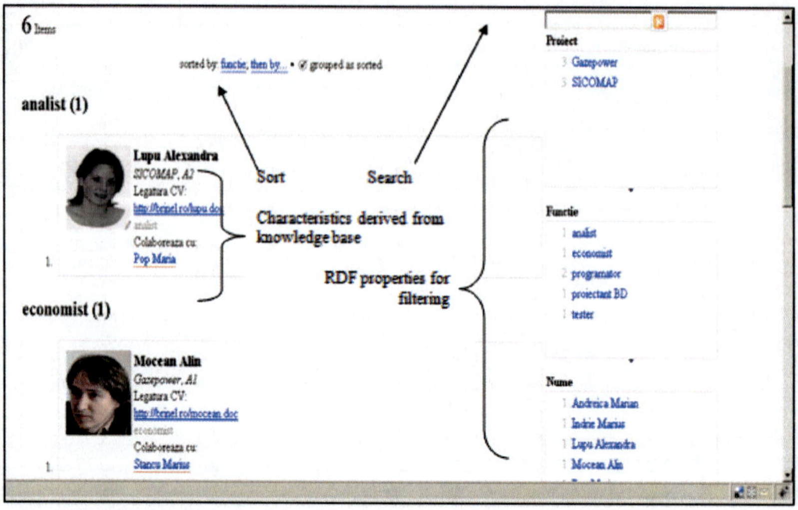

Fig. 6 The EXHIBIT interface for the EMPLOYEE class

5 Conclusions and Future Work

The proposed approach represents a practical model for integrating the semantic or knowledge oriented applications with the data-oriented or legacy systems, at minimum cost, and without sacrificing the tools already in place and well accepted by user community. Despite the limitations of the classical paradigm, these tools have won the confidence of users and provide a highly user experience of inertia and familiarity. The development of the systems is a relevant requirement in the context of knowledge society. The solution provides added value for ERP systems, expanding their ability to support all levels of management decision, but especially the executive ones.

The solution is based on a loose coupling of the knowledge base and ERP. It is well known accepted that the most robust solutions are based on a tight coupling and not on a loose coupling approach. The proposed solution provides flexibility in ERP usage, but not flexibility in the system reengineering. The semantic extension was provided in an incremental manner, with multiple points of conversion and interoperability, increasing the heterogeneity of extended ERP system. Some popular semantic technologies, but still during on the experimental stage are involved in the proposed solution. These technologies include Web services to generate Exhibit reports and to convert Babel knowledge structures. Integrating these functionalities in the system and not the online would be preferable.

The semantic emerging paradigm for data-oriented systems, even if it was launched some time ago, has still a limited visibility and spreading inside the companies. However, the Web 3.0 is emerging and the major producers seek to propose standards and provide technology solutions for emerging semantic technologies markets.

As for the future work, the authors will consider to use an evaluation framework, for the assessment of the proposed solution. After that, the implementation process will be decided. In parallel, a tight coupling of the will be considered for the CMS development.

References

1. Biesalski, E., Abecker, A.: Ansätze zum ontologiebasierten Human Resource Management. In: Workshop on IT Tools for Knowledge Management Systems: Applicability, Usability, and Benefits @ WM 2005, Kaiserslautern (April 2005)
2. Bizer, C., Heese, R., Mochol, M., Oldakowski, R., Tolksdorf, R., Eckstein, R.: The impact of semantic web technologies on job recruitment process. In: Proceedings of International Conference on Economical Informatics, Germany (2005)
3. Bodea, C.N., Elmas, C., Tanasescu, A., Dascalu, M.: An ontological-based model for competences in sustainable development projects: a case study for project's commercial activities. Amfiteatru Economic 27, 177–189 (2010) ISSN: 1582-9146
4. Brabrand, C., Dahl, B.: Using the solo taxonomy to analyze competence progression of university science curricula. Higher Education 58, 531–549 (2009)
5. The Shape of Business: the next 10 years (2010), `http://www.cbi.org.uk/pdf/20091123-cbi-shape-of-business.pdf`, CBI Report (accessed September 5, 2011)
6. Cheetham, G., Chives, G.: Professions, competence, and informal learning. Edward Edgar Publishing, Cheltenham (2005)
7. Dittmann, L., Zelewski, S.: Ontology-based Skills Management. In: Proceedings of the 8th World Multi-conference on Systemics, Cybernetics and Informatics, vol. IV, pp. 190–195 (2004)
8. Draganidis, F., Chamopoulou, P., Mentzas, G.: An ontology based tool for competency management and learning paths. In: 6th International Conference on Knowledge Management (I-KNOW 2006), Special track on Integrating Working and Learning, Graz (September 6, 2006)
9. Dubois, D., Rothwell, W., Stern, D., Kemp, L.: Competency-Based Human Resource Management. Davies-Black Publishing, CPP Inc., Palo Alto, CA (2004)
10. Gruber, T.: Ontolingua: A Mechanism to Support Portable Ontologies. Knowledge Systems Laboratory, Stanford University, California (1992)
11. Harzallah, M., Leclere, M., Trichet, F.: CommOnCv: Modelling the Competencies Underlying a Curriculum Vitae. In: ACM Proceedings of the 14th Int. Conference on Software Engineering and Knowledge Engineering, Italy, pp. 65–71 (2002)
12. HR-XML Consortium: HR-XML Competency (2009), `http://ns.hr-xml.org/schemas/org_hr-xml/3_0/Documentation/indexes/index.php` (accessed September 5 2011)
13. ICEL: Determining the future demand for ICT skills in Europe (2001), `http://duda.imag.fr/edu/supplydemand.pdf` (accessed September 5, 2011)
14. International Project Management Association: ICB - IPMA Competence Baseline, Nijkerk, Netherlands (2006)
15. Ley, T.: Organizational Competence Management - A Competence Performance Approach, Ph. D. University of Graz, Graz (2006)

16. Lindgren, R., Stenmark, D., Ljungberg, J.: Rethinking competence systems for knowledge-based organizations. European Journal of Information Systems 12(1), 18–29 (2003)
17. Maedche, A., Motik, B., Stojanovic, L.: Managing multiple and distributed ontologies in the semantic web. VLDB Journal 12, 286–302 (2003)
18. Niculescu, C., Trăuşan-Matu, S.: An Ontology-centred Approach for Designing an Interactive Competence Management System for IT Companies. Revista de Informatică Economică (4), 159–167 (2009) ISSN: 1453-1305
19. Noda, K.: Towards a representational model of evaluation ontology. In: Proceedings of the International Symposium on Large Scale Knowledge Resources, pp. 159–160 (2006)
20. Prins, F., Nadolski, R., Berlanga, A., Drachler, H., Hummel, H., Koper, R.: Competence Description for Personal Recommendations: The importance of identifying the complexity of learning and performance situations. Educational Technology & Society 11(3), 141–152 (2008)
21. Schmidt, A.: Context-steered learning: The Learning in Process approach. In: IEEE International Conference on Advanced Learning Technologies (ICALT 2004), pp. 684–686. IEEE Computer Society, Joensuu (2004)
22. Schmidt, A., Kunzmann, C.: Towards a Human Resource Development Ontology for Combining Competence Management and Technology-Enhanced Workplace Learning. In: Meersman, R., Tari, Z., Herrero, P. (eds.) OTM 2006 Workshops. LNCS, vol. 4278, pp. 1078–1087. Springer, Heidelberg (2006)
23. Sicilia, M.A.: Ontology-based competency management: infrastructures for the knowledge-intensive learning organization. In: Lytras, M.D., Naeve, A. (eds.) Intelligent Learning Infrastructures in Knowledge Intensive Organizations: A Semantic Web Perspective, pp. 302–324. Idea Group, Hershey (2005)
24. Trichet, F., Leclère, M.: A Framework for Building Competency-Based Systems Dedicated to Human Resource Management. In: Zhong, N., Raś, Z.W., Tsumoto, S., Suzuki, E. (eds.) ISMIS 2003. LNCS (LNAI), vol. 2871, pp. 633–639. Springer, Heidelberg (2003)
25. Vasconcelos, J., Kimble, C., Miranda, H., Henriques, V.: A knowledge-engine architecture for a competence management information system. In: UKAIS 2009 Conference Oxford, England (2009)

Knowledge-Based Adaptive Machining Concept for Service Oriented Architectures

Alexandru Dumitrache[1], Theodor Borangiu[1], Sylvain Pateloup[2], and Grigore Gogu[2]

[1] University Politehnica of Bucharest, Romania,
Centre for Research & Training in Robotics and CIM (CIMR)
{alexandru.dumitrache,theodor.borangiu}@cimr.pub.ro
[2] Clermont Université, IFMA, EA 3867,
Laboratoire de Mécanique et Ingénierie,
BP 10448, F-63000, Clermont-Ferrand, France
{sylvain.pateloup,grigore.gogu}@ifma.fr

Abstract. This chapter presents a new architecture for intelligent CNC machining in the context of a service-oriented manufacturing system involving material processing and robotized assembly operations. The proposed approach is a knowledge-based adaptive machining concept (KBAMC), which encompasses four modules: management, observation, prediction and strategy.

Keywords: Service-oriented manufacturing, CNC, adaptive cutting, knowledge-based machining.

1 Introduction

Some of the problems that discrete, repetitive manufacturing industry faces are: resource availability (unexpected failure or recovery of a resource) and treatment of rush orders. As a result, several manufacturing paradigms were proposed for addressing these problems, and a possible classification is [12]:

- *Mass production systems*, which were focused on the reduction of product cost;
- *Lean manufacturing*, which places emphasis on continuous improvement in product quality while decreasing product costs; according to [10], *agile manufacturing* is a natural development from this concept;
- *Flexible manufacturing system*, which makes possible the manufacture of a variety of products on the same system; they have a major disadvantage of being expensive, since in many cases they include more functions than needed;
- *Reconfigurable manufacturing systems*, which allow flexibility not only in producing a variety of parts, but also in changing the system itself.

Service-oriented methodology attempts to compensate for the deficiencies of current control / supervision systems, which are not agile because any shop floor

T. Borangiu, A. Thomas, and D. Trentesaux (Eds.): SOHOMA 2011, SCI 402, pp. 335–347.
springerlink.com

change requires programming modifications, implying the need for qualified programmers, usually not available in manufacturing environments. Even small changes (e.g. rush orders) might affect the global system architecture, which inevitably increases the programming effort and the potential for side-effect errors [2].

Service-oriented manufacturing is a new product pattern and manufacturing paradigm, which incorporates consumer services, producer services and manufacturing industries [8].

2 Overview of the Knowledge-Based Adaptive Machining Concept

The KBAMC (Knowledge-Based Adaptive Machining Concept) is proposed in order to realize an effective, reliable and superior manufacturing system, and encompasses four modules (Fig. 1):

- *Management*, which represents the interface to the holonic manufacturing system and handles production orders. This module exposes machining services in a heterarchical control system [11] where each job is negotiated between the resource units and the central management module of the manufacturing system. An implementation framework for this system was presented in detail in [5].
- *Observation* of milling process by sensor fusion, acquiring data about cutting forces, process temperature and machine load, and store the results of data analysis into a central knowledge base. This module also employs 3D vision for measuring the position, size and exact shape of the raw stock part for automating the initial machine setup stage.
- *Prediction* of cutting forces (short-term) or tool wear, milling accuracy figures, or machine failures (long-term), using mathematical models of the milling process with parameters estimated from observation data.
- *Strategy*, which generates the machining program and optimizes it according to the knowledge base data, at planning level (reordering cutting operations to minimize total time), at cutting process level (tool load control, feed rate and spindle speed optimization) and at dynamics level (ensuring smoothness, i.e. low acceleration and jerk values, and also exploiting kinematical redundancy when possible).

This chapter will focus on techniques for *observation* of the milling process, in order to gain knowledge about the underlying mathematical model, and also to be able to automate the machine setup for small batches of products, and *adaptive machining* which can be used either with or without external sensors (by simulating the milling process) and which can take advantage of prediction data and of observed process models which are stored in a central knowledge base.

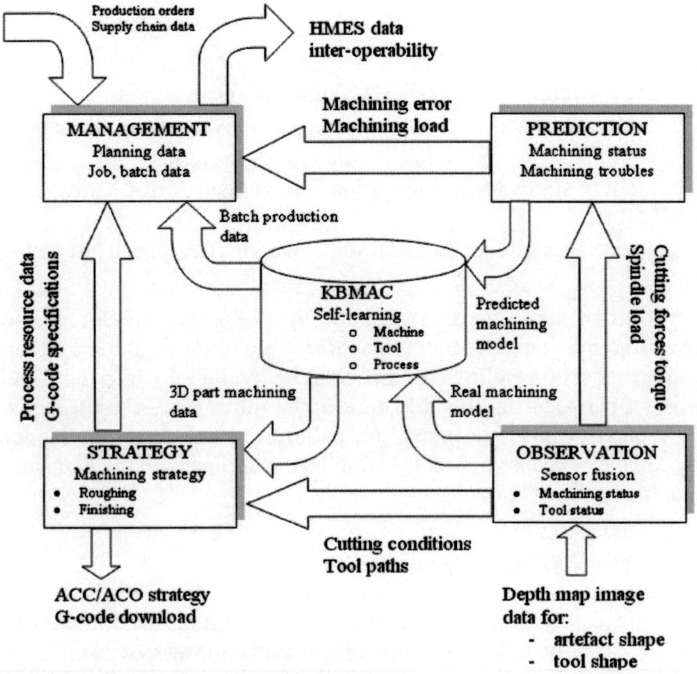

Fig. 1 Overview of the KBAMC system

3 Management Module

This module represents the interface between the CNC machine and the holonic manufacturing system into which the machine will be integrated. In this context, the milling machine is seen as a *resource holon*, which is able to offer *machining services.*

The following operations are defined for the milling machine and its associated tending robot for integration in the system presented in [5]:

- *Machine loading*: robot places the raw stock part in CNC clamping device;
- *Milling operation*: has a CNC program associated with it;
- *Machine unloading*: robot takes processed part from CNC;
- Local stock management: robot handles local storage buffers with raw and finished parts.

In our implementation, robot - machine communication is performed using the DNC protocol over Ethernet, where the robot controller is the master controller of the robot - CNC holon, and the interface between the controller and the PLC is only implemented on the robot side. The milling machine acts as a slave device of the robot, which is simply able to execute orders and report status.

The following operations were implemented via DNC between the robot and the machine, and can be started from a robot program:

- loading an arbitrary machining program (G-Code) from a text file;
- controlling the automatic machine door (opening and closing) from the robot;
- controlling the clamping vice of the milling machine from the robot;
- reading CNC machine status (program status: idle or busy; error status; machine and vice status;
- setting work piece origin and calibration data, which is required for automatic calibration.

From a scheduling point of view, we split the milling operation into four basic items (machine loading, milling operation, machine unloading and local stock management) because this scheme allows the robot to be available while the CNC machine is processing a part. During this idle period, the robot is able to fill its local buffers with raw parts (which are placed on pallets arriving from the feeder system of the work cell), to execute orders (placing a finished part on a pallet) and to do other tasks, such as quality control.

4 Observation of the Milling Process

Monitoring of milling process is necessary in any implementation of autonomous operation in a shop floor, especially in the context of agile manufacturing systems [1].

Monitoring is accomplished by detecting a significant change in the machining process by using some kind of sensor. For milling process, commonly used sensors measure cutting forces, acceleration and acoustic signals [6], but other variables like temperature, spindle current or part surface measurements are possible. It is generally agreed that a sensor for process monitoring must be easy to use, mount and maintain, must not interfere with the machine tool, must not be affected by the machining process, and it should be possible to mount the sensor close to the machining point to have reliable output [6]. Spindle current measurement is practical because it can be obtained with low cost and without affecting the milling process, but it may not offer enough information about cutting forces and vibration. Acoustic signal monitoring is also inexpensive and can serve to replace the traditional ability of the operator, based on his experience and senses (mainly vision and hearing) [14], but it may require complex signal processing in order to extract meaningful information [16]. 3D vision devices are expensive, but can acquire detailed information about part location and shape, and can be used for both fine-tuning of the work piece reference frame, and for tolerance checking of processed part; they are affected by chips and coolant and therefore their application in real-time process monitoring is limited.

In [13], the authors used a Leica laser tracker to analyze machine tool movements in a comparison between a serial machine (Huron KX-15) and a parallel machine (Triptéor X7). Laser tracker measurements can be compared with NC set point information and machine-tool encoder information in order to monitor the trajectory of the cutting tool, which in the case of parallel-kinematics machines (PKM) can deviate from the ideal trajectory to a higher extent than in the case of a serial kinematics machine (SKM), due to lower rigidity of the former structures. However, PKM machine offer a significant productivity gain over SKM due to superior dynamics; a monitoring solution can be using for reducing machining errors on a PKM while maintaining productivity gains.

A parameter frequently used to characterize 2D milling operations is the *stepover percentage*; a related parameter is the *tool engagement angle* (TEA), defined by [9] as the amount of sweep subtended by each cutting edge as it engages and leaves the stock (Fig. 2).

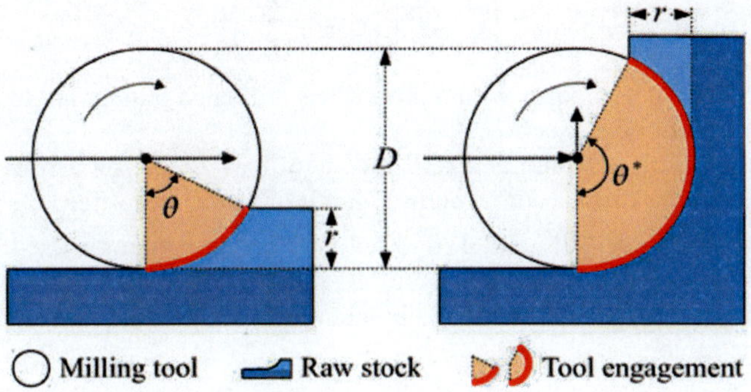

Fig. 2 Stepover and tool engagement angle, in straight line motion (left) and 90 degrees corner (right)

For straight line motions, there is a direct, nonlinear relationship between tool engagement angle and stepover (Fig. 2 b):

$$s = \frac{1+\sin\left(\theta - 90^o\right)}{2}, \quad 0 \le \theta \le 180^o$$

$$\theta = 90^o + \arcsin\left(2s-1\right), \quad 0 \le s \le 1$$

In [17], the authors defined a set of quantifiable metrics for evaluating a tool path quality, which included two main criteria:

- *Path curvature*, which determines the acceleration and jerk required to track the trajectory;
- Instantaneous cutter engagement, which can be evaluated by geometrical process simulation.

The authors proposed a method for evaluating the instantaneous cutter engagement by discrete process simulation. Both the in-process geometry of the raw work piece and the cutter shape are discretized, by encoding them as bitmap images. Pixels on the raw image can have two states: *empty* and *material_present*, while the pixels on the tool image can be *outside*, *inside* or on the *circumference*.

Another parameter useful for evaluating tool path quality is the material removal rate, which can also be determined by pixel simulation, even if the geometry of the parts have arbitrary complexity.

Cutting forces should be controlled during the milling process, since excessive values will lead to premature tool wear, overheating and other unwanted side effects. In a closed-loop scenario, these parameters can be obtained from force sensors mounted on the milling machine, for example on the work piece holder.

Fig. 3 shows a test trajectory for establishing a relationship between measured cutting forces in *XY* plane and *Z* direction, and parameters which can be computed by discrete simulation: stepover, tool engagement angle and material removal rate. This tool path is designed to have engagement value oscillating between $TEA_{\min}$ and $TEA_{\max}$ with a period p, which covers both very high and very low stepover values:

$$TEA_{ref} = TEA_{\min} + (TEA_{\max} - TEA_{\min}) \cdot \left(1 - \cos\frac{2\pi t}{p}\right) \cdot 0.5;$$

The method used for generating toolpaths with a prescribed tool engagement angle is the one which will be described in Section 5.

The test was performed with a 10mm end mill tool with two flutes, on a raw part having dimensions of 100x40mm, with a cutting depth of 1mm, and reference engagement angle varying between 20° and 170°. Actual engagement angle used for plotting was computed by pixel simulation from NC position data, and cutting forces were measured in X, Y and Z directions by clamping the test work piece on a fixed dynamometric table.

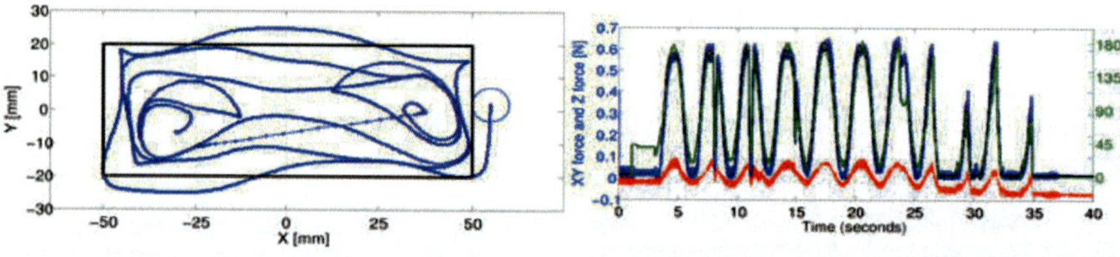

Fig. 3 Test tool path (a) with variable engagement angle, used for measuring the cutting forces in XY plane and Z direction (b)

There is a strong correlation between tool engagement angle and cutting forces in *XY* plane, and, to a lesser extent, between tool engagement angle and cutting force in Z direction; however, the magnitude of cutting forces in Z direction is much smaller in 2D milling, as it can be observed from Fig. 3b).

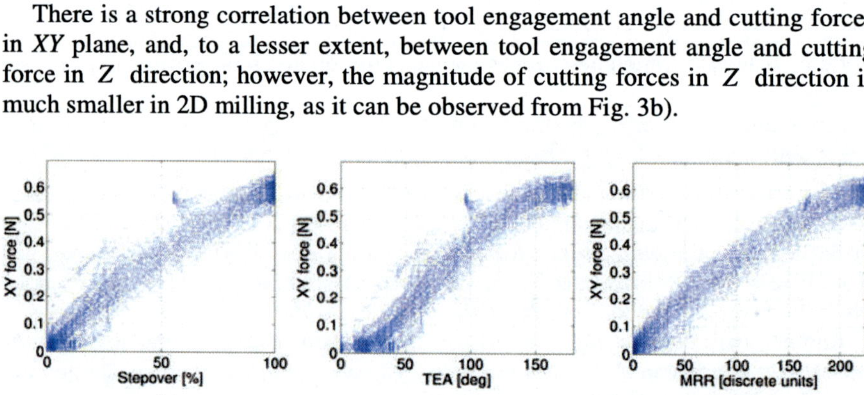

(a) (b) (c)

Fig. 4 Dependence of cutting force in XY plane on stepover (a), tool engagement angle (b) and material removal rate (c)

Fig. 4 shows the similarity between the 3 parameters, material removal rate (a), stepover (b) and tool engagement angle (c), in determining the cutting force only by means of pixel simulation, under the assumption that the material processed is homogeneous.

The observation module can be also used for automating the initial machine setup. Usually, CNC machines require a great deal of preparation before being able to execute the effective machining task. This includes aligning the coordinate system of the machine with the raw part, and calibrating the tool offsets, which are currently performed manually. This is not a problem when processing large batches of identical products, but for small batches, setting up the machine can take more time than the machining operation itself, which greatly limits system's agility.

Fig. 5 3D rendering of a raw stock part mounted in a CNC clamping vice, computed from laser scan data

Depending on the raw stock shape used, there are a few mathematical models which can be used:

- *Point cloud from 3D scan data*, which is a starting point for other models;
- *Depth map*: is suitable for modelling any part and/or raw stock shape which can be described by $z = f(x, y)$, i.e. for a 3-axis milling machine;
- *Cylindrical depth map*: is suitable for modelling relatively simple raw stock without concave regions, which can be described by $\rho = f(x, \theta)$, especially parts for 4-axis milling. This model was used for rendering the part shown in Fig. 5.
- *Ideal geometric models* like cylinder or prism. These models are commonly used by current CAM software and may include position and orientation data, which allow compensation for less accurate fixturing.

5 Adaptive Machining Strategy

Traditionally, CNC machining is performed by employing CAD/CAM software to design and generate milling toolpaths, which are eventually exported as G-Code (DIN 66025) programs. The G-Code program encodes the machine motion, which should be followed no matter what happens during milling, and has to be adjusted whenever the tool dimensions change due to mechanical wear. With this restriction, adaptive machining could be performed by adjusting feed rate and spindle speed at values different than the prescribed ones; however, spindle speed and feed rate have to be correlated in order to maintain a constant chip size, so the usefulness of this method is limited.

Conventional tool-path generation strategies generated by current CAM software employ either direction-parallel or contour-parallel toolpaths; while the resulting part is correct from a geometrical point of view, the approaches seldom take into account the milling process conditions. Advanced CAM programs use techniques such as trochoidal steps for controlling the tool load [15].

In [7] a technique is presented for generating adaptive 2D toolpaths which adjust to the geometry of the raw stock and design part, being parameterized by the *tool engagement angle* (TEA). This parameter is proportional to the cutting force, and limiting its value has benefits such as reducing chatter, cutting forces and cooling requirements and increasing tool life. Other parameters which can be chosen for this purpose are stepover and material removal rate (Fig. 4); the main advantage of choosing tool engagement angle is that the direction for achieving a prescribed value is straightforward to compute from a geometric intersection between tool circumference and raw stock geometry.

5.1 Algorithm Overview

In a 2D milling step, the raw stock and the part can be represented by two binary depth maps. There are four combinations possible, which allow encoding the geometry of part, stock and clamping device:

- *white part, white stock*: design part (raw material which must not be removed);
- white part, black stock: forbidden areas (e.g. for modelling the clamping fixtures);
- *black part, white stock*: material which has to be removed (where the milling paths will be generated);
- *black part, black stock*: free space (no material and no restrictions; useful for performing lateral entries into the raw stock in order to avoid plunging)

The milling tool path generator can be regarded as a state machine (Fig. 6a), with the following states:

1. Find Starting Point
2. Constant Engagement Milling
3. Contour Milling

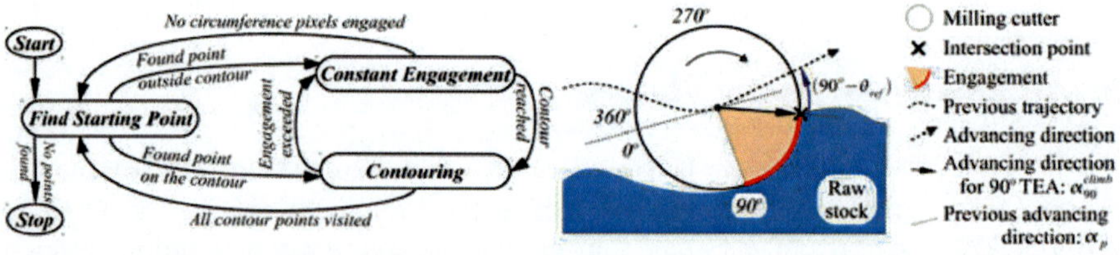

Fig. 6 (a) State diagram for tool path generation strategy; (b) Advancing direction for constant TEA, in climb milling

The second state will advance the cutter into the raw material by maintaining a constant tool engagement value. When the tool touches the design part, it is not possible to continue with the same engagement, because the tool path would remove material which must be kept in place. When this happens, the tool path generator will switch to the third state, contour milling.

In contour milling, the tool moves along the offset contour, being always tangent to the design part. It is necessary for the tool to move on the entire offset contour in order to obtain the correct geometry of the part; however, the maximum engagement angle can be exceeded, reaching values up to 180°.

Therefore, whenever the maximum engagement angle is exceeded by more than a given threshold, let's say 20°, the generator will switch back to constant engagement state.

5.2 Core Algorithm: Advancing with Constant Engagement

The core of the tool path generation algorithm is in the second state, which generates a constant engagement tool path, considering only the starting point and raw stock geometry. When the tool touches the part geometry, the core algorithm terminates. This prevents the cutting tool from removing more material than necessary.

The advancing direction α which maintains constant engagement angle can be computed from the intersection point between the tool circumference and the raw part geometry.

The direction of the intersection point between the tool circumference and the part edge (Fig. 6b) indicates the advancing direction for a 50% stepover, or 90° engagement; for other TEA values, it is necessary to apply a constant angular offset.

Let α_p be the direction of the tool path at the previous step. At the first step, α_p should point from the tool centre towards the raw stock. Also, let (x_k^O, y_k^O) be the coordinates of pixels from the tool circumference, with respect to the tool centre. Then:

$$\phi_k = \left(\operatorname{atan2}(y_k^O, x_k^O) + \alpha_p + 180^o\right) \bmod 360^o$$

For *climb milling*, the *white* (non-engaged) circumference pixel with the *largest* ϕ value will indicate the intersection between the tool circumference and the raw material edge. In this case, the advancing direction α^{climb} is (Fig. 6 b):

$$\alpha^{climb} = \alpha_{90}^{climb} + 90^o - \theta_{ref}$$

For *conventional milling*, the *smallest* ϕ will be used to compute the advancing direction α^{conv}. Therefore:

$$\alpha_{90}^{conv} = \min(\phi_k^{white}) - \alpha_p - 180^o$$

$$\alpha_{conv} = \alpha_{90}^{conv} - 90^o + \theta_{ref}$$

The algorithm can finish its loop in two ways. The stop reason will indicate the next state of the tool path generator (Fig. 6 a):

- The tool touched the design part (i.e. it tried to remove white pixels from the part image)
- There are no more white pixels on the tool circumference.

In the second case, the computation for the advancing direction can be repeated after artificially increasing the tool radius, up to a preset value (for example, 5 pixels or 10% of the nominal tool radius). The approach helps when the shape of the raw part exhibits external sharp corners.

Algorithm 1: Constant engagement milling

Input:
- Discrete tool diameter: D_D
- Reference engagement angle: θ_{ref}
- Raw stock: $\mathcal{R}$
- Part shape: $\mathcal{P}$
- Entry point: (x_0, y_0)
- Advancing step: Δd

Output:
- Trajectory points: $(x_1, y_1) \dots (x_n, y_n)$
- Raw stock after the generated milling step: $\mathcal{R}$
- Reason why the algorithm finished: *stop_reason*

```
Initialization:
(x, y) = (x0, y0)
i = 0
repeat
    i++
    if engaged (white) pixels on tool circumference
        α = direction for constant TEA (eq. 4-8)
        (x, y) = (x, y) + Δd · (cos α, sin α)
        if tool at (x, y) does not touch P
            (xi, yi) = (x, y)
            Update R
        else
            stop_reason = "touched part"
    else
        stop_reason = "no material around the tool"
until stop_reason ;
```

5.3 Examples of Generated Toolpaths

Fig. 7 shows toolpaths generated for milling a test part with simple geometry, which presents a concave region where tool engagement would have a sharp peak when using conventional contour-parallel toolpaths.

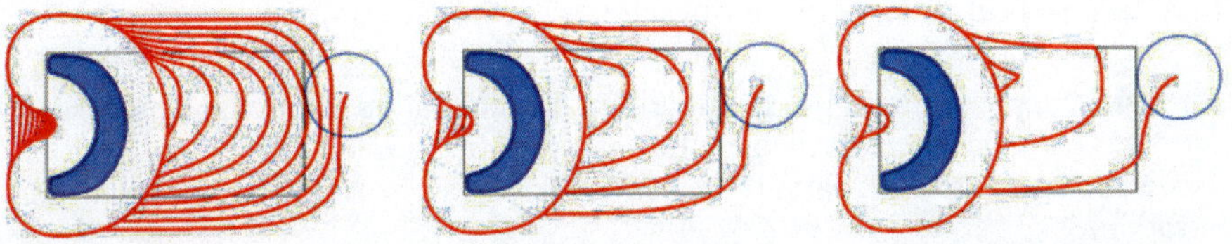

(a) TEA = 37° (b) TEA = 60° (c) TEA = 90°

Fig. 7 Milling toolpaths generated with the proposed algorithm for a simple test part

Fig. 8 shows examples of generated toolpaths for different values of reference engagement angle (a-c), and a graph showing the evolution of tool engagement angle along the milling process (d), showing the ability of the algorithm to limit the maximum value of the tool engagement, and therefore the cutting forces.

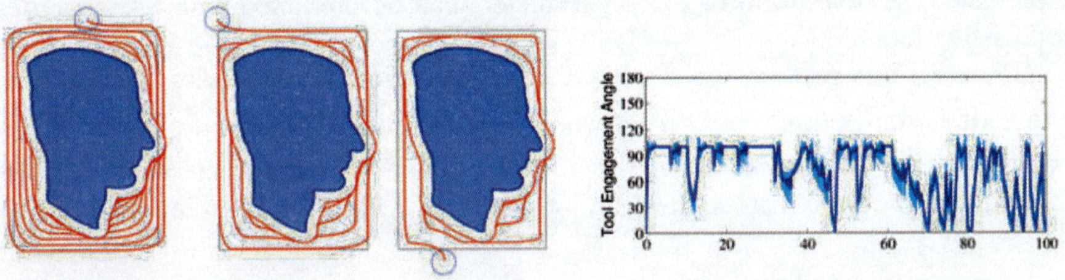

(a) TEA=60° (b) TEA=90° (c) TEA=120° (d) TEA evolution for (b)

Fig. 8 Toolpaths generated with the proposed algorithm for a part with complex geometry

6 The Prediction Module

Under typical usage, the milling machine will become less accurate due to wear, and requires costly recalibration sessions. However, adding extra sensors will make possible to predict the change in calibration parameters. We will study the possibility of using the 3D scanning sensor for inspecting finite parts and estimate the calibration parameters and their evolution.

We will consider the following parameters for machine calibration:

- Tool length and radius (for each milling tool)
  - $L_i, i = \overline{1,n}$
  - $R_i, i = \overline{1,n}$
- Lack of orthogonality between X, Y and Z axes of the machine
  - 3×3 matrix (in the ideal case: I_3)
- Lack of parallelism between rotary axis and X axis (for the particular machine in our study)
  - yaw / pitch correction angles

- Rotary axis offset in *YZ* plane (for our particular machine)
    - *dy*, *dz*

A possible approach is to use an algorithm for simulating the milling process under perturbations, where we can apply small variations to the parameters described above and compute a global error function which can be used for nonlinear minimization.

As shown in [10], we can model the part surface, stock surface and tool profile as depth map images, and apply the following generalized morphological image processing operators:

- Gray scale dilation: for generating part offset (both in 2D and 3D), which is used for generating milling toolpaths;
- Gray scale erosion: this simulates the result of the milling process from a given 2.5D milling tool path.

Once a method of online measurement of machine calibration parameters has been established, the evolution of those parameters can be predicted with a time series forecasting technique.

However, this method has a very high computational complexity, as only one simulation step is done in $O(n^4)$, where n is the mm-to-pixel ratio used for geometry discretization (modelling the milling surfaces as depth maps) and in order to obtain an accurate results, the search time is very high for any practical use.

7 Conclusions

This chapter proposes an architecture for integrating intelligent CNC machining under the service oriented manufacturing paradigm: a knowledge-based framework for adaptive machining, which addresses the main limitations of current CNC machine controllers, i.e. the lack of flexibility in a machining program (G-Code), and difficulty to setup the machine for small batches of parts. A method for 2D trajectory generation parameterized by tool engagement angle, which can be applied for arbitrarily complex part and raw stock geometry, is presented; when coupled with a 3D measurement system, this technique allows optimizing of milling trajectories for every unique raw stock part, assuming their exact shape and size may vary.

References

1. Bagci, E.: Monitoring and analysis of MRR-based feedrate optimization approach and effects of cutting conditions using acoustic sound pressure level in free-form surface milling. Scientific Research and Essays 6(2), 256–277 (2011)
2. Borangiu, T.: A service-orientated architecture for holonic manufacturing control, Report University Politehnica of Bucharest (2009)
3. Borangiu, T., Dogar, A., Dumitrache, A.: Integrating a short range laser probe with a 6-dof vertical robot arm and a rotary table. In: RAAD 2008 - The 17th Int'l Workshop on Robotics in Alpe-Adria-Danube Region, Ancona, Italy (2008)

4. Borangiu, T., Dogar, A., Dumitrache, A.: Calibration of wrist-mounted profile laser scanning probe using a tool transformation approach. In: RAAD 2009 – The 18th International Workshop on Robotics in Alpe-Adria-Danube Region, Brasov, Romania (2009)
5. Borangiu, T., Gilbert, P., Ivanescu, N.A., Rosu, A.: An implementing framework for holonic manufacturing control with multiple robot-vision stations. Eng. Appl. Artif. Intell. 22, 505–521 (2009)
6. Cho, D.W., Lee, S.J., Chu, C.N.: The state of machining process monitoring research in Korea. International Journal of Machine Tools and Manufacture 39(11), 1697–1715 (1999)
7. Dumitrache, A., Borangiu, T., Dogar, A.: Automatic generation of milling tool-paths with tool engagement control for complex part geometry. In: The 10th IFAC Workshop on Intelligent Manufacturing Systems - IMS 2010, Lisbon, Portugal (July 2010)
8. Gao, J., Yao, Y., Zhu, V.C.Y., Sun, L., Lin, L.: Service-oriented manufacturing: a new product pattern and manufacturing paradigm. Journal of Intelligent Manufacturing (2009)
9. Gasparraj, E.: Constant material removal: The key to hard milling. Modern Machine Shop (2005)
10. Gunasekaran, A., Yusuf, Y.Y.: Agile manufacturing: a taxonomy of strategic and technological imperatives. International Journal of Production Research 40(6), 1357–1385 (2002)
11. Leitao, P.: Agent-based distributed manufacturing control: A state-of-the-art survey. Engineering Applications of Artificial Intelligence 22(7), 979–991 (2009)
12. Mehrabi, M.G., Ulsoy, A.G., Koren, Y.: Reconfigurable manufacturing systems: Key to future manufacturing. Journal of Intelligent Manufacturing 11(4), 403–419 (2000)
13. Pateloup, S., Chanal, H., Duc, E.: Comparison of pkm and skm performances for aeronautics part machining. In: 8th International Conference in High Speed Machining HSM 2010, Metz (2010)
14. Rubio, E.M., Teti, R.: Process monitoring systems for machining using audible sound energy sensors. Future Manufacturing Systems, InTech (2010) ISBN: 978-953-307-128-2
15. Sprut Technology: Sprutcam v7 documentation (2009)
16. Tnshoff, H.K., Li, X., Lapp, C.: A fast Haar transform and concurrent learning-based monitoring approach and its application to tool breakage detection. IEEE Trans. Mechatronics, MT01-068R (2007)
17. Wang, H., Jang, P., Stori, J.A.: A metric-based approach to 2D tool path optimization for high speed machining. Journal of Manufacturing Science and Engineering (JMSE) 127(1), 33–48 (2005)

Author Index

Aissani, Nassima 115
Ali, Osman 1
Aly, Amir 183
Anton, Florin Daniel 75
Anton, Silvia 75

Ballot, Eric 279
Bekrar, Abdelghani 115
Berger, Thierry 21
Berraf, Sana Belmokhtar 169
Bodea, Constanta-Nicoleta 321
Borangiu, Theodor 21, 75, 265, 293, 335
Buchmann, Robert 321
Burlacu, George 129

Caramihai, Simona 255
Carbone, Giuseppe 75
Cardin, Olivier 197
Castagna, Pierre 197
Ceccarelli, Marco 75

Derigent, William 89
Dobrescu, Radu 307
Douche, Sidali Ad 143
Dubromelle, Yves 155
Dumitrache, Alexandru 335
Dupont, Daniel 49
Duta, Luminita 143

Enyedi, Szilárd 103

Germain, Bart Saint 1
Gobet, Olivier 279
Gogu, Grigore 335

Herrera, Carlos 169

Ivanescu, Nick Andrei 241

Kubler, Sylvain 89

Merzouki, Rochdi 49
Miclea, Liviu 103
Moiş, George 103
Moisescu, Mihnea Alexandru 129, 255
Montreuil, Benoit 279
Morariu, Cristina 293
Morariu, Octavian 293
Morvan, Gildas 49
Munteanu, Calin 255

Neagu, Gabriel 211

Ounnar, Fouzia 155

Panescu, Doru 61
Parlea, Mihai 241, 265
Pascal, Carlos 61
Pateloup, Sylvain 335
Philips, Johan 1
Prabhu, Vittaldas V. 227
Pujo, Patrick 155

Raileanu, Silviu 21, 265
Rey, Gabriel Zambrano 115
Rondeau, Éric 89
Rosu, Andrei 241

Sacala, Ioan Stefan 129, 255
Sallez, Yves 35
Soyez, Jean-Baptiste 49

Stanescu, Aurelian Mihai 129
Ştefan, Iulia 103
Stocklosa, Octavian 21, 265

Tahon, Christian 21
Tapus, Adriana 183

Thomas, André 89, 169
Trentesaux, Damien 21, 115

Valckenaers, Paul 1
Van Belle, Jan 1
Van Brussel, Hendrik 1

Subject Index

A

active entity 23
active entity system 25
active holon 26
active holon entity 268
active product 35
actual conflict 257
adaptive machining 335
advanced metering infrastructure 229
agent 103
agent-based approach 211
aggregation process 29
allocation process 121
ambient control entity 159
ambient services 157
analytic hierarchy process 165
analytic network process 165
architecture modelling 95
architecture performance 98
augmentation 35
autonomy 178

B

bar code 247
BayesiaLab 150
bayesian network 144, 145
behavioural model 30

C

closed-loop PLM 36
cloud computing 238
collaborative enterprises 129
coloured petri nets 65
communicating material 90
communication paths 82
competence description 324
competency management system 324, 325
competency modelling 325
concurrent enterprising 136
conflict resolution 257
continuous petri nets 255
contract-net 41, 62, 270
control 241, 243
control system 266
coordination 67, 212, 224
coordinator agent 109
cross-docking 15
cutting forces 339

D

data security 112
data-oriented systems 332
DCPN diagram 217
decision control petri nets 214
decision making 143
decision support 164, 216, 308
decomposition 126
decryption algorithm 112
delegate MAS 4
depth map 341
description logics 329
DEVS 163
diagnosis 43
Directory Facilitator 108
disassembly planning 143
disassembly policies 144
discrete-event observer 197
dismantling 143
distributed architecture 227
distributed control 202
distributed database systems 92
distributed private keys 112
distributed production scheduling 227, 229

driving strategies 268
dynamic approach 145

E

eEnterprise 136
eHealth 307
electric grid 229
electronic health records 308
electronic prescribing
(e-prescribing) 308
Enterprise Resource Planning 326
EPC 140
ERP 228, 232
Ethernet network 82
Euler angles 186
evolution graph 259
expected occupancy 121
Extended ERP system 327

F

fault tolerant 85, 86
features extraction 184
flexibility 103
flexible automation solution 22
flow holon 29
flow shop 6
FMS 123
framework 293, 305
functional view 38

G

generalized active entity 24
generalized active holon 27
generic control architecture 32
generic prototyping approach 214
global planner 244
GridMOSI 224

H

head pose estimation 185
Health Information Technology 308
Health Level Seven 310
Healthcare Information Network 309
Healthcare Services Specification
Project 310
high availability 81
high available shared storage 79
HL7 Development Framework 313
HMES 2, 61, 63
HMS 3, 78, 200, 202, 283
holonic agent model 66
holonic supply chain 279
holonification process 26
HPN model 261
human resources management 322
human-robot interaction 183
hybrid petri nets 259

I

inference algorithms 151
information technology systems 307
instantaneous firing speed 256
integration process 216
intelligent object 140
intelligent product 2, 5, 35, 173, 249
intelligent resource 2, 5
interface 295, 296, 312
Internet of Things 89, 132
Interoperability 129
IT competence ontology 329

J

JADE 275
job shop 8, 9
job-shop scheduling 211

K

knowledge base 327

L

load-balancing 103, 107
logistics 14, 93
loose coupling 332

M

machine calibration 346
maintenance 39
manufacturing 6, 35, 104, 197
manufacturing control 214
Manufacturing Execution System 1
manufacturing planning and control
systems 169
manufacturing systems 85, 293–295

Markov chains 146
Medicaid Information Technology Architecture 317
Message Development Framework 316
modelling holons 119
modelling resources 120
monitoring of milling process 338
multi-agent society 103
myopic behaviour 117, 202

N

Nao robot 190
network failure 85
network simulator 95
networked enterprise 130
networked production 10
non-active entity 23
non-active holon 26

O

Object Management Group 311
offer request management 294–296, 305
offline planner 242
OGSA 223
ontology 325
ontology-centric approach 326
OPC client 251
Open Distributed Processing 311
open hub network 283
open supply web 282

P

PEGAF platform 222
petri nets 62, 64, 73, 95, 255
Physical Internet 280, 281
planning 61, 68
platform independent model 312
platform specific model 312
PLC 242, 247
PLC supervisor 267
point cloud 341
potential field 41
product decomposition 77
product holon 64, 244
product lifecycle 36
product-driven production 227
product-driven system 170
production management 104
project management competence 328
project-oriented organization 322
PROSA 3, 61, 67, 198, 200, 201
prototyping process 215

Q

Quality of Service 312

R

reachability graph 71, 73
reactivity 122
real time decision making 154
reconfiguration 103
recursiveness 118
resource allocation 212
resource scheduling 31
RFID 41, 139, 158, 162, 166, 176, 230, 247
robotics 12
route planning 272
routing algorithm 251

S

scheduling 105
search algorithm 295, 301
search criteria 296, 299, 300, 303
search engine 293, 295, 297, 298
self-adaptation 176
semantic extension 332
Sequential Function Chart 244
service 2, 5
service agents 107
Service Functional Models 318
service orientation 212
Service Oriented Architecture 156, 161, 202, 227, 309
service reception 31
service supplier 79
Service-oriented Architecture 309
smart grid 227, 237
social interaction 194
speech temporal segmentation 189
staff holon 62, 67, 203
static structure 28
stochastic variables 147

structural conflict 257
Supply Chain 141, 280
Supply Chain Management 89
supply network 282
switch strategy 269
system characterized by a guided flow 21
System of Systems 130

T

tool engagement angle 339
traceability 36, 39
transport system 242

U

Unified Service Action Model 314

V

viable system model 169
Virtual Enterprise 134
virtual organization 131
vision systems 83
volunteer based search 293, 295, 297

W

Web 3.0 332